Methods in Molecular Biology™

Series Editor
John M. Walker
School of Life Sciences
University of Hertfordshire
Hatfield, Hertfordshire, AL10 9AB, UK

For further volumes:
http://www.springer.com/series/7651

Data Mining for Systems Biology

Methods and Protocols

Edited by

Hiroshi Mamitsuka

Bioinformatics Center,
Institute for Chemical Research, Kyoto University, Uji, Kyoto, Japan

Charles DeLisi

Department of Biomedical Engineering, Boston University,
Boston, MA, USA

Minoru Kanehisa

Bioinformatics Center,
Institute for Chemical Research, Kyoto University, Uji, Kyoto, Japan

Editors
Hiroshi Mamitsuka
Bioinformatics Center
Institute for Chemical Research
Kyoto University
Uji, Kyoto, Japan

Charles DeLisi, Ph.D.
Department of Biomedical Engineering
Boston University
Boston, MA, USA

Minoru Kanehisa
Bioinformatics Center
Institute for Chemical Research
Kyoto University
Uji, Kyoto, Japan

ISSN 1064-3745 ISSN 1940-6029 (electronic)
ISBN 978-1-62703-106-6 ISBN 978-1-62703-107-3 (eBook)
DOI 10.1007/978-1-62703-107-3
Springer New York Heidelberg Dordrecht London

Library of Congress Control Number: 2012947383

Printed on acid-free paper

Humana Press is a brand of Springer
Springer is part of Springer Science+Business Media (www.springer.com)

Preface

The post-genomic revolution is witnessing the generation of petabytes of information annually, with deep implications ranging across evolutionary theory, developmental biology, agriculture, and disease processes. The great challenge during the coming decades is not so much in generating the data, for that will continue at an accelerating pace, but in converting it into the information and knowledge that will improve the human condition and deepen our understanding of the world around us. A first step in meeting that challenge is to structure data so that it is easily accessed, integrated, and assimilated. Data Mining in Systems Biology surveys and demonstrates the science and technology of this important initial step in the data-to-knowledge conversion. The volume is organized around two overlapping themes, network inference and functional inference.

Network Inference

Tsuda and Georgii (*Dense Module Enumeration in Biological Networks*) discuss a rigorous, robust, and inclusive approach to inferring a particular type of network; viz, networks defined by databases that record physical interactions between proteins. Willy, Sung, and Ng (*Discovering Interacting Domains and Motifs in Protein–Protein Interactions*) discuss a method for discovering interactions between protein domains and short linear sequences, which are fundamental to multiple cellular processes. In particular, they discuss and demonstrate how to exploit the surge in structural data to infer such interactions. Mongiovì and Sharan (*Global Alignment of Protein–Protein Interaction Networks*) describe a novel method for identifying proteins that are orthologous across species. Their method is based on alignment of protein–protein interaction networks. This paper and that of Tsuda and Georgii represent a good example of the knowledge amplification that can be achieved by research on different but potentially complementary projects carried out by different labs. These three papers illustrate important directions in the discovery and analysis of *protein–protein* interactions.

While protein–protein interactions define the repertoire of cellular processes, *protein–DNA interactions* regulate those processes. In general, gene/protein networks defined by such interactions can be inferred from experimental data by various multivariate statistical methods. One of the widely used forms of inference is Bayesian probabilistic modeling. Larjo, Shmulevich, and Lähdesmäki (*Structure Learning for Bayesian Networks as Models of Biological Networks*) review recent progress in the development and application of these methods. Mordelet and Vert (*Supervised Inference of Gene Regulatory Networks from Positive and Unlabeled Examples*) discuss SIRENE, a machine learning method for inferring networks of transcriptional regulators and their targets from expression data and known regulatory relationships. Honkela, Rattray, and Lawrence (*Mining Regulatory Network Connections by Ranking Transcription Factor Target Genes Using Time Series Expression Data*) developed a reverse engineering approach to infer regulator target interactions and applied it to candidate targets of the p53 tumor suppressor promoter.

Historically, molecular biology has focused on proteins and nucleic acids. One of the major changes in the past decade has been a dramatic increase in understanding metabolism; this, of course, is also stimulated by the availability of whole genome sequence data. This constitutes the subject of *Protein–Chemical Substance Interactions*. Hancock, Takigawa, and Mamitsuka (*Identifying Pathways of Co-ordinated Gene Expression*) present a tutorial for the use of gene expression data to identify metabolic networks associated with a given condition.

More direct approaches to metabolism include an increased emphasis on the structure of complex carbohydrates. Aoki-Kinoshita (*Mining Frequent Subtrees in Glycan Data Using the RINGS Glycan Miner Tool*) describes an algorithmic method for finding frequently occurring tree structures with glycan databases, which are relevant to the binding of particular proteins. This can be thought of as the metabolic analogue to approaches that identify protein–protein and protein–DNA binding sites.

The chapter by Yamanishi (*Chemogenomic Approaches to Infer Drug–Target Interaction Networks*) discusses another kind of network, those formed by drug–target interactions. In this case, sequence and chemical structure databases provide the information that enable statistical classification methods to identify plausible drug–target interactions.

Functional Inference

The ability to predicatively localize proteins to one or another cellular compartment can generate important clues about their possible function. Imai, Hayat, Sakiyama, Fujita, Tomii, Elofsson, and Horton (*Localization Prediction and Structure-Based In Silico Analysis of Bacterial Proteins: With Emphasis on Outer Membrane Proteins*) evaluate localization prediction tools against a known dataset, and illustrate with an application to β-barrel outer membrane proteins in *E. coli*. For biological interpretation of large-scale datasets, visualization tools play key roles. Hu (*Analysis Strategy of Protein–Protein Interaction Networks*) explains how to use the multiple data sources and analytical tools in VisANT to identify and analyze networks of various kinds. Karp, Paley, and Altman (*Data Mining in the MetaCyc Family of Pathway Databases*) present an introduction to the contributions made by Karp and his colleagues over many years. The chapter is a rich source of tools and methods for mining this extensive, well-curated, and extremely important set of databases.

Approaches to genotype–phenotype correlations have evolved continuously over the past several decades. With the advent of whole genome sequencing, the search for correlations between genes and Mendelian traits accelerated enormously, but complex phenotypes, whether normal traits or diseases, find their genetic basis in sets of genes, and in particular combinations of alleles. Various procedures have been developed to infer such sets from variations in transcriptional variation. Hung (*Gene Set/Pathway Enrichment Analysis*) describes in detail how the so-called gene set enrichment analysis can be used to draw functional inferences from such transcriptional datasets. The method has been applied to identify processes that distinguish disease phenotypes from normal phenotypes. This leads to the final four chapters of the volume, which are all disease related.

Linghu, Franzosa, and Xia (*Construction of Functional Linkage Gene Networks by Data Integration*) discuss an approach to combining heterogeneous datasets in order to construct full genome networks in which each gene is surrounded by functionally related neighbors, with the relationships specified by evidence-weighted links. Such functional linkage networks (FLNs) of human genes can uncover surprising genetic associations between phenotypically unrelated diseases and suggest that our current disease nosology may need to be reformulated.

The chapter by Yang, Kon, and DeLisi (*Genome-Wide Association Studies*) presents an overview of genome-wide association methods and explains how multiple data sources, including databases generated by high-throughput genotyping technologies, can be used to identify disease-associated chromosomal locations.

Kuiken, Yoon, Abfalterer, Gaschen, Lo, and Korber (*Viral Genome Analysis and Knowledge Management*) discuss three of the major infectious disease sequence-function databases—those for the human immunodeficiency, hepatitis C, and hemorrhagic fever viruses. The challenge here again is combining information from different sources, but in this case, integration and quality control are achieved by a continually upgraded community-developed infrastructure.

Kanehisa (*Molecular Network Analysis of Diseases and Drugs in KEGG*) presents another integrated approach where known disease genes and drug targets are integrated into the KEGG molecular network database and explains how to make use of this resource with the KEGG Mapper tool in large-scale data analysis.

We expect this book to be of interest to cell biologists and biotechnologists, as well as to the scientists and engineers developing the databases and mining and visualization systems that are central to the paradigm-altering discoveries being made with increasing frequency.

Uji, Kyoto, Japan — *Hiroshi Mamitsuka*
Boston, MA, USA — *Charles DeLisi*
Uji, Kyoto, Japan — *Minoru Kanehisa*

Contents

List of Contributors

WERNER ABFALTERER • *Los Alamos National Laboratory, Theoretical Biology and Biophysics (MS K710), Los Alamos, NM, USA*

TOMER ALTMAN • *Bioinformatics Research Group, SRI International, Menlo Park, CA, USA*

KIYOKO FLORA AOKI-KINOSHITA • *Department of Bioinformatics, Faculty of Engineering, Soka University, Hachioji, Tokyo, Japan*

CHARLES DELISI • *College of Engineering, Boston University, Boston, MA, USA*

ARNE ELOFSSON • *Science for life laboratory, Department of Biochemistry and Biophysics, Stockholm Bioinformatics Center, Center for Biomembrane Research, Swedish E-science Research Center, Stockholm University, Stockholm, Sweden*

ERIC A. FRANZOSA • *Bioinformatics Program, Boston University, Boston, MA, USA*

NAOYA FUJITA • *AIST, Computational Biology Research Center, Tokyo, Japan, Taiho Pharmaceutical Company, Ibaraki, Japan*

BRIAN GASCHEN • *Los Alamos National Laboratory, Theoretical Biology and Biophysics (MS K710), Los Alamos, NM, USA*

ELISABETH GEORGII • *Helsinki Institute for Information Technology HIIT Aalto University, School of Science, Aalto, Finland*

TIMOTHY HANCOCK • *Bioinformatics Center, Institute for Chemical Research, Kyoto University, Uji, Japan*

SIKANDER HAYAT • *Science for life laboratory, Department of Biochemistry and Biophysics, Stockholm Bioinformatics Center, Center for Biomembrane Research, Swedish E-science Research Center, Stockholm University, Stockholm, Sweden*

ANTTI HONKELA • *Department of Computer Science, Helsinki Institute for Information Technology HIIT, University of Helsinki, Helsinki, Finland*

PAUL HORTON • *AIST, Computational Biology Research Center, Tokyo, Japan*

ZHENJUN HU • *Bioinformatics Program, Boston University, Boston, MA, USA*

WILLY HUGO • *School of Computing, National University of Singapore, Singapore, Singapore*

JUI-HUNG HUNG • *Program in Bioinformatics and Integrative Biology, Worcester, MA, USA*

KENICHIRO IMAI • *AIST, Computational Biology Research Center, Tokyo, Japan, Japan Society for the Promotion of Science, Chiyoda, Tokyo, Japan*

MINORU KANEHISA • *Bioinformatics Center, Institute for Chemical Research, Kyoto University, Uji, Japan*

PETER KARP • *Bioinformatics Research Group, SRI International, Menlo Park, CA, USA*

MARK KON • *Department of Mathematics and Statistics, Boston University, Boston, MA, USA*

BETTE KORBER • *Los Alamos National Laboratory, Theoretical Biology and Biophysics (MS K710), Los Alamos, NM, USA*

CARLA KUIKEN • *Los Alamos National Laboratory, Theoretical Biology and Biophysics (MS K710), Los Alamos, NM, USA*

HARRI LÄHDESMÄKI • *Department of Information and Computer Science, School of Science, Aalto University, Aalto, Finland*
ANTTI LARJO • *Department of Signal Processing, Tampere University of Technology, Tampere, Finland*
NEIL D. LAWRENCE • *Department of Computer Science, Regent Court, University of Sheffield, Sheffield, UK The Sheffield Institute for Translational Neuroscience, University of Sheffield, Sheffield, UK*
BOLAN LINGHU • *Biomarker Development Group, Translational Sciences Department, Novartis Institutes for BioMedical Research, Cambridge, MA, USA*
CHIENCHI LO • *Los Alamos National Laboratory, Theoretical Biology and Biophysics (MS K710), Los Alamos, NM, USA*
HIROSHI MAMITSUKA • *Bioinformatics Center, Institute for Chemical Research, Kyoto University, Uji, Japan*
MISAEL MONGIOVI • *Computer Science Department, University of California Santa Barbara, Santa Barbara, CA, USA*
FANTINE MORDELET • *Department of Computer Science, Duke University, NC, USA*
SEE-KIONG NG • *Institute for Infocomm Research, Connexis, Singapore*
SUZANNE PALEY • *Bioinformatics Research Group, SRI International, Menlo Park, CA, USA*
MAGNUS RATTRAY • *Department of Computer Science, Regent Court, University of Sheffield, Sheffield, UK The Sheffield Institute for Translational Neuroscience, University of Sheffield, Sheffield, UK*
NORIYUKI SAKIYAMA • *AIST, Computational Biology Research Center, Tokyo, Japan*
RODED SHARAN • *Blavatnik School of Computer Science, Tel Aviv University, Tel Aviv, Israel*
ILYA SHMULEVICH • *Institute for Systems Biology, Seattle, WA, USA*
WING-KIN SUNG • *School of Computing, National University of Singapore, Singapore*
ICHIGAKU TAKIGAWA • *Bioinformatics Center, Institute for Chemical Research, Kyoto University, Uji, Japan*
KENTARO TOMII • *AIST, Computational Biology Research Center, Tokyo, Japan*
KOJI TSUDA • *AIST Computational Biology Research Center, Tokyo, Japan, JST ERATO Minato Project, Sapporo, Japan*
JEAN-PHILIPPE VERT • *Mines ParisTech, Centre for Computational Biology, Fontainebleau, France*
YU XIA • *Bioinformatics Program, Boston University, Boston, MA, USA*
YOSHIHIRO YAMANISHI • *Institut Curie, Centre de recherche Biologie du developpement, U900 Unit of Bioinformatics and Computational Systems Biology of Cancer, Paris, France*
TUN-HSIANG YANG • *Bioinformatics program, College of Engineering, Boston University, Boston, MA, USA*
HYEJIN YOON • *Los Alamos National Laboratory, Theoretical Biology and Biophysics (MS K710), Los Alamos, NM, USA*

Chapter 1

Dense Module Enumeration in Biological Networks

Koji Tsuda and Elisabeth Georgii

Abstract

Automatic discovery of functional complexes from protein interaction data is a rewarding but challenging problem. While previous approaches use approximations to extract dense modules, our approach exactly solves the problem of dense module enumeration. Furthermore, constraints from additional information sources such as gene expression and phenotype data can be integrated, so we can systematically detect dense modules with interesting profiles. Given a weighted protein interaction network, our method discovers all protein sets that satisfy a user-defined minimum density threshold. We employ a reverse search strategy, which allows us to exploit the density criterion in an efficient way.

Key words: Protein complex, Dense module enumeration, Reverse search, Gene expression, Protein interaction

1. Introduction

Today, a large number of databases provide access to experimentally observed protein–protein interactions. The analysis of the corresponding protein interaction networks can be useful for functional annotation of previously uncharacterized genes as well as for revealing additional functionality of known genes. Often, function prediction involves an intermediate step where clusters of densely interacting proteins, called modules, are extracted from the network; the dense subgraphs are likely to represent functional protein complexes (1). However, the experimental methods are not always reliable, which means that the interaction network may contain false positive edges. Therefore, confidence weights of interactions should be taken into account.

A natural criterion that combines these two aspects is the average pairwise interaction weight within a module (assuming a weight of zero for unobserved interactions, cf. (2)). We call this the module *density*, in analogy to unweighted networks (3). We present a method

Hiroshi Mamitsuka et al. (eds.), *Data Mining for Systems Biology: Methods and Protocols*, Methods in Molecular Biology, vol. 939, DOI 10.1007/978-1-62703-107-3_1, © Springer Science+Business Media New York 2013

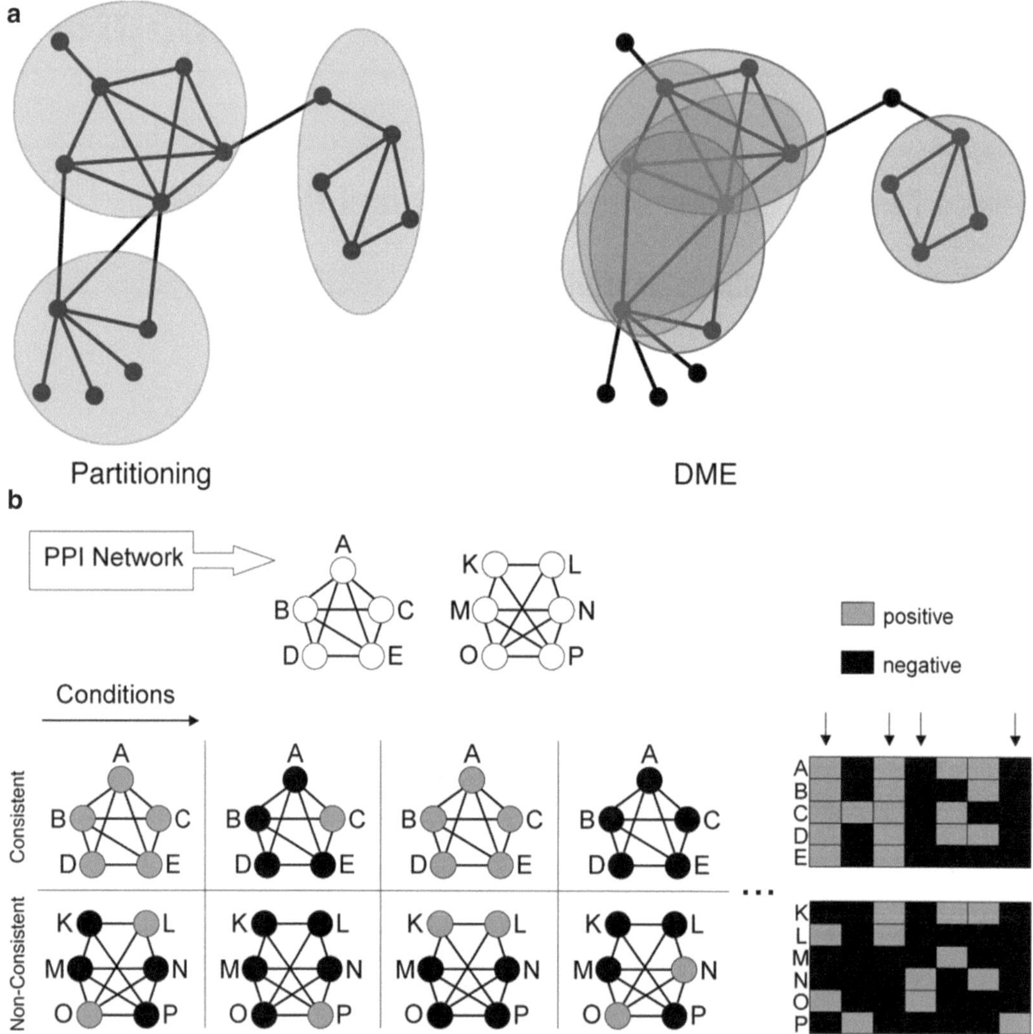

Fig. 1. Dense module enumeration approach. (**a**) DME versus partitioning. While partitioning methods return one clustering of the network, DME discovers all modules that satisfy a minimum density threshold. (**b**) Combination with profile data. Integration of protein–protein interaction (PPI) and external profile data allows to focus on modules with consistent behavior of all member proteins in a subset of conditions. The top module has two conditions where all nodes are positive and one condition where all nodes are negative. The arrows in the profile show such consistent conditions. On the other hand, the bottom module does not have such consistency.

to enumerate all modules that exceed a given density threshold. It solves the problem efficiently via a simple and elegant reverse search algorithm, extending the unweighted network approach in (4).

There is a large variety of related work on module discovery in networks. The most common group are graph partitioning methods (5–7). They divide the network into a set of modules, so their approach is substantially different from dense module enumeration (DME), which provides an explicit density criterion for modules (Fig. 1a). Another group of methods define explicit module criteria, but employ heuristic search techniques to find the modules (3, 8). This contrasts with complete enumeration algorithms, which form the third line of research: they give explicit criteria and return all modules that satisfy them. For example, clique search has been frequently applied (9, 10). The enumeration of cliques can be considered as a special case of our approach, restricting it to unweighted graphs and a density threshold of one. Further enumerative approaches use different module criteria assuming unweighted graphs (11).

In recent years, many module finding approaches which integrate protein–protein interaction networks with other gene-related data have been published. One strategy, often used in the context of partitioning methods, is to build a new network whose edge weights are determined by multiple data sources (12). Tanay et al. (13) also create one single network to analyze multiple genomic data at once; however, they use a bipartite network where each edge corresponds to one data type only. In both cases, the different data sets have to be normalized appropriately before they can be integrated. In contrast to that, other approaches keep the data sources separate and define individual constraints for each of them. Consequently, arbitrarily many data sets can be jointly analyzed without the need to take care of appropriate scaling or normalization. Within this class of approaches, there exist two main strategies to deal with profile data like gene expression measurements. In the first case, the profile information is transformed into a gene similarity network, where the strength of a link between two genes represents the global similarity of their profiles (2, 14, 15). In the second case, the condition-specific information is kept to perform a context-dependent module analysis (16–18). Our approach follows along this line, searching for modules in the protein interaction network that have consistent profiles with respect to a subset of conditions. In contrast to the previous methods, our algorithm systematically identifies all modules satisfying a density criterion and optional consistency constraints.

2. Materials

1. A protein interaction network: It can be downloaded, e.g., from the following Web sites, IntAct (19), MINT (20), and BIND (21).
2. Gene expression data: For example, global human gene expression profiles across different tissues can be obtained from the supplementary information of (22).

3. Methods

We describe the basic idea of DME using the examplar graph shown in Fig. 2. First, we discuss how to enumerate dense modules in a network, and then proceed to explain how gene expression data can be involved.

3.1. Enumeration of Dense Modules

Our method is based on the reverse search paradigm (23), which is quite popular in the algorithm community, but only in a limited degree known in the data mining community. A weighted graph is represented as a symmetric association matrix (edges that are not shown have zero weight). We denote by w_{ij} the weight between two nodes, and define the density of a node subset U as

$$\rho(U) = \sum_{i,j \in U, i<j} w_{ij} \Big/ \frac{|U|(|U|-1)}{2}.$$

We would like to enumerate all subsets U with $\rho(U) \geq \theta$, where θ is a prespecified constant.

All subsets form a natural graph-shaped search space, where one can move downwards or upwards by adding or removing a node, respectively (Fig. 3a). Here, the root node corresponds to the empty set. For efficient traversal, however, one needs a spanning tree, not a graph. When a tree is made by lexicographical ordering (Fig. 3b), the search tree is not *anti-monotonic* with respect to the density. Namely, the density is not monotonically decreasing when the tree is traversed from the root to a leaf. This property disallows early pruning and makes the enumeration difficult. However, there exists indeed a search tree which is anti-monotonic (Fig. 3c). It can be constructed by reverse search.

In reverse search, the search tree is specified by defining a *reduction map* $f(U)$ which transforms a child to its parent. In our case, the parent is created by removing the node with minimum degree from the child. Here, the degree of a node is defined as the sum of weights of all adjacent edges within U. If there are multiple

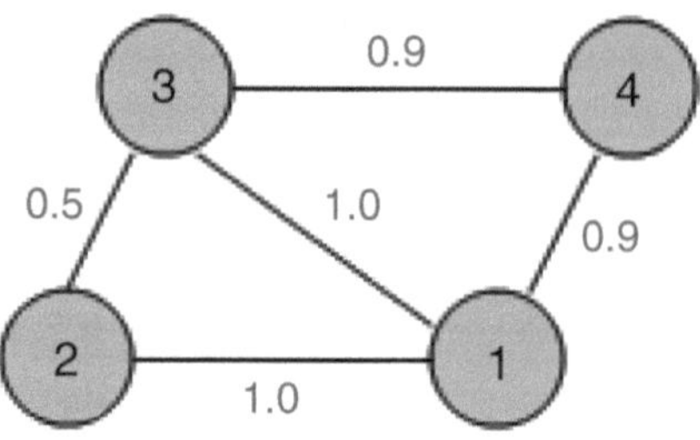

Fig. 2. An examplar graph for dense module enumeration.

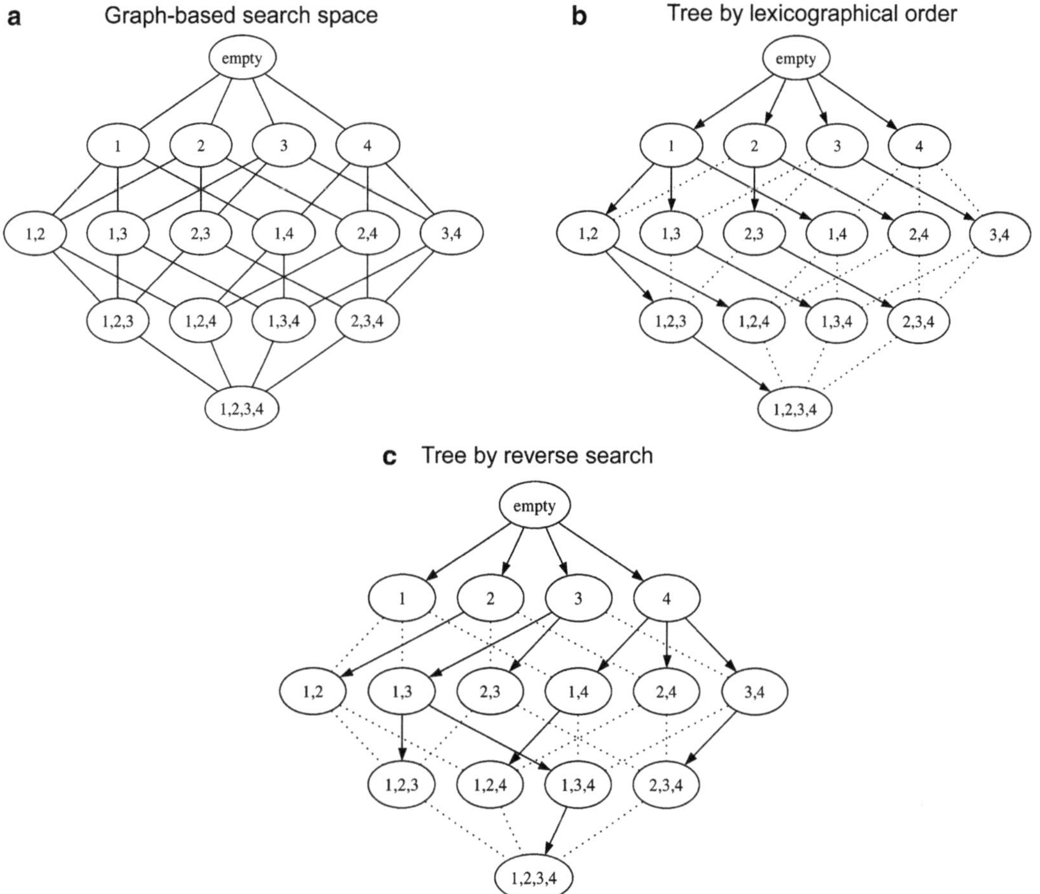

Fig. 3. Illustration of reverse search.

nodes with minimum degree, the one with the smallest index is removed. It is proven that the density of a parent is at least as high as the maximum density among the children, ensuring that the search tree induced by the reduction map is anti-monotonic.

In addition to the anti-monotonicity property, a valid reduction map must satisfy the following reachability condition (23): starting from any node of the search tree, we can reach a root node after applying the reduction map a finite number of times. This condition ensures that the induced search tree is indeed spanning. For the reduction map stated above, it is trivial to show that the reachability condition is satisfied, because any cluster shrinks to the empty set by removing nodes repeatedly.

To enumerate all clusters with density $\geq \theta$, one has to traverse this implicitly defined search tree in a depth-first or a breadth-first manner. During traversal, children are generated *on demand*. As the reduction map defines how to get from children to parents and not vice versa, we cannot directly derive the children from a given parent. Instead, to generate the children of a cluster U, we have to consider all candidates $U \cup \{i\},\ i \notin U$ and apply the reduction map to every candidate (reverse search principle). Qualified candidates with $f(U \cup \{i\}) = U$ are then taken as children. A naive implementation of this child generation process can make the algorithm very slow. Thus, it is important to engineer this process well. As the search tree is anti-monotonic, one can prune the tree whenever the density goes below θ.

The definition of a search tree is not an issue in the context of frequent pattern mining (24), because frequency is anti-monotonic in any tree. Reverse search is interesting because it provides a systematic way of defining an anti-monotonic tree. Notice, however, that it is not applicable to all score functions. Cluster density is an example where reverse search can be applied most effectively.

3.2. Integration of Additional Constraints

The DME framework makes it easy to incorporate and systematically exploit constraints from additional data sources. For illustration, consider the case where we have an additional data set which provides profiles of proteins or genes across different conditions (Fig. 1.1b). For simplicity, let us assume binary profiles being 1 if the protein is positively associated with the corresponding condition, and 0 otherwise. Then, dense modules where all member proteins share the same profile across a certain number of conditions are of particular interest; we call these modules *consistent*. The problem of DME with consistency constraints is formalized as follows.

Definition 1: Given a graph with node set V and weight matrix W, a density threshold $\theta > 0$, a profile matrix $(m_{ij})_{i \in V, j \in C}$, and nonnegative integers n_0 and n_1, find all modules $U \subset V$ with $\rho_W(U) \geq \theta$ s.t. there exist at least n_0 conditions $c \in C$ with $m_{uc} = 0, \quad \forall u \in U$ and there exist at least n_1 $c \in C$ with $m_{uc} = 1, \forall u \in U$.

Given such a *consistency constraint*, we can stop the module extension during the dense module mining as soon as the constraint is violated. This is due to the fact that the number of consistent profile conditions cannot increase while extending the module; more generally, this property is called *anti-monotonicity*. So we simply add to the module enumeration algorithm a condition which checks for the consistency requirements. These are then automatically taken into account in the check for local maximality. The use of additional constraints can restrict the search space considerably, so it accelerates the computation and helps to focus on biologically interesting solutions.

We have described a method for enumerating dense modules in a network. Methodological details and experimental results are available in (25). Our framework can be extended to module detection from multiple networks. see ref. 26 for detailed explanation.

4. Notes

1. If one starts from a low density threshold, our algorithm often takes too much time. One should start from very large threshold first, and gradually reduce the threshold to meet one's requirement.

References

1. Sharan R, Ulitsky I, Shamir R (2007) Network-based prediction of protein function. Mol Syst Biol 3:88
2. Ulitsky I, Shamir R (2007) Identification of functional modules using network topology and high-throughput data. BMC Syst Biol 1:8
3. Bader GD, Hogue CW (2003) An automated method for finding molecular complexes in large protein interaction networks. BMC Bioinformatics 4:2
4. Uno T (2007) An efficient algorithm for enumerating pseudo cliques. In: Proceedings of ISAAC 2007, pp. 402–414
5. Chen J, Yuan B (2006) Detecting functional modules in the yeast protein-protein interaction network. Bioinformatics 22(18):2283–2290
6. van Dongen S (2000) Graph clustering by flow simulation. PhD thesis, University of Utrecht
7. Newman ME (2006) Modularity and community structure in networks. Proc Natl Acad Sci USA 103(23):8577–8582
8. Everett L, Wang LS, Hannenhalli S (2006) Dense subgraph computation via stochastic search: application to detect transcriptional modules. Bioinformatics 22(14):e117–e123
9. Palla G, Derenyi I, Farkas I, Vicsek T (2005) Uncovering the overlapping community structure of complex networks in nature and society. Nature 435(7043):814–818
10. Spirin V, Mirny LA (2003) Protein complexes and functional modules in molecular networks. Proc Natl Acad Sci USA 100(21):12123–12128
11. Zeng Z, Wang J, Zhou L, Karypis G (2006) Coherent closed quasi-clique discovery from large dense graph databases. KDD '06: proceedings of the 12th ACM SIGKDD international conference on knowledge discovery and data mining. ACM, New York, pp 797–802
12. Hanisch D, Zien A, Zimmer R, Lengauer T (2002) Co-clustering of biological networks and gene expression data. Bioinformatics 18 (suppl 1):S145–S154
13. Tanay A, Sharan R, Kupiec M, Shamir R (2004) Revealing modularity and organization in the yeast molecular network by integrated analysis of highly heterogeneous genomewide data. Proc Natl Acad Sci USA 101(9):2981–2986
14. Segal E, Wang H, Koller D (2003) Discovering molecular pathways from protein interaction and gene expression data. Bioinformatics 19 (suppl 1):i264–i271
15. Pei J, Jiang D, Zhang A (2005) Mining cross-graph quasi-cliques in gene expression and protein interaction data. ICDE '05: proceedings of the 21st international conference on data engineering (ICDE'05). IEEE Computer Society, Washington, DC, pp 353–354
16. Ideker T, Ozier O, Schwikowski B, Siegel AF (2002) Discovering regulatory and signalling circuits in molecular interaction networks. Bioinformatics 18(suppl 1):S233–S240
17. Huang Y, Li H, Hu H, Yan X, Waterman MS, Huang H, Zhou XJ (2007) Systematic discovery of functional modules and context-specific functional annotation of human genome. Bioinformatics 23(13):i222–i229
18. Yan X, Mehan MR, Huang Y, Waterman MS, Yu PS, Zhou XJ (2007) A graph-based approach to systematically reconstruct human transcriptional regulatory modules. Bioinformatics 23(13):i577–i586
19. Hermjakob H, Montecchi-Palazzi L, Lewington C, Mudali S, Kerrien S, Orchard S, Vingron M, Roechert B, Roepstorff P, Valencia A, Margalit H, Armstrong J, Bairoch A, Cesareni G, Sherman D, Apweiler R (2004) IntAct: an open source molecular interaction database. Nucleic Acids Res 32(suppl 1): D452–D455
20. Chatr-aryamontri A, Ceol A, Palazzi LM, Nardelli G, Schneider MV, Castagnoli L, Cesareni G (2007) MINT: the Molecular INTeraction database. Nucleic Acids Res 35 (suppl 1):D572–D574
21. Bader GD, Betel D, Hogue CWV (2003) BIND: the Biomolecular Interaction Network Database. Nucleic Acids Res 31(1):248–250
22. Su AI, Wiltshire T, Batalov S, Lapp H, Ching KA, Block D, Zhang J, Soden R, Hayakawa M,

Kreiman G, Cooke MP, Walker JR, Hogenesch JB (2004) A gene atlas of the mouse and human protein-encoding transcriptomes. Proc Natl Acad Sci U S A 101(16):6062–6067

23. Avis D, Fukuda K (1996) Reverse search for enumeration. Discrete Appl Math 65:21–46
24. Han J, Kamber M (2006) Data mining: concepts and techniques of the Morgan Kaufmann series in data management systems, 2nd edn. Morgan Kaufmann Publishers, San Francisco
25. Georgii E, Dietmann S, Uno T, Pagel P, Tsuda K (2009) Enumeration of condition-dependent dense modules in protein interaction networks. Bioinformatics 25:933–940
26. Georgii E, Tsuda K, Schölkopf B (2011) Multi-way set enumeration in weight tensors. Mach Learn 82:123–155

Chapter 2

Discovering Interacting Domains and Motifs in Protein–Protein Interactions

Willy Hugo, Wing-Kin Sung, and See-Kiong Ng

Abstract

Many important biological processes, such as the signaling pathways, require protein–protein interactions (PPIs) that are designed for fast response to stimuli. These interactions are usually transient, easily formed, and disrupted, yet specific. Many of these transient interactions involve the binding of a protein domain to a short stretch (3–10) of amino acid residues, which can be characterized by a sequence pattern, i.e., *a short linear motif (SLiM)*. We call these interacting domains and motifs *domain–SLiM* interactions. Existing methods have focused on discovering SLiMs in the interacting proteins' sequence data. With the recent increase in protein structures, we have a new opportunity to detect SLiMs directly from the proteins' 3D structures instead of their linear sequences. In this chapter, we describe a computational method called SLiMDIet to directly detect SLiMs on domain interfaces extracted from 3D structures of PPIs. SLiMDIet comprises two steps: (1) interaction interfaces belonging to the same domain are extracted and grouped together using structural clustering and (2) the extracted interaction interfaces in each cluster are structurally aligned to extract the corresponding SLiM. Using SLiMDIet, de novo SLiMs interacting with protein domains can be computationally detected from structurally clustered domain–SLiM interactions for PFAM domains which have available 3D structures in the PDB database.

Key words: Short linear motifs, Protein structural mining, Domain–SLiM interactions, Protein–protein interactions

1. Introduction

Many protein–protein interactions (PPIs), such as those in signal transductions pathways, require fast response to stimuli. These interactions, also known as transient interactions, are designed to be easily formed and disrupted, and specific. While other PPIs are mediated by the binding of two large globular domain interfaces (domain–domain interactions), these transient interactions typically involve the binding of a protein domain to a short stretch (3–10) of amino acids (domain–motif interactions).

Hiroshi Mamitsuka et al. (eds.), *Data Mining for Systems Biology: Methods and Protocols*, Methods in Molecular Biology, vol. 939, DOI 10.1007/978-1-62703-107-3_2, © Springer Science+Business Media New York 2013

Many bioinformatics researchers have worked on discovering domain–domain interactions computationally. One of the earlier works is the InterDom database (1) created by detecting interacting domains from overabundant domain pairs in the protein sequence data of PPIs. With the increase in the availability of protein 3D structure data, researchers are able to detect domain–domain interactions directly from the PDB structural database (2); the databases in this line include iPFAM (3), 3DID (4), SCOPPI (5), and SNAPPI-DB (6).

Recently, researchers have found that in addition to domain-–domain interactions, protein domains can recognize a second type of interaction motifs on other proteins called *short linear motifs (SLiMs)* (7–12). The SLiMs are short and degenerate, typically only 3–20 residues long containing just a few conserved positions. The SLiMs are often found to mediate PPIs that are specific but, at the same time, easily formed and disrupted. This type of interaction is found in many key biological pathways such as the signal transduction. Because of their small sizes, SLiM-based PPIs are good targets for small-molecule drug therapy, for it is easier to design drugs to mimic the SLiMs (13) than the larger structural motifs like domains. The listing of all known SLiMs to date can be found in databases like ELM (14) and MiniMotif (MnM) (15).

Experimental methods to find SLiMs include site-directed mutagenesis and phage display. These are tedious and expensive methods to apply on whole interactomes. As such, bioinformatics researchers have developed a number of computational methods for predicting SLiMs from other biological data. The current methods can be broadly classified into two approaches. The first approach mines motifs from a given set of *related* protein sequences, with the relations being established by prior biological knowledge such as similarity in known biological functions, similar localization to a certain cell compartment, or sharing of interaction partners. Methods in this class include DILIMOT (16), SLiMDisc (17), and SLiMFinder (18). The second approach is to mine SLiMs that are overrepresented in the available PPI data. Methods in this class include D-STAR (19), MotifCluster (20), and SLIDER (21). There are several drawbacks with these two approaches. First, the motifs identified via these sequence-based approaches are not guaranteed to occur on the binding interface. Such atomic level of details can only come from high-resolution 3D structures (22). Second, because SLiMs are highly degenerative, most of these algorithms masked conserved structured regions (which are assumed not to have many SLiMs) such as globular domains to reduce false positives. However, it was found that such filtering has caused true motifs to be missed (18). Third, the algorithms are highly dependent on the accuracy of the interaction identification experiments, but high-throughput PPI data are well known to be noisy (23).

Just as the development of domain–domain interaction detection methods has progressed from sequence-based into structure-based approaches, the rapid increase of protein structure data in the PDB database also offers an excellent opportunity for detecting SLiMs directly from 3D structures instead of the proteins' sequences. The atomic level of details available in the high-resolution 3D structures are much richer than the linear protein sequences for discovering the weak signals of SLiMs, and the detected SLiMs are guaranteed to occur on the binding surfaces. In this chapter, we describe a method called SLiMDIet (24) to find SLiMs solely from 3D structure data. From the protein structure dataset downloaded from PDB on August 24, 2009, SLiMDIet was able to detect 452 distinct SLiMs on the domain interaction interfaces. One hundred and fifty-five of them were validated using the literatures, available structures, or statistical enrichment in the high-throughput PPI data. In addition, 198 SLiMs have been detected on domain–domain interaction interfaces (we call these *domain–domain SLiMs*), suggesting that the common belief that SLiMs occur outside the globular domain regions is not completely accurate, and that some of the apparent domain–domain interactions could in fact be mediated by domain–SLiM interactions.

2. Materials

1. *Protein 3D structure data.* The protein structure dataset can be downloaded from public databases such as the PDB. In the running example of this chapter, we used a protein 3D structure dataset downloaded from PDB on August 24, 2009, containing 57,559 structures. We chose structures with at least one protein chain and whose crystallographic resolution is 3.0 Å or better. We also included all NMR structures. In total, we have a working dataset of 54,981 structures with 130,488 protein chains.
2. *Protein domain annotations.* We compute PFAM domain annotations on each PDB chain using the *hmmpfam* program from the HMMER library version 2.3.2 (25) with the PFAM 23.0 library (26). We use PFAM as our choice of protein domain definition instead of the structurally defined SCOP (27) or CATH (28) because of the relatively better coverage of the former (see Note 1). However, PFAM domain does have its own limitation. It currently does not define structural domains that are formed by multiple protein chains. Nevertheless, SLiMDIet can also be applied on SCOP/CATH domain definitions if needed.
3. *PPIs.* To compute the statistical significance of the SLiMs detected by SLiMDIet, we compute their enrichment within a large nonhomologous PPI dataset which can be downloaded from online databases such as the BioGRID (29) (see Note 2).

3. Methods

SLiMDIet consists of two main steps: a DIet step (*D*omain *I*nterface *ext*raction and clustering step), followed by a SLiM step (SLiM extraction step):

1. The DIet step takes a set of protein structures from PDB as input, finds all known domains within the input structures, and extracts the domain interfaces associated with each of them. A domain interface comprises two sets of amino acid residues: one found within a protein domain (the set is called the *domain face*) while the other on a partner chain (the *partner face*) that are in close vicinity of each other. The interaction interfaces of each domain are then clustered based on their structural similarity.
2. In the SLiM step, we conduct an approximate structural multiple alignment to align the domain faces and the partner faces in each cluster. We then check if the alignment of the partner faces contains any conserved linear region (called a "block") of an appropriate length. If so, we construct a (linear) gapped position-specific scoring matrix (PSSM) from the block to represent the detected SLiM.

The details of each step are given as follows (see also Fig. 1).

3.1. DIet Step: Domain Interface Extraction and Clustering

3.1.1. Domain Interface Extraction

1. For each PDB structure, we use the HMMER program to identify the PFAM domains in its chains. Regions with overlapping domain annotations are resolved by choosing the PFAM domain with the best HMMER *E*-value.
2. For all possible domain interfaces, we retain those in which each amino acid on the domain face is within a distance threshold of 5 Å (as done in PSIMAP (30)) from some amino acid on the partner face and vice versa. We define the distance between two amino acid residues to be the nearest distance between any pair of non-hydrogen atoms in the two residues.
3. To curb possible nonbiological (crystal) interfaces, which generally have smaller interface area, we set a threshold of having domain interfaces involving a minimum of *eight amino acids on the domain face* and *four amino acids on the partner face*. This lower bound corresponds to a binding area larger than 800 Å^2—the average size of a domain interface as given in (5).
4. For intrachain domain interfaces, we also require that the residues on the partner face are at least *ten residues* from the ends of the domain (see Note 3).

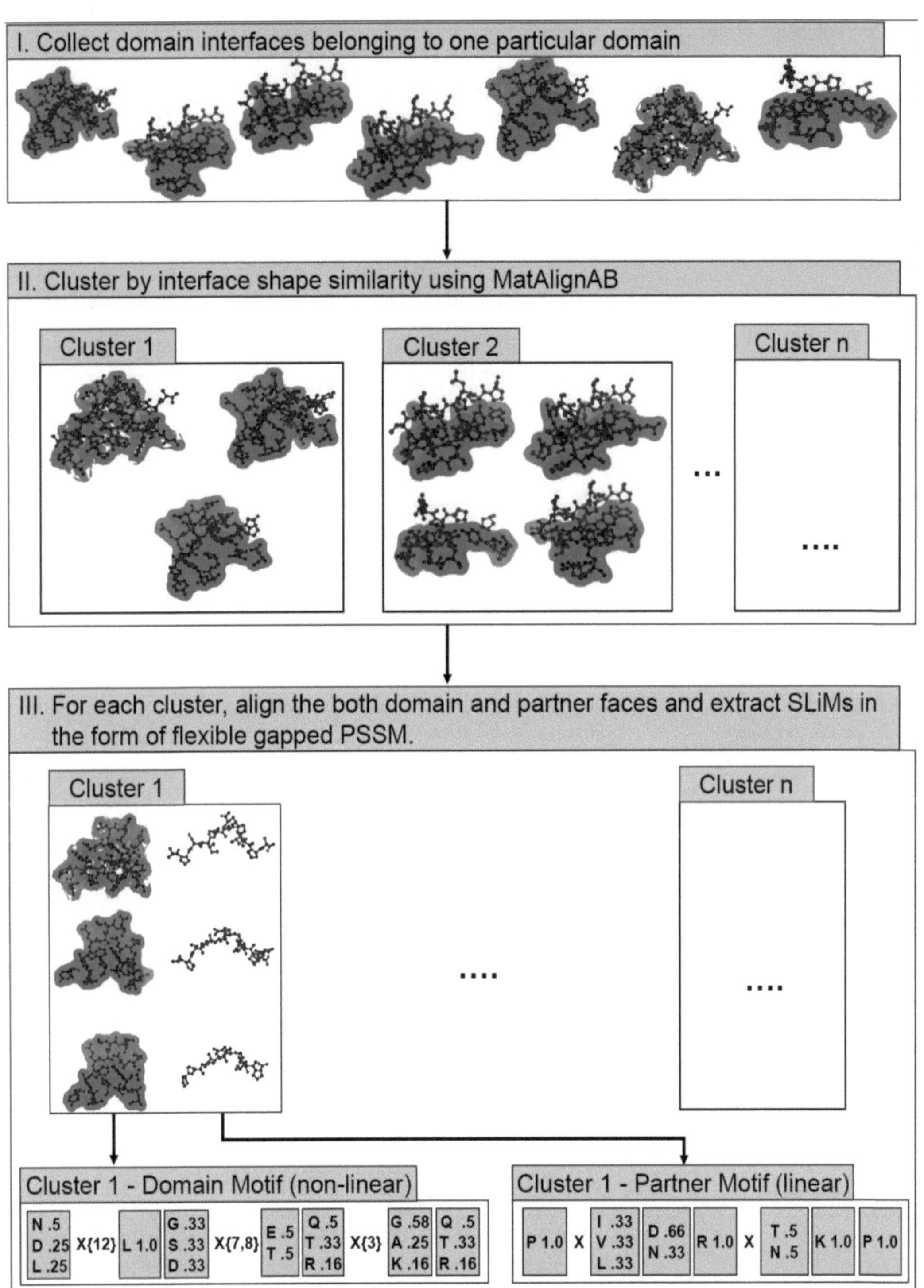

Fig. 1. SLiMDlet's overview. The domain interfaces of each PFAM domain are clustered by their structural similarity. Next, from each cluster, the domain and partner faces are structurally aligned and we build a gapped PSSM based on the contacts on the partner faces. The gapped PSSM has flexible gaps defined by the minimum and maximum gaps observed between two PSSM positions. We define a gapped PSSM as linear when the total length of its non-gap positions is 3–20 residues with gaps of at most four residues between any consecutive residue positions. To detect domain–SLiM interfaces, we collect domain interface clusters whose partner faces are covered by a linear gapped PSSM.

3.1.2. Pairwise Structural Alignment

1. Next, we compute the similarity scores and pairwise alignments among all pairs of domain interfaces of each PFAM domain in our dataset.
2. Alignment of two domain interfaces is done by treating each interface (both the domain and partner face) as one rigid body. Moreover, we enforce the alignment of the domain face residues on one interface against the domain face residues on the other, and do the same for the partner face residues (see Note 4).
3. We define the similarity of two interfaces using the modified *S*-score function by Alexandrov and Fischer (31) as follows: $S_{\text{norm}} = \frac{1}{(1+\Delta)} \frac{N}{\min(|A|,|B|)}$ where Δ is the root mean square distance (RMSD) between the two structures being aligned, N is the number of aligned residues between the two interfaces, and $|A|$ and $|B|$ are the sizes of the aligned interfaces, respectively. The similarity of two interfaces is measured using both the backbone and side chain conformation of the residues on each interface (see Note 5). To this end, we designed MatAlignAB for comparing domain interfaces' both Cα and Cβ atoms, based on the MatAlign algorithm (32).

3.1.3. Hierarchical Agglomerative Clustering of the Domain Interfaces

1. For every domain, we cluster its interfaces using a hierarchical agglomerative clustering algorithm using average linkage (see Note 6) as follows.
2. We start by setting every domain interface as a trivial cluster with one member.
3. Next, we pick the pair of clusters which has the highest similarity and combine them into a new cluster. We compute the average similarity of the newly combined cluster.
4. We repeat the above step until the similarity score between every possible pair of the clusters is below a certain threshold (for threshold setting, see Note 7).

3.2. SLiM Step: SLiM Extraction

3.2.1. Multiple Alignment of the Partner Faces in a Cluster

1. For each resulting domain interface cluster, we choose the interface with the least average distance to all other cluster member as the cluster center.
2. To generate an approximate multiple alignment of the partner faces, we align the partner faces from the interfaces in the cluster to the cluster center's partner face. We keep only the alignments that contain at least four nonhomologous partner faces. A face f_a is defined as homologous to f_b when (1) f_a's and f_b's aligned residues in the alignment are exactly the same and (2) their full protein chains share more than 50% sequence similarity.
3. We also make sure that each alignment column has at least 50% occupancy, i.e., the number of nonempty residues aligned in each column (see Note 8) must be at least half of the number of nonhomologous interfaces aligned.

3.2.2. SLiM Extraction from the Longest Linear Block

1. We identify the longest linear block in each of the above alignments of the nonhomologous faces (see also Fig. 2 for an example). A linear block is defined as a set of 3–12 consecutive alignment positions with gaps of at most four residues.
2. We also require that the block must cover at least half of the partner faces in the alignment. A block is said to cover a partner face f_a when it includes at least half of the contact residues in f_a.
3. From the longest linear block thus identified, we construct a gapped PSSM (i.e., a PSSM with flexible gaps) to represent the SLiM recognized in the particular domain interface cluster. The (flexible) gap in between each alignment column is computed by taking the minimum and maximum gap observed between two residue positions.
4. Given that our multiple alignment is derived from limited structural data, we do not directly score a residue with its observed frequency in the alignment. Instead, we define the score of a residue X on the alignment column i by

$$\text{GappedPSSM}(i, X) = \ln\left(\sum_{\text{AA}\in \text{Res}(i)} \text{freq}_i(\text{AA}) \cdot e^{\text{BLOSUM}(X,\text{AA})}\right),$$

where Res(i) is the set of amino acids seen in the column i of the alignment and freq$_i$(AA) is the frequency of residue AA in column i. Basically, the equation computes the weighted combination of the BLOSUM62 substitution score (33) of any residue X against the residues observed in the alignment—it extrapolates the feasibility of having other residues in that position based on the BLOSUM62 substitution matrix. An illustration of gapped PSSM construction can be seen in Fig. 3.

3.2.3. Computing the Statistical Significance of the SLiM Using PPI Data

1. For each SLiM extracted from an interface cluster, we also verify whether the motif occurs significantly more in the interaction partners of the domain as compared to random PPIs.
2. We define the gapped PSSM score of a particular position j in a given protein sequence S as the maximum sum of the gapped PSSM's residue scores starting at j over all possible gap value in the PSSM (see Note 9 for an example).
3. We define a position j in a protein with a gapped PSSM score s as an *occurrence* of the PSSM if the probability of scoring j with s or higher in a set of random protein sequence is at most 0.0001 (see Note 10).
4. Given a SLiM's gapped PSSM and a set of PPI data, the probability of observing a certain number of occurrences in the interaction partners of a protein domain by random can be

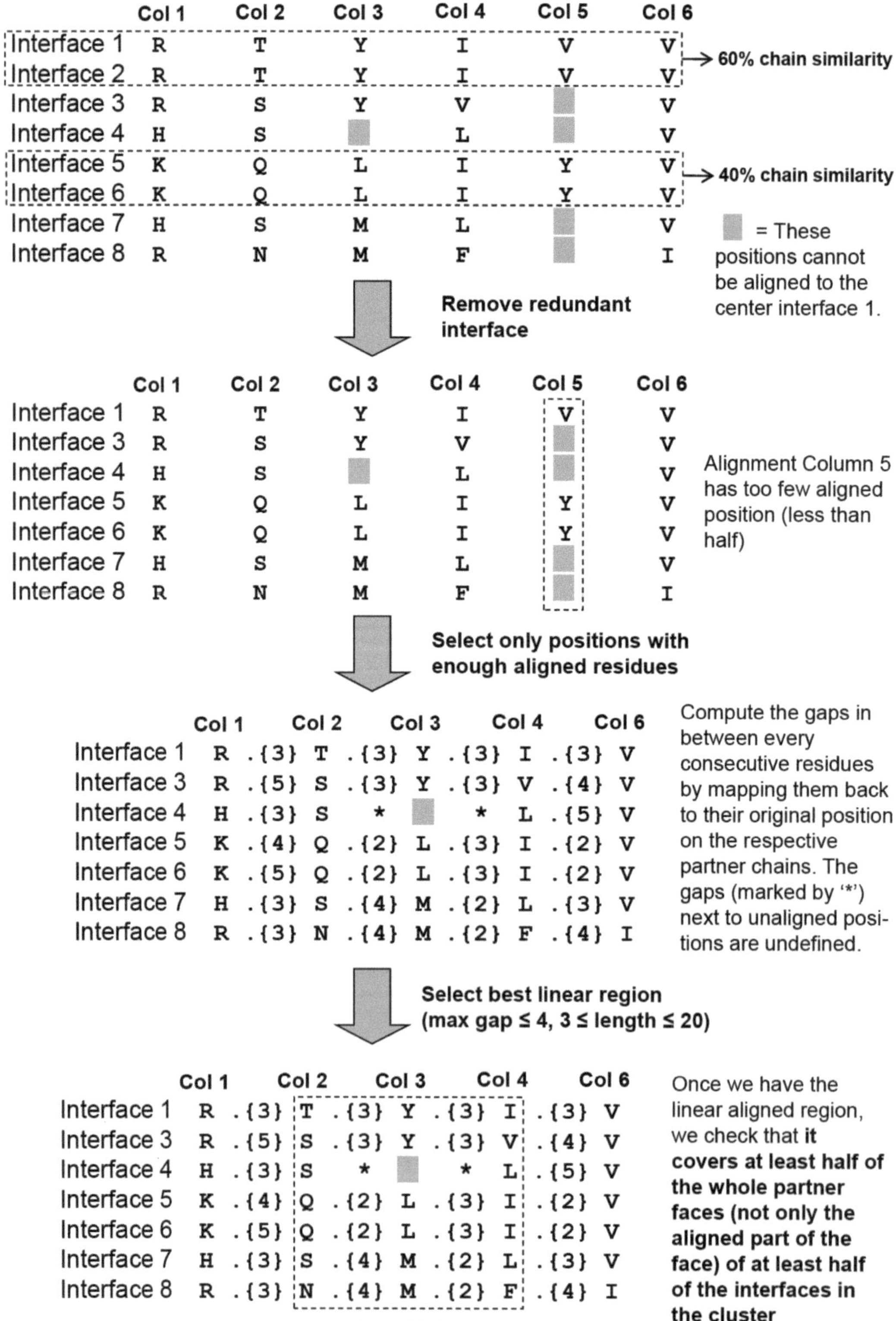

Fig. 2. Partner face alignment steps for finding the longest linear block. The latter is where we extract the SLiM from.

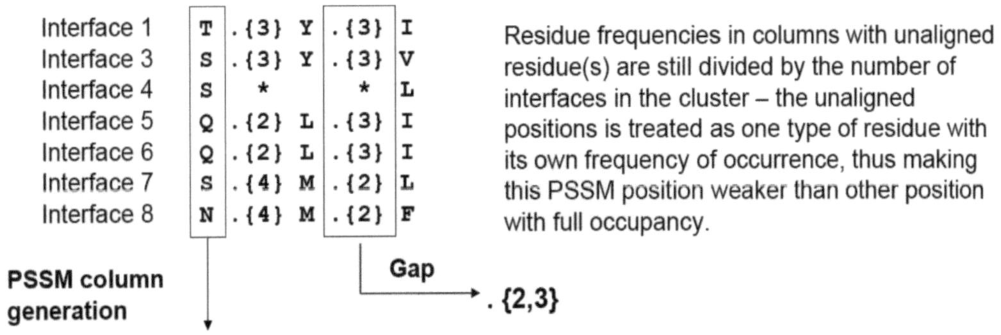

```
Interface 1   T .{3} Y .{3} I
Interface 3   S .{3} Y .{3} V
Interface 4   S   *      *  L
Interface 5   Q .{2} L .{3} I
Interface 6   Q .{2} L .{3} I
Interface 7   S .{4} M .{2} L
Interface 8   N .{4} M .{2} F
```

Residue frequencies in columns with unaligned residue(s) are still divided by the number of interfaces in the cluster – the unaligned positions is treated as one type of residue with its own frequency of occurrence, thus making this PSSM position weaker than other position with full occupancy.

PSSM column generation

Gap → **. {2,3}**

```
A = ln(3/7.e^BLOSUM62(S,A)+2/7.e^BLOSUM62(Q,A)+1/7.e^BLOSUM62(T,A)+1/7.e^BLOSUM62(N,A))
  = ln(3/7.e^1 + 2/7.e^-1 + 1/7.e^-1 + 1/7.e^-1)
  = 0.318

V = ln(3/7.e^BLOSUM62(S,V)+2/7.e^BLOSUM62(Q,V)+1/6.e^BLOSUM62(T,V)+1/7.e^BLOSUM62(N,V))
  = ln(3/7.e^-2 + 2/7.e^-2 + 1/7.e^-2+ 1/7.e^-3)
  = -2.094
..
..
(The same calculation is repeated for all 20 AA)
..
..
```

Fig. 3. An illustration of SLiMDIet's gapped PSSM generation from a linear block computed from the multiple interface alignment.

computed by the standard hypergeometric distribution function:

$$P\text{ - Value}(I, I_D, I_M, I_{DM}) = \frac{\binom{|I_M|}{|I_{DM}|}\binom{(|I|-|I_M|)}{(|I_D|-|I_{DM}|)}}{\binom{|I|}{|I_D|}},$$

where I is the whole set of the high-throughput PPI data, I_M is the subset of I which contains an occurrence of the gapped PSSM M, I_D is the subset of I containing the domain D, and I_{DM} is the subset of I_D which contains an instance of M in the domain D's interaction partners. SLiMs with P-value ≤ 0.05 are considered to be enriched in the PPI data and reported as detected SLiMs.

4. Notes

1. PFAM has higher PDB chain coverage on the current dataset [it covers 112,424 chains (86.16% coverage)] as compared to SCOP [version 1.75, dated June 2009, covering 87,064 chains (66.72% coverage)] and CATH [version 3.2.0, dated July 2008, covering 86,105 chains (65.99% coverage)].

2. We collected a set of 181,997 nonhomologous PPI data from the BioGRID interaction database version 2.0.58. We removed genetic (nonphysical) interactions (as defined by BioGRID) and those derived directly from structural data (to avoid self-discovery). Non-homology is enforced by keeping only one interaction among those whose both proteins are at least 70% homologous to another pair(s) of interacting proteins. The PPIs are collected as ordered tuples, i.e., for a given pair of interacting protein A and B, the tuple (A, B) is distinct from (B, A). From each tuple, we collect the domain face from the left element and the partner face from the right one.
3. The requirement is important in order to avoid recognizing local secondary structures at the end of a domain as interface contacts. A similar filtering is also used by Stein and Aloy (34) (published shortly after SLiMDIet).
4. We do not align the structures of entire domains as done in SCOPPI (5) and SNAPPI-DB (6), which greatly reduces computation time. By considering both domain and partner face as one rigid body, we also avoid the need of considering the relative orientation between the domain and partner face.
5. Usually, the RMSD between two proteins is approximated only by the RMSD of their backbone's Cα atoms. Since SLiMDIet's domain interfaces only consist of the contact residues (instead of the whole domain), the Cα representation is rather inadequate. We use the Cβ atom position as a first-order approximation of the side chain with respect to its backbone Cα (a similar Cβ approximation was mentioned in (35)).
6. The similarity of two clusters is the average pairwise similarity between all the members of the two clusters (as done in (5)).
7. We use the multiple thresholds, 0.15, 0.2, 0.25, and 0.3, to generate different sets of (possibly overlapping) domain interface clusters. Those clusters which originate from different thresholds but have more than 70% overlapping cluster member are grouped and SLiMDIet only reports the one with the most stringent cutoff as the representative of the group.
8. Some alignment column may have empty residues because the pairwise structural alignment may not align a residue from a particular partner face to the cluster center's residue when these residues' 3D positions are too different.
9. For example, the best score of position 0 in the string *F*SDTK based on the gapped PSSM $\begin{bmatrix} L: 4.62 \\ F: 1.38 \end{bmatrix} \cdot \{1,2\} \begin{bmatrix} T: 2.4 \\ D: -0.12 \end{bmatrix}$ would be $\max\begin{cases} 1.38 + (-0.12) & (\text{gap} = 1), \\ 1.38 + 2.4 & (\text{gap} = 2) \end{cases}.$

Note that this is a mini-version of a gapped PSSM for exemplary purpose; the real gapped PSSM would have entries for all 20 amino acids.

10. We created a set of 10,000 random protein sequences, each of length 500, whose amino acid distribution follows the distribution observed in our PPI data (BioGRID 2.0.58). For each gapped PSSM, we compute its scores on all protein positions in the random dataset (of approximately five million positions) and sort the scores in nonincreasing order. The 500th score on the sorted score list would have an empirical *P*-value of 0.0001 and is chosen as the cutoff score for the gapped PSSM's occurrence.

References

1. Ng SK et al (2003) InterDom: a database of putative interacting protein domains for validating predicted protein interactions and complexes. Nucleic Acids Res 31:251–254
2. Berman HM et al (2000) The Protein Data Bank. Nucleic Acids Res 28(1):235–242
3. Finn RD, Marshall M, Bateman A (2005) iPfam: visualization of protein–protein interactions in PDB at domain and amino acid resolutions. Bioinformatics 21(3):410–412
4. Stein A, Céol A, Aloy P (2011) 3did: identification and classification of domain-based interactions of known three-dimensional structure. Nucleic Acids Res 39:D718–D723
5. Kim WK et al (2006) The many faces of protein-protein interactions: a compendium of interface geometry. PLoS Comput Biol 2(9): e124
6. Jefferson ER et al (2007) SNAPPI-DB: a database and API of structures, iNterfaces and alignments for protein-protein interactions. Nucleic Acids Res 35(Database Issue): D580–D589
7. Pawson T, Scott JD (1997) Signaling through scaffold, anchoring, and adaptor proteins. Science 278(5346):2075–2080
8. Sudol M (1998) From Src homology domains to other signaling modules: proposal of the 'protein recognition code'. Oncogene 17:1469–1474
9. Neduva V, Russell RB (2005) Linear motifs: evolutionary interaction switches. FEBS Lett 579(15):3342–3345
10. Neduva V, Russell RB (2006) Peptides mediating interaction networks: new leads at last. Curr Opin Biotechnol 17(5):465–471
11. Diella F et al (2008) Understanding eukaryotic linear motifs and their role in cell signaling and regulation. Front Biosci 13:6580–6603
12. Fox-Erlich S, Schiller MR, Gryk MR (2009) Structural conservation of a short, functional, peptide-sequence motif. Front Biosci 14:1143–1151
13. Vagner J, Qu H, Hruby VJ (2008) Peptidomimetics, a synthetic tool of drug discovery. Curr Opin Chem Biol 12:1–5
14. Puntervoll P et al (2003) ELM server: a new resource for investigating short functional sites in modular eukaryotic proteins. Nucleic Acids Res 31(13):3625–3630
15. Rajasekaran S et al (2009) Minimotif miner 2nd release: a database and web system for motif search. Nucleic Acids Res 37(Database Issue):D185–D190
16. Neduva V et al (2005) Systematic discovery of new recognition peptides mediating protein interaction networks. PLoS Biol 3(12):e405
17. Davey NE, Shields DC, Edwards RJ (2006) SLiMDisc: short, linear motif discovery, correcting for common evolutionary descent. Nucleic Acids Res 34(12):3546–3554
18. Edwards RJ, Davey NE, Shields DC (2007) SLiMFinder: a probabilistic method for identifying over-represented, convergently evolved, short linear motifs in proteins. PLoS One 2(10): e967
19. Tan SH et al (2006) A correlated motif approach for finding short linear motifs from protein interaction networks. BMC Bioinformatics 7:502
20. Leung HC et al (2009) Clustering-based approach for predicting motif pairs from protein interaction data. J Bioinform Comput Biol 7(4):701–716
21. Boyen P et al (2009) SLIDER: mining correlated motifs in protein-protein interaction networks. In: Proceedings of the 2009 ninth IEEE

international conference on data mining (ICDM) Miami, FL, USA, on December 6–9, pp. 716–721

22. Aloy P, Russell RB (2006) Structural systems biology: modelling protein interactions. Nat Rev Mol Cell Biol 7:188–197
23. von Mering C et al (2002) Comparative assessment of large-scale data sets of protein-protein interactions. Nature 417(6887):399–403
24. Hugo W et al (2010) SLiM on Diet: finding short linear motifs on domain interaction interfaces in Protein Data Bank. Bioinformatics 26 (8):1036–1042
25. Eddy SR (1998) Profile hidden Markov models. Bioinformatics 14:755–763
26. Finn RD et al (2008) The Pfam protein families database. Nucleic Acids Res 36(Database Issue):D281–D288
27. Andreeva A et al (2008) Data growth and its impact on the SCOP database: new developments. Nucleic Acids Res 36(Database Issue):419–425
28. Cuff AL et al (2009) The CATH classification revisited–architectures reviewed and new ways to characterize structural divergence in superfamilies. Nucleic Acids Res 37(Database Issue): D310–D314
29. Stark C et al (2011) The BioGRID Interaction Database: 2011 update. Nucleic Acids Res 39 (Database Issue):D698–D704
30. Dafas P et al (2004) Using convex hulls to extract interaction interfaces from known structures. Bioinformatics 20(10):1486–1490
31. Alexandrov NN, Fischer D (1996) Analysis of topological and nontopological structural similarities in the PDB: new examples with old structures. Proteins 25(3):354–365
32. Aung Z, Tan K (2006) MatAlign: precise protein structure comparison by matrix alignment. J Bioinform Comput Biol 4 (6):1197–1216
33. Henikoff S, Henikoff JG (2005) Amino acid substitution matrices from protein blocks. Proc Natl Acad Sci USA 89 (22):10915–10919
34. Stein A, Aloy P (2010) Novel peptide-mediated interactions derived from high-resolution 3-dimensional structures. PLoS Comput Biol 6(5):e1000789
35. Torrance JW et al (2005) Using a library of structural templates to recognise catalytic sites and explore their evolution in homologous families. J Mol Biol 347 (3):565–581

Chapter 3

Global Alignment of Protein–Protein Interaction Networks

Misael Mongiovì and Roded Sharan

Abstract

Sequence-based comparisons have been the workhorse of bioinformatics for the past four decades, furthering our understanding of gene function and evolution. Over the last decade, a plethora of technologies have matured for measuring Protein–protein interactions (PPIs) at large scale, yielding comprehensive PPI networks for over ten species. In this chapter, we review methods for harnessing PPI networks to improve the detection of orthologous proteins across species. In particular, we focus on pairwise global network alignment methods that aim to find a mapping between the networks of two species that maximizes the sequence and interaction similarities between matched nodes. We further suggest a novel evolutionary-based global alignment algorithm. We then compare the different methods on a yeast-fly-worm benchmark, discuss their performance differences, and conclude with open directions for future research.

Key words: Network alignment, Protein–protein interaction, Functional orthology, Network evolution

1. Introduction

Over the last decade, high-throughput techniques such as yeast two-hybrid assays (1) and co-immunoprecipitation experiments (2), have allowed the construction of large-scale networks of Protein–protein interactions (PPIs) for multiple species. Comparative analyses of these networks have greatly enhanced our understanding of protein function and evolution.

Analogously to the sequence comparison domain, two main concepts have been introduced in the network comparison context: *local* network alignment and *global* network alignment. The first considers local regions of the network, aiming to identify small subnetworks that are conserved across two or more species (where conservation is measured in terms of both sequence and interaction patterns). Local alignment algorithms have been utilized to detect

Hiroshi Mamitsuka et al. (eds.), *Data Mining for Systems Biology: Methods and Protocols*, Methods in Molecular Biology, vol. 939, DOI 10.1007/978-1-62703-107-3_3, © Springer Science+Business Media New York 2013

protein pathways (3) and complexes that are conserved across multiple species (4–6), to predict protein function, and to infer novel PPIs (4).

In global network alignment (GNA), the goal is to associate proteins from two or more species in a global manner so as to maximize the rate of sequence and interaction conservation across the aligned networks. In its simplest form, the problem calls for identifying a 1-1 mapping between the proteins of two species so as to optimize some conservation criterion. Extensions of the problem consider multiple networks and many-to-many (rather than 1-1) mappings. Such analyses assist in identifying (functional) orthologous proteins and orthology families (7) with applications to predicting protein function and interaction. They aim to improve upon sequence-only methods that partition proteins into orthologous groups based on sequence-similarity computations (8–10).

GNA methods can be classified into two main categories. The first category contains matching methods that explicitly search for a one-to-one mapping that maximizes a suitable scoring function. The scoring function favors mappings that conserve sequence and interaction. Methods in this category include the integer linear programming (ILP) method of (11) and a greedy gradient ascent method of (12). The second category includes ranking methods that consider all possible pairs of interspecies proteins that are sufficiently sequence-similar, and rank them according to their sequence and topological similarity. These ranks are then used to derive a 1-1 mapping. Methods in this category include a Markov random field (MRF) approach (13), the IsoRank method that is based on Google's Page Rank (7), and a diffusion-based method—hybrid RankProp (14). In addition, there are several very recent ranking approaches that do not use sequence-similarity information at all (15, 16).

Here, we aim to propose a third, evolutionary perspective on global alignment by designing a GNA algorithm that is based on a probabilistic model of network evolution. The evolution of a network is described in terms of four basic events: gene duplication, gene loss, edge attachment, and edge detachment. This model allows the computation of the probability of observing extant networks given the ancestral network they originated from; by maximizing this probability, one obtains the most likely ancestor–descendant relations, which naturally translate into a network alignment.

This chapter is organized as follows: Subheading 3 reviews GNA methods that are based on graph matching. Subheading 4 presents the ranking-based methods. Subheading 5 describes in detail the probabilistic model of evolution and the proposed alignment method. The different approaches are compared in Subheading 6. Finally, Subheading 7 gives a brief summary and discusses future research directions.

2. Preliminaries and Problem Definition

We focus the presentation on methods for pairwise global alignment, where the input consists of two networks and possibly sequence-similarity information between their nodes, and the output is a correspondence, commonly one-to-one, between the nodes of the two networks.

A protein network $G=(V, E)$ has a set V of nodes, corresponding to proteins, and a set E of edges, corresponding to PPIs. For a node $i \in V$, we denote its set of (direct) neighbors by $N(i)$. Let $G_1 = (V_1, E_1)$ and $G_2 = (V_2, E_2)$ be the two networks to be aligned. Let $R \subseteq V_1 \times V_2$ be a compatibility relation between proteins of the two networks, representing pairs of proteins that are sufficiently sequence-similar. A many-to-many correspondence that is *consistent* with R is any subset $R^* \subseteq R$. Under such a correspondence, we say that an edge (u, v) in one of the networks is *conserved* if there exists an edge (u', v') in the other network such that $(u, u'), (v, v') \in R^*$ or $(u', u), (v', v) \in R^*$. We let $T(G_1, G_2) = \{(u, u', v, v'): (u, v), (u', v') \in R, (u, u') \in E_1, (v, v') \in E_2\}$ denote the set of all quadruples of nodes that induce a conserved interaction.

In its simplest formulation, the alignment problem is defined as the problem of finding an injective function (one-to-one mapping) $\varphi: V_1 \rightarrow V_2$ such that (i) it is consistent with R and (ii) it maximizes the number of conserved interactions. More elaborate formulations of the problem can relax the 1-1 mapping to a many-to-many mapping and possibly define an alignment score to be optimized that combines the amount of interaction conservation and the sequence similarity of the matched nodes. The definition of a conserved interaction can also be made more elaborate by taking into account the reliability of the pertaining interactions and by allowing "gapped" interactions, i.e., a directed interaction in one network is matched to two nodes that are of distance 2 in the other network. We defer the discussion of these extensions and the specific scoring functions used to the next sections, where the different GNA approaches are described.

The problem of finding the optimal one-to-one alignment between two networks, as defined above, can be shown to be NP-hard by reduction from maximum common subgraph (11). Consequently, an efficient algorithm cannot be designed for the general case. However, under certain relaxations the problem can be solved optimally on current data sets in acceptable time.

3. Graph Matching Methods

In this section, we describe GNA methods that look for an explicit 1-1 correspondence between the two compared networks. The first method, by Klau, is based on reformulating the alignment problem as an ILP (11). The variables of the program represent the 1-1 mapping sought. Specifically, for each pair $(u, v) \in R$, the author defines a binary variable x_{uv} denoting whether u and v are matched ($x_{u,v} = 1$) in the alignment or not ($x_{u,v} = 0$). The ILP formulation is as follows:

$$\max \sum_{(u,u',\upsilon,\upsilon')\in T(G_1,G_2)} x_{u,\upsilon} \cdot x_{u',\upsilon'} + \sum_{(u,\upsilon)\in R} \sigma(u,v) \cdot x_{u,\upsilon}$$

s.t.

$$\sum_{u\in V_1} x_{u,\upsilon} \leq 1 \quad \forall \upsilon \in V_2$$

$$\sum_{\upsilon\in V_2} x_{u,\upsilon} \leq 1 \quad \forall u \in V_1,$$

where $\sigma(u, v)$ denotes the sequence similarity of u and v. The objective function can be linearized in an obvious way by introducing binary variables $t_{u,u',\upsilon,\upsilon'} = x_{u,\upsilon} \cdot x_{u',\upsilon'}$ (for $(u, u', \upsilon, \upsilon') \in T(G_1, G_2)$) with appropriate constraints.

While the author uses optimization techniques, such as Lagrangian decomposition and Lagrangian relaxation, to solve this problem, an optimum solution for restricted instances can be found in reasonable time as we report in Subheading 6. We note that if $V_1 \cap V_2$ is first partitioned into sufficiently small orthology clusters (using, e.g., the Inparanoid algorithm (8)) and if the graph of potential conserved interactions across clusters has no loops, then the optimum alignment can be found in polynomial time via a dynamic programming algorithm (12).

In the general case, the computation of optimal solutions is too costly, hence the use of heuristics is necessary. Vert et al. (12) suggested a gradient ascent approach. It starts from a feasible solution and computes a sequence of moves in the direction of the objective's gradient until converging to a local maximum. Denoting the adjacency matrices of the two graphs by A_1 and A_2, respectively, and assuming that $|V_1| = |V_2| = n$ (otherwise, add dummy vertices), the goal of the optimization is to find a permutation matrix P that maximizes a weighted sum of the number $J(P)$ of conserved interactions and a sequence similarity term $S(P)$. In matrix notation, $J(P) = \frac{1}{2} tr(A_1^T P A_2 P^T)$ and its gradient is $A_1^T P A_2$; $SP = tr(PC)$ where C, the matrix of sequence-similarity scores, is its gradient.

The initial solution P_0 is given by sequence similarity alone, using a maximum matching algorithm. At each step, the algorithm employs a maximum matching computation to update the current permutation in the direction of the gradient:

$$P_{n+1} = \arg\max_P tr([\lambda A_1^T P_n A_2 + (1 - \lambda)C]P),$$

where $0 \leq \lambda \leq 1$ is a weighting constant.

4. Methods Based on Ranking

A second class of methods is based on assigning a score to each pair of compatible nodes and only at a second step choosing a global pairing of the nodes. The latter pairing is effectively ***disambiguating*** the compatibility relations, pinpointing the "best" 1-1 mapping. The disambiguation can be achieved by computing a maximum weighted bipartite matching or via simple greedy strategies. The difference between the various methods lies mainly in the first, scoring phase.

The first method for GNA has been proposed by Bandyopadhyay et al. (13) and uses a ranking that is based on a MRF model. It starts by building an alignment graph, where the nodes represent candidate pairs of (sequence-similar) proteins and the edges represent potentially conserved interactions. Each node in the alignment graph is associated with a binary state z indicating if that node represents a true orthology relation or not. The state values are modeled using a MRF. The MRF model assumes that for each node of the alignment graph $j = (u, v)$, the probability that j represents a true pair of orthologs ($z_j = 1$) depends only on the states of its neighbors ($N(j)$), and the dependence is through a logistic function:

$$P(z_j|z_{N(j)}) = \frac{1}{1 + e^{-\alpha-\beta \cdot c(j)}},$$

where α and β are parameters and $c(j)$ is the conservation index of j, defined as twice the number of conserved interactions between j and neighbors of j whose states are pre-assigned with value 1 (true orthologs), divided by the total number of interactions involving u and v across the two species. The inference of the states of the nodes is conducted using Gibbs sampling (17), yielding orthology probabilities for every node. These estimated probabilities are used to disambiguate the pairing.

Singh et al. (7) proposed an alignment method (IsoRank) that is based on Google's PageRank algorithm. As for MRF, the method first computes a score for each candidate pair of orthologs and then uses the scores for disambiguating the pairing. The score $R_{(i,j)}$ of the pair $(i, j) \in V_1 \times V_2$ is a weighted average of the scores of its neighboring pairs (assuming that all node pairings are allowed):

$$R_{(i,j)} = \sum_{u \in N(i)} \sum_{v \in N(j)} \frac{R_{(u,v)}}{|N(u)||N(v)|}.$$

The authors translate the problem of finding R into an eigenvector problem by expressing it in matrix form as $R = AR$ where A is defined as:

$$A_{(i,j)(u,v)} = \begin{cases} \frac{1}{|N(u)||N(v)|} & \text{if } (i,u) \in E_1, (j,v) \in E_2 \\ 0 & \text{otherwise.} \end{cases}$$

Under this formulation, the problem reduces to finding the dominant eigenvector of A, which is efficiently solved using the power method. To account for sequence similarity, the objective is modified as $R = [\alpha A + (1 - \alpha)B1^T]R$ where B is the vector of normalized bit scores and 1^T is an all-1 row vector.

Yosef et al. (14) devised the hybrid RankProp algorithm. It considers one "query node" of the first network at a time and ranks the nodes of the second network with respect to it by using a diffusion procedure. To this end, they constructed a composite network with two types of edges: PPI and sequence similarity. The query node is assigned a score of 1.0 that is continually pumped to the other nodes through the network's edges. The scores that the nodes assume after the diffusion process converges induce a ranked list of candidates for matching the query node. In detail, at step $t + 1$, the score of a node i with respect to a query q is given by:

$$S_i(t+1) = W_{qi} + \alpha \sum_{j \in N(i) \setminus \{q\}} W_{ji} S_j(t),$$

where α is a parameter controlling the diffusion rate and W is a weight matrix that represents the composite network—it is the normalized confidence of an interaction for PPI edges and a normalized sequence similarity for sequence-similarity edges. Finally, to make the score symmetric, proteins from both networks are queried and each pair is assigned the average score of its two associated queries.

5. Network Evolution-Based Alignment

In this section, we present a new alignment method, called PME, that is based on a probabilistic model of evolution. PME aims to reconstruct the most probable ancestral network that gave rise to the observed extant networks. Such a network induces a many-to-many alignment in the descending networks by associating groups of proteins in the two input networks with the corresponding ancestral proteins. The method is based on a probabilistic model of the evolutionary dynamics of a network, that supports four kinds of evolutionary events: link attachment, link detachment, gene duplication and gene loss (18).

An alignment between two networks G_1 and G_2 is defined by an ancestral network $G_0 = (V_0, E_0)$ and two functions $f_1 : V_1 \rightarrow V_0$ and

$f_2 : V_2 \rightarrow V_0$ which map the nodes of G_1 and G_2 into the nodes of G_0 (ancestral proteins). The score of an alignment $A = (G_0, f_1, f_2)$ is the product of the prior probability for A and the likelihood of observing G_1 and G_2 given A. We describe the probability computations in detail below.

The probability $P(A)$ is the product of two terms that consider the prior probability of observing G_0 and the probability of the pattern of gene duplications and losses implied by f_1 and f_2. For the former, we adopt a simple Erdős–Rényi model where edges occur independently with some constant probability P_E. For the latter, we focus on gene duplications (as in (18)), assuming that gene duplication events occur independently with some fixed probability P_d. For computational efficiency, we disallow gene losses, although those could be easily incorporated to the model in a similar manner. Formally, the two terms are as follows:

- A priori ancestral network probability:

$$\prod_{(u,v)\notin E_0} (1 - P_E) \cdot \prod_{(u,v)\in E_0} P_E.$$

- Gene duplication ($i \in \{1, 2\}$):

$$\prod_{\substack{v\in V_0 \\ f_i^{-1}(v)\neq\phi}} P_d^{|f_i^{-1}(v)|-1} \cdot \prod_{\substack{v\in V_0 \\ |f_i^{-1}(v)|\leq 1}} (1 - P_d).$$

The probability $P(G_i|A)$ of observing the network G_i, $i \in \{1, 2\}$ is given by the product of two factors that consider edge attachment and edge detachment events, assuming these events occur independently with probabilities P_A and P_D, respectively.

- Edge attachment:

$$\prod_{(u,v)\notin E_0} \left(\prod_{\substack{(u',v')\notin E_i \\ f_i(u')=u, f_i(v')=v}} (1 - P_A) \cdot \prod_{\substack{(u',v')\in E_i \\ f_i(u')=u, f_i(v')=v}} P_A \right).$$

- Edge detachment:

$$\prod_{(u,v)\in E_0} \left(\prod_{\substack{(u',v')\in E_i \\ f_i(u')=u, f_i(v')=v}} (1 - P_D) \cdot \prod_{\substack{(u',v')\notin E_i \\ f_i(u')=u, f_i(v')=v}} P_D \right).$$

Our goal is to find an alignment that maximizes $P(G_1, G_2, A) = P(A) \cdot P(G_1|A) \cdot P(G_2|A)$. In the following, we provide an ILP formulation of the problem. Consider a set of n hypothetical nodes of the ancestral network, where $n = |V_1| + |V_2|$ is the maximal number of nodes in the ancestral network. With each node, we associate a binary variable z_i which is 1 if and only if node i has some descendant node in the extant networks. With each vertex pair (i, j), we associate a binary variable t_{ij} which is 1 if and only if nodes i and j interact with each other in the ancestral network. To model the mappings f_1 and f_2, we define binary variables x_{iu} and y_{iv}, where $x_{iu} = 1$ ($y_{iv} = 1$) if and only if $f_1(u) = i$ ($f_2(v) = i$). Finally, in order to consider gene duplications, we add binary variables d_i^j, $j \in \{1, 2\}$ such that $d_i^j = 0$ if and only if i has more than one descendant in G_j.

5.1. The ILP Formulation

The constraints of the ILP are defined as follows:

$$t_{ij} \leq z_i, z_j, \quad 1 \leq i < j \leq n$$

to allow edges only between "true" vertices of the ancestral network.

$$\sum_{i=1}^{n} x_{iu} = 1, \quad u \in V_1,$$
$$\sum_{i=1}^{n} y_{iv} = 1, \quad \in V_2$$

to model the fact that each protein descends from a single ancestor.

$$\sum_{u \in V_1} x_{iu} \geq z_i, \quad 1 \leq i \leq n,$$
$$\sum_{v \in V_2} y_{iv} \geq z_i, \quad 1 \leq i \leq n,$$
$$x_{iu} \leq z_i, \quad 1 \leq i \leq n, \ u \in V_1,$$
$$y_{iu} \leq z_i, \quad 1 \leq i, \leq n, \ u \in V_2$$

to model the fact that each true node of the ancestral network ($z_i = 1$) must have at least one descendant in each network and each dummy node of the ancestral network ($z_i = 0$) cannot have any descendants.

$$d_i^1 \leq 1 + z_i - x_{iu} - x_{iv}, \quad 1 \leq i \leq n, \ u, v \in V_1,$$
$$d_i^1 \geq 1 + z_i - \sum_{u \in V_1} x_{iu}, \quad 1 \leq i \leq n,$$
$$d_i^2 \leq 1 + z_i - y_{iu} - y_{iv}, \quad 1 \leq i \leq n, \ u, v \in V_2,$$
$$d_i^2 \geq 1 + z_i - \sum_{u \in V_2} y_{iu}, \quad 1 \leq i \leq n$$

to impose that nodes that have only one descendant have not undergone a duplication event. Finally, we add the integer constraints:

$$x_{iu}, y_{iv}, z_i, t_{ij}, d_i^1, d_i^2 \in \{0, 1\} \ 1 \leq i, j \leq n, u \in V_1, v \in V_2.$$

The objective is to maximize $P(G_1, G_2, A)$ or, equivalently, to maximize $\log P(G_1, G_2, A)$. The latter is a sum of four terms:

- A priori ancestral network probability:

$$\varphi_E = \sum_{i<j} \left(\log(P_E) \cdot t_{ij} + \log(1 - P_E) \cdot (1 - t_{ij})\right).$$

- Gene duplication (for simplicity, we specify only the sub-term involving G_1):

$$\varphi_d = \sum_{i=1}^{n} \left(\sum_{u \in V_1} x_{iu} - z_i \right) \cdot \log(P_d) + \sum_{i=1}^{n} \log(1 - P_d) \cdot d_i^1.$$

- Edge attachment (for simplicity, we specify only the sub-term involving G_1):

$$\varphi_A = \sum_{i<j} (1 - t_{ij}) \cdot \left(\sum_{(u,v) \notin E_1} x_{iu} \cdot x_{jv} \cdot \log(1 - P_A) + \sum_{(u,v) \in E_1} x_{iu} \cdot x_{jv} \cdot \log(P_A) \right).$$

- Edge detachment (for simplicity, we specify only the sub-term involving G_1):

$$\varphi_D = \sum_{i<j} t_{ij} \cdot \left(\sum_{(u,v) \in E_1} x_{iu} \cdot x_{jv} \cdot \log(1 - P_D) + \sum_{(u,v) \notin E_1} x_{iu} \cdot x_{jv} \cdot \log(P_D) \right).$$

In order to make the problem linear, we introduce the following additional binary variables with appropriate constraints: $p_{ijuv} = t_{ij} \cdot x_{iu} \cdot x_{jv}$ and $q_{ijuv} = (1 - t_{ij}) \cdot x_{iu} \cdot x_{jv}$.

5.2. Refinements and Variable Reduction

In some cases, there are not enough interactions to support a match. To avoid an arbitrary choice among identically scored solutions, we choose the solution that agrees best with the sequence-similarity information. To this end, we add a small penalty to each ancestral-descendant connection whose value is $10^{-8} \cdot \log(S + 1)$, where S is the bit score of the two proteins.

Although PME naturally produces a many-to-many correspondence between orthologous proteins, we focus here on its reduction to a one-to-one mapping to facilitate its comparison to other methods. To this end, we rank all pairs of inter-species proteins that are predicted to descend from the same common ancestor. For any potentially matched pair (u, v), with $f(u) = f(v) = i$, the score of (u, v) is given by the score of the global alignment after removing all the nodes that descend from i except for u and v (i.e., forcing the alignment to match u and v). These scores are then fed to a maximum bipartite matching computation to construct a 1-1 alignment.

The sequence-similarity information allows us to greatly reduce the number of variables considered. We start with a set $V = V_1 \cap V_2$ of hypothetical ancestral nodes. We build two relations $R_1 \subseteq V \times V_1$ and $R_2 \subseteq V \times V_2$ as follows: For each $i \in V$, we add to R_1 all pairs (i, u) with $u \in V_1$ such as u is sequence-similar to i and $u \leq i$. Analogously, we add to R_2 all pairs (i, v) with $v \in V_2$ such that u is sequence-similar to i and $u \leq i$. The search is then restricted to alignments whose ancestor–descendant pairs are in $R_1 \cap R_2$.

The relations R_1 and R_2 also allow us to reduce the number of possible edges of the ancestral network. Consider a pair of nodes (u, v) of the ancestral network such that all possible pairs of descendants of these nodes span non-edges. Clearly, in the optimal solution, (u, v) will be a non-edge. Since the networks are usually very sparse, this simple rule greatly reduces the number of variables required to model the topology of the ancestral network and, consequently, greatly saves in variables introduced by the linearization. Although non-edges contribute to the objective function, we can modify the latter so that the contribution of non-edges is zero (by adding $-\log(1 - P_E)$ to all ancestral vertex pairs). In a similar manner, we can reduce the number of ancestor–descendant pairs that are considered in the computation of edge attachment events.

6. Experimental Results

To compare the different GNA methods, we used the benchmark in (13), which focuses on the pairwise global alignment of the PPI networks of yeast and fly, starting from an initial clustering of the proteins into orthology families formed by the Inparanoid algorithm (8). In addition, we compared, under the same setting, the alignments of each of these networks to a PPI network of worm. The worm network was constructed by collecting data from recently published papers and public databases (19–21) and spanned 2,967 proteins and 4,852 interactions. The yeast network contained 4,393 proteins and 14,318 interactions; the fly network contained 7,042 proteins and 20,719 interactions. We considered 2,244 Inparanoid groups between yeast and fly, 1,833 groups between yeast and worm, and 4,228 groups between worm and fly.

We included in the comparison the following methods: ILP (11), MRF (13), IsoRank (7) and PME (Subheading 5). We did not consider gradient ascent (12) and hybrid RankProp (14) in our tests. Gradient ascent tries to approximate the same objective as the ILP method, hence the latter should be superior to it. Hybrid RankProp was shown by its authors to be equivalent in performance to the original RankProp method, which is based on sequence only.

We implemented ILP, IsoRank, and PME in Matlab and used ILOG CPLEX as an ILP solver. For MRF, we report on the results published in the original paper (13). The parameter that balance topology versus sequence similarity was set as $c = 0.01$ for both IsoRank and ILP in order to give higher weight to topology. For PME, we used the following settings: The probability of attachment and detachment was set so as to obtain the same global rate of attachments and detachments estimated from the unambiguous clusters of Inparanoid ($P_A = 0.0026$; $P_D = 0.9617$). The probability of an edge in the ancestral network was estimated from the density of the two networks ($P_E = 3.32e^{-4}$). The probability of duplication was set to $P_d = 0.03$ with the results being robust to a wide range of values for this parameter (in the range 10^{-4}–0.5). All the experiments were executed on a DELL server with eight processors Quad-Core AMD Opteron and 16 GB RAM, OS Ubuntu 9.04.

To evaluate the functional coherency of the aligned proteins, we considered two measures: (1) the number of pairs that are classified as orthologs by HomoloGene (22), considered as a gold standard; and (2) a score based on the gene ontology (GO) (23), focusing on the biological process and molecular function branches. To evaluate the significance of the number of HomoloGene pairs that were matched, we computed a hypergeometric p-value, which measures the probability that a random set of matches (of the same size as our alignment and constrained to the Inparanoid clusters) would yield the observed overlap or higher. The GO score is computed as the average GO similarity of all matched pairs. We employed the Resnik similarity among terms and considered as a similarity between proteins the value of the best matching between their terms (24). We restricted our analysis to the set of ambiguous clusters, i.e., clusters that contain more than one protein for at least one of the species.

The results for yeast-fly, yeast-worm and fly-worm are reported in Table 1. Evidently, all methods perform similarly. ILP and IsoRank always attain the maximum number of conserved interactions. This is expected for ILP and suggests that IsoRank is a good heuristic for maximizing the number of conserved interactions. ILP also achieves the maximum number of HomoloGene pairs, except in the yeast-worm alignment, where it is outperformed by IsoRank. With respect to the GO measures, ILP attains the highest scores in most cases, with PME performing better on the molecular function measure.

Table 1
A comparison of GNA methods on a yeast-fly-worm benchmark

Dataset	Method	Total pairs	Conserved interactions	HomoloGene			GO Sim	
				Pairs	%	P-value	(MF)	(BP)
Yeast-fly	ILP	545	91	134	0.246	4.33e − 09	3.32	1.87
	MRF	535	87	133	0.248	2.17e − 09	3.26	1.85
	IsoRank	545	91	133	0.244	1.01e − 08	3.28	1.88
	PME	545	86	132	0.242	2.33e − 08	3.25	1.83
Yeast-worm	ILP	194	48	72	0.371	0.059	2.95	2.23
	IsoRank	194	48	74	0.381	0.021	2.97	2.22
	PME	194	47	72	0.371	0.059	2.98	2.22
Fly-worm	ILP	209	38	93	0.445	0.004	2.32	1.62
	IsoRank	209	38	87	0.416	0.084	2.32	1.50
	PME	209	36	92	0.440	0.007	2.34	1.61

7. Conclusions

In this chapter, we present the GNA problem and discuss extant methods for solving it. A guiding principle in most of these methods is the maximization of conserved interactions across the two aligned networks. We further present a novel strategy to the problem that is based on a probabilistic model of protein network evolution. We test the methods on a yeast-fly-worm benchmark and find that all methods perform similarly on current networks when starting from a defined set of orthology groups.

We believe that future research in this domain should cover both the development of better alignment methods and the benchmarking of such methods. While current methods do reasonably well with respect to maximizing the number of conserved interactions, evolutionary considerations are still scarcely used and could potentially guide the alignment in a more refined way, particularly when comparing species that are less distant apart. Additional developments could include going beyond 1-1 alignments and pairwise comparisons (25). An orthogonal axis is the development of gold standard alignments. Current benchmarks such as the Homologene collection are mostly sequence-driven and, thus, potentially lead to biased assessment of methods. In summary, we expect GNA methods to have greater impact as protein networks and orthology information continue to accumulate.

Acknowledgements

M.M. was partially supported by the Army Research Laboratory, under Cooperative Agreement Number W911NF-09-2-0053. The views and conclusions contained in this document are those of the authors and should not be interpreted as representing the official policies, either expressed or implied, of the Army Research Laboratory or the US Government. The US Government is authorized to reproduce and distribute reprints for government purposes notwithstanding any copyright notation here on. R.S. was supported by a research grant from the Israel Science Foundation (grant no. 385/06).

References

1. Fields S, Song O (1989) A novel genetic system to detect Protein–protein interactions. Nature 340(6230):245–246
2. Aebersold R, Mann M (2003) Mass spectrometry-based proteomics. Nature 422:198–207
3. Kelley BP, Yuan B, Lewitter F, Sharan R, Stockwell BR, Ideker T (2004) PathBLAST: a tool for alignment of protein interaction networks. Nucl Acids Res 32(Suppl 2): W83–W88
4. Sharan R, Suthram S, Kelley R, Kuhn T, McCuine S, Uetz P, Sittler T, Karp R, Ideker T (2005) Conserved patterns of protein interaction in multiple species. Proc Natl Acad Sci USA 102(6):1974–1979
5. Kalaev M, Bafna V, Sharan R (2009) Fast and accurate alignment of multiple protein networks. J Comput Biol 16(8):989–999
6. Koyuturk M, Kim Y, Topkara U, Subramaniam S, Szpankowski W, Grama A (2006) Pairwise alignment of protein interaction networks. J Comput Biol 13(2):182–199
7. Singh R, Xu J, Berger B (2008) Global alignment of multiple protein interaction networks with application to functional orthology detection. Proc Natl Acad Sci USA 105 (35):12763–12768
8. Remm M, Storm CE, Sonnhammer EL (2001) Automatic clustering of orthologs and in-paralogs from pairwise species comparisons. J Mol Biol 314(5):1041–1052
9. Tatusov R et al (2003) The COG database: an updated version includes eukaryotes. BMC Bioinformatics 1(4):41
10. Datta RS, Meacham C, Samad B, Neyer C, Sjolander K (2009) Berkeley PHOG: phylo-Facts orthology group prediction web server. Nucl Acids Res 37(Suppl 2):84–89
11. Klau G (2009) A new graph-based method for pairwise global network alignment. BMC Bioinformatics 10(Suppl 1):S59
12. Zaslavskiy M, Bach F, Vert JP (2009) Global alignment of Protein–protein interaction networks by graph matching methods. Bioinformatics 25(12):i259–1267
13. Bandyopadhyay S, Sharan R, Ideker T (2006) Systematic identification of functional orthologs based on protein network comparison. Genome Res 16(3):428–435
14. Yosef N, Sharan R, Noble WS (2008) Improved network-based identification of protein orthologs. Bioinformatics 24(16):i200–i206
15. Milenkovic T, Wong W, Hayes W, Przulj N (2010) Optimal network alignment with graphlet degree vectors. Canc Inform 9:121–137
16. Kuchaiev O, Milenkovic T, Memisevic V, Hayes W, Przulj N (2010) Topological network alignment uncovers biological function and phylogeny. J R Soc Interface 7(50):1341–1354
17. Smith AFM, Roberts GO (1993) Bayesian computation via the gibbs sampler and related markov chain monte carlo methods. J Roy Stat Soc B Stat Meth 55(1):3–23
18. Berg J, Lassig M, Wagner A (2004) Structure and evolution of protein interaction networks: a statistical model for link dynamics and gene duplications. BMC Evol Biol 4(1):51
19. Li S et al (2004) A map of the interactome network of the Metazoan C. elegans. Science 303(5657):540–543
20. Xenarios I, Salwinski L, Duan XJ, Higney P, Kim SM, Eisenberg D (2002) DIP, the database of interacting proteins: a research tool for studying cellular networks of protein interactions. Nucl Acids Res 30(1):303–305
21. Chen N et al (2005) WormBase: a comprehensive data resource for Caenorhabditis biology

and genomics. Nucl Acids Res 33(Suppl 1): D383–389

22. Wheeler DL et al (2005) Database resources of the national center for biotechnology information. Nucl Acids Res 33(Suppl 1):D39–D45
23. Ashburner M (2000) Gene ontology: tool for the unification of biology. Nat Genet 25:25–29
24. Schlicker A, Albrecht M (2007) FunSimMat: a comprehensive functional similarity database. Nucl Acids Res 36(Suppl 1):D434–439
25. Liao CS, Lu K, Baym M, Singh R, Berger, B (2009) IsoRankN: spectral methods for global alignment of multiple protein networks. Bioinformatics 25(12):i253–258

Chapter 4

Structure Learning for Bayesian Networks as Models of Biological Networks

Antti Larjo, Ilya Shmulevich, and Harri Lähdesmäki

Abstract

Bayesian networks are probabilistic graphical models suitable for modeling several kinds of biological systems. In many cases, the structure of a Bayesian network represents causal molecular mechanisms or statistical associations of the underlying system. Bayesian networks have been applied, for example, for inferring the structure of many biological networks from experimental data. We present some recent progress in learning the structure of static and dynamic Bayesian networks from data.

Key words: Static Bayesian networks, Dynamic Bayesian networks, Structure learning, Active learning

1. Introduction

Many biological systems are naturally modeled as networks. In many applications, one would often like to know the (network) structure of a measured system. Knowing the network structure also allows the prediction of future states or effects of interventions. Consider a set $X = X_1, \ldots, X_N$ of N random variables describing the state of a system. These nodes may be states of genes (expressed/not expressed, available/not available for transcription), proteins (phosphorylation state, protein level), or other biomolecules, and the states may be discrete, continuous, or both. Causal mechanisms or statistical associations between variables in X can be represented in a form of graphical network structure (see Fig. 1a for example). Often, the structure of (at least part of) the system is unknown and needs to be inferred from measurement data. Bayesian networks are well suited for this task, and several inference methods have been developed (1, 2).

1.1. Bayesian Networks

A Bayesian network represents a joint probability distribution by decomposing it into several local distributions

Hiroshi Mamitsuka et al. (eds.), *Data Mining for Systems Biology: Methods and Protocols*, Methods in Molecular Biology, vol. 939, DOI 10.1007/978-1-62703-107-3_4, © Springer Science+Business Media New York 2013

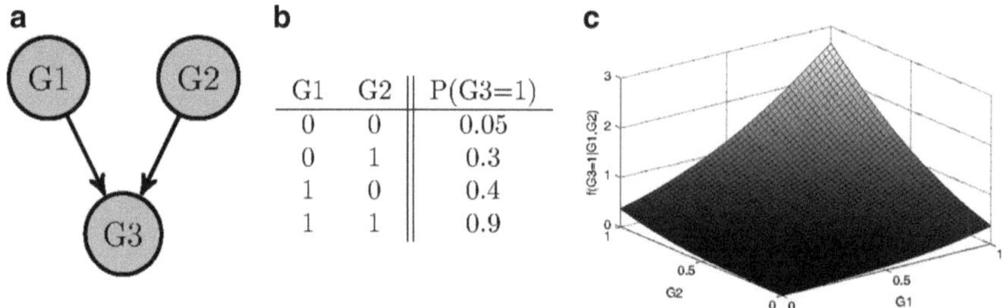

G1	G2	P(G3=1)
0	0	0.05
0	1	0.3
1	0	0.4
1	1	0.9

Fig. 1. (**a**) An example of a small Bayesian network, consisting of three nodes with G3 having G1 and G2 as parents. (**b**) shows an example of the parameters of node G3 when the BN is discrete valued and all nodes are binary, which can be interpreted, for example, as being on/off or present/absent. Thus, considering the example figure in a biological setting, it could present a case where the presence of transcription factors G1 and G2 (i.e., G1=1 and G2=1) indicates a probability of 0.9 that gene G3 is expressed. In (**c**), an example where the nodes in (a) are allowed to have continuous values from [0, 1] is considered and the plotted function is the value of probability distribution function f for G3=1. The parameters used in this case were $\mu = 0$, $W = [0.5, 0.5]$, $\sigma = 0.3$ (see text).

$$P(X|G,\theta) = \prod_{i=1}^{N} P(X_i = x_i | Pa_G(X_i), \theta_i) \,, \tag{1}$$

where G is a directed acyclic graph (DAG) describing the independence relations between the random variables, X_i denotes the ith variable and x_i its value, $\mathrm{Pa}_G(X_i)$ denotes the parents of X_i as defined in G, and $\theta = \theta_1, \ldots, \theta_N$ is the set of parameters defining the probability distributions. A Bayesian network is formally said to consist of the pair G, $\boldsymbol{\theta}$.

When node X_i and its parents are discrete valued, its conditional probability distribution $P(X_i = x_i | \mathrm{Pa}_G(X_i), \boldsymbol{\theta}_i)$ can be defined as a conditional probability table (see Fig. 1b) for an example). The probability of each configuration in the table can be calculated as the frequency of incidences of that configuration in the training data. For continuous valued nodes perhaps the most used distribution is Gaussian. For example, the distribution for node X_i with continuous nodes can be $\mathrm{Norm}(\mu + W\mathrm{pa}_G(X_i), \sigma)$, where μ is the mean of the normal distribution, σ its covariance, $\mathrm{pa}_G(X_i)$ denotes the values of the parents of X_i (as a column vector), and W is the weight (regression) coefficients (see Fig. 1c) for an example). The network can also be a mix of both discrete and continuous nodes.

Given a dataset D that is measured from a system consisting of N random variables X, one would like to obtain the underlying structure of the system. However, often the measurements are noisy, the models do not capture all relevant aspects of the system, or the system itself is inherently noisy. So, instead of just one structure, it is often more sensible to consider a set of network structures (or ideally all potential structures) and weight them based on the posterior distribution

$$P(G|D) = \frac{P(D|G)P(G)}{P(D)} \,, \tag{2}$$

where $P(G)$ is the prior probability of G, $P(D) = \Sigma_{G'} P(D|G')P(G')$ is the marginal probability of data, and $P(D|G)$ is the marginal likelihood of G that can be calculated by integrating over the parameter space, i.e., $P(D|G) = \int P(D|G, \theta_G)P(\theta_G|G)d\theta_G$.

Assuming complete data (i.e., no missing data points) and mutual independence of parameters θ_G, then with certain choices of parameter priors, the marginal likelihood $P(D|G)$ can be solved in closed form. In these cases, it is required that the prior is the conjugate prior of the posterior distribution, the two main cases being multinomial distributions with Dirichlet priors (3) and Gaussian distributions with normal-Wishart priors (4). For incomplete data or other priors that are not conjugate, it is necessary to utilize approximations. Here, we concentrate on BNs with multinomial distributions and Dirichlet priors.

The marginal likelihood can be used to score different network structures as shown in Eq. 2. Additionally, the posterior can be used, for example, to predict next measurements:

$$P(X|D) = \sum_G P(G|D) \int P(X|\theta_G, G)P(\theta_G|D, G)\, d\theta_G \,, \tag{3}$$

which can also be used when the network has been intervened with. Predictive behavior will be discussed more in Subheading 2.2. Posterior probabilities of network features can also be calculated

$$P(f|D) = \sum_G P(f, G|D) = \sum_G f(G)P(G|D) \,, \tag{4}$$

where f is an indicator function, i.e., $f(G) = 1$ if graph G contains the wanted feature and $f(G) = 0$ otherwise.

The prior probability of a network structure can be used to include additional knowledge about the domain into learning. This knowledge can include expert knowledge (transferred to subjective probabilities) or measurement data from other sources, which allows combining heterogeneous data types.

Relevant to inference of biological network structure is the causal interpretation of BN structure, meaning that each edge $A \rightarrow B$ in the BN denotes A is a direct cause of B. Thus, by learning the structure of a BN, one is also learning the causal structure.

1.2. Dynamic Bayesian Networks

Dynamic Bayesian networks are created as temporal extensions of static BNs. That is, each node $i = 1, \ldots, N$ becomes a random variable $X_i[t]$ which also depends on time $t \in \{1, 2, \ldots\}$. DBNs can overcome one drawback of static BNs, namely, the acyclicity requirement, which, for example, prohibits feedback loops that are frequently found in biological systems. This is because it is legal to have in the same network the edges $X_i[t-1] \rightarrow X_j[t]$ and $X_j[t-1] \rightarrow X_i[t]$, which in the "time rolled" network become $X_i \rightleftharpoons X_j$. For example, see Fig. 2. If we assume the random variables depend only on the

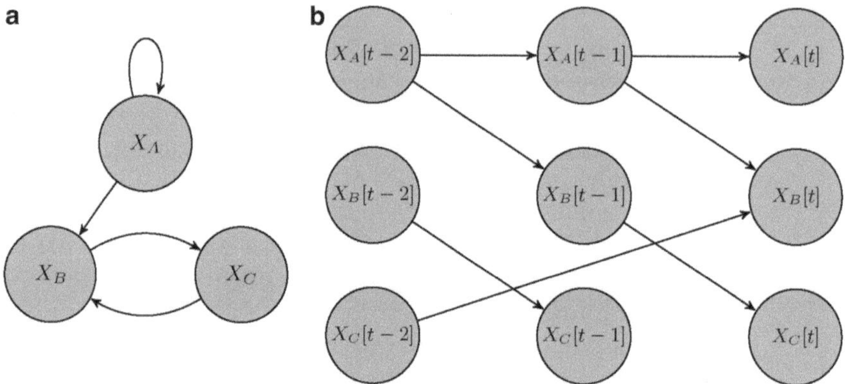

Fig. 2. The "time unrolled" DBN network in the right can be seen to represent the left-hand side network. Note that it is also possible to include influences with slower effects, such as the effect of C on B.

values of the previous timestep (i.e., make the first-order Markov assumption), then $P(X[t]|X[t-1],\ldots,X[1]) = P(X[t]|X[t-1])$, and the joint distribution decomposes as

$$P(X[1],X[2],\ldots,X[T]) = P(X[1])\prod_{t=2}^{T}P(X[t]|X[t-1]) \tag{5}$$

$$= P(X[1])\prod_{t=2}^{T}\prod_{i=1}^{N}P(X_i[t]|Pa(X_i[t]))\,, \tag{6}$$

although in general the parents $\text{Pa}(X_i[t])$ can be from both $X[t]$ and $X[t-1]$.

2. Structure Learning

For selecting among all the possible model structures one or more that best describe the biological system, a way to score the networks is needed. An evident scoring criterion is Eq. 2, whose denominator we are unable to evaluate but as it is independent of G, it suffices to evaluate the nominator to find the relative posterior probabilities of different structures. Thus, we can use, for example, $\log p(D,G) = \log p(g) + \log p(D|G)$.

The number of different Bayesian network structures grows superexponentially as a function of the number of nodes. This makes exhaustive evaluations prohibitive for all but smallest numbers of nodes (about $N < 9$), and heuristic search algorithms must be used instead.

Search methods are commonly iterative and move in the space of DAGs by making allowed (i.e., fulfilling acyclicity constraint) one-edge modifications (addition, deletion, reversal) to a DAG. Perhaps the simplest one, called greedy search, moves by selecting the modification giving the highest improvement in score. An issue with this approach is its inability to escape local maxima; however, this can be overcome with random restarts of the algorithm or using some other application-dependent heuristics.

In addition to only passively observing the states of the system, exact identifiability of the network structure requires that the system is intervened with or perturbed in some ways. Intuitively, the reason is simple: consider a simple network of two nodes A and B. Just by observing their states, one is not able to distinguish causation (either $A \rightarrow B$ or $B \leftarrow A$) from mere correlation. This results from the fact that $P(A)P(B|A) = P(B)P(A|B)$, i.e., both structures produce the same probability distribution.[1] Such inseparable DAGs are said to belong to the same equivalence class (6) and the scoring function should give equal scores for DAGs in the same equivalence class (7). By including interventional data (e.g., by measuring the system when A or B have been forced to certain values), it is possible to break these equivalence classes into smaller, and, with enough properly selected interventions, the size of the most probable class should reduce to one, thus allowing to differentiate the causal structure. Note that the effect of intervention on calculating the score is such that the contribution of each intervened node i is $P(X_i|Pa_G(X_i), \hat{D}) = 1$, where $\hat{D}$ consists of the measurements in dataset D that have node i intervened (8).

Selection of interventions in a beneficial way is discussed in Subheading 2.2. It has also been noted that the interventions are not always perfect, which can also be modeled while learning the structure (9).

The search methods may aim at returning a single network structure, preferably the one giving the highest-scoring (i.e., maximum a posteriori, MAP) model. Often the posterior landscape can be very peaky, perhaps due to sparseness of data or because the underlying system may not be accurately modeled with BNs. In such cases, finding a single MAP model can be very challenging or no single network can describe the system data well. It would be desirable to have the whole posterior distribution available, but in practice, one can only sample the posterior with methods such as Markov chain Monte Carlo (MCMC).

In order to sample from the posterior of structures, a Markov chain is set up so that its target distribution is $P(G|D)$ (10). This is done using the Metropolis-Hastings algorithm which consists of proposing a move from structure G to G' with probability $Q(G'|G)$ and accepting the move with probability

[1] In more exact terms, given causal interpretation, there is a concordance between independence equivalence (given by the v-structure method) and likelihood equivalence (5).

$$\min\left\{1, \frac{P(G'|D)Q(G|G')}{P(G|D)Q(G'|G)}\right\} = \min\left\{1, \frac{P(D|G')P(G')Q(G|G')}{P(D|G)P(G)Q(G'|G)}\right\}, \tag{7}$$

where the probability distribution $Q()$ is called the proposal distribution. In the case of BN structure learning, the proposal distribution is most often defined as

$$Q(G'|G) = \begin{cases} \frac{1}{|N_Q(G)|}, & \text{if } G' \in N_Q(G) \\ 0, & \text{if } G' \notin N_Q(G) \end{cases}, \tag{8}$$

where $N_Q(G)$ is the neighborhood of G reachable by $Q()$, for example, the set of DAGs that are the result of a single-edge-modification (addition, deletion, reversal) to G, and $|N_Q(G)|$ is the cardinality of this set. This algorithm is run for long enough (called burn-in period), after which the actual sample is taken (sampling period) and this sample should be relatively large in order to correctly represent the posterior distribution.

Note that both sets $N_Q(G)$ and $N_Q(G')$ are needed in calculating Eq. 7. Forming these neighborhood sets can get cumbersome for greater than one consecutive edge modifications, which is why normally the proposal distributions are restricted to single-edge-modifications. However, the MCMC chains using such proposals can be very slow in converging to the steady-state distribution, especially in cases of peaky posteriors. Improvements have been achieved by using more versatile proposals, such as in (11, 12).

Alternatively, it is possible to learn in the space of node orders instead of structure space (13), which considerably enhances convergence of MCMC chains. Notable improvements in BN structure learning also include (14), who utilize dynamic programming to calculate the posterior probabilities of all BNs in exponential time, and variations of this in (15, 16).

Assessment of convergence of the MCMC chains is critical to ensure that the results are from the steady-state distribution. Yet, without evaluating all the structures, it is impossible to say whether a chain is converged, and one is most often left to check only some necessary (not sufficient) convergence criteria. Among the most used ones is running two or more chains in parallel and calculating the posterior probabilities of edges for the samples from each chain:

$$P(e(G)|D) \approx \frac{1}{|\mathcal{G}|} \sum_{G \in \mathcal{G}} e(G), \tag{9}$$

where e is a function so that $e(G) = 1$ if graph G contains a given edge and $e(g) = 0$ otherwise, and $\mathcal{G}$ is a sample from an MCMC chain, having $|\mathcal{G}|$ sampled graphs. The edge posterior probabilities can then be examined, for example, by plotting them pairwise in a scatter-plot, where any marked deviations from the diagonal indicate partial convergence or convergence to different distributions. Another simple method is to plot the Bayesian scores ($\log p(D, G)$) for the samples from each chain, in which case lack of overlap denotes the chains have not converged and "wandering" of a chain in the score-plot may suggest a longer burn-in period would be needed.

2.1. Learning DBN Structure

The same methods as the ones used for BN structure learning can be used for DBNs since DBNs are in effect BNs where all nodes are duplicated for each time-slice. If the parents of $X_i[t]$ are constrained to only $X_i[t-1]$ (i.e., within-time-slice edges disallowed), then the DBN structure is guaranteed to be acyclic, thus facilitating movement of search methods in the structure space.

The data used for DBN structure inference must consist of time-series measurements. In many systems biology applications, however, it is not possible to carry out time-course experiments. There are also methods for inferring DBN structure from either static measurements alone or a combination of static and time-series measurements (17). DBNs are essentially vector-valued Markov chains, which under relatively mild technical assumptions possess a stationary distribution. It is natural to assume that static measurements are then sampled from this stationary distribution. For a given discrete-valued DBN, likelihood of the static data can be computed in a straightforward manner by solving for the steady-state distribution of the DBN. Frequentist methods, however, are not well-suited for this problem because, for example, several DBNs can have the same steady-state distribution. Bayesian inference in turn is computationally challenging because the marginal likelihood cannot be computed in a closed form. An efficient reversible jump MCMC (RJMCMC) method is proposed in (17) to sample from the full posterior of DBNs, including both G and θ. The RJMCMC structure learning method is similar in nature with the standard MCMC algorithms for BNs or DBNs except that the RJMCMC for DBNs must propose two different moves: (1) a "jump" into a new network structure in the neighborhood of the current network structure, in which case the parameters are also sampled from a proposal distribution, and (2) a "null" move where only new parameter values are proposed. To achieve the detailed balance and convergence to the desired posterior distribution, care needs to be taken that a bijective mapping between DBN models with different dimensionality is obtained. For full details, the reader is referred to (17).

2.2. Active Learning of BN Structure

Not all measurements are equally informative for the task of learning a model structure, and therefore it is of interest to select (e.g., biological) experiments to be done in order to gain maximal information from them. This is a highly nontrivial problem, and methods implemented for such experimental design are called active learning methods. In the case of BN (and other model classes) learning, they can be divided into ones aiming at inferring either the parameters or the structure of the system. For the latter category, the methods are used to propose the nodes of a BN that should be intervened or clamped for maximum benefit.

Two closely related approaches for selecting the perturbations have been presented: those that break equivalence classes (18) and decision theoretic methods that aim to diminish uncertainty (or increase information maximally) about some edges (19, 20). The similarity of the methods is due to the fact that within an equivalence class, the inability to say which direction an edge takes is, in other words, uncertainty about that edge.

As an example of an active learning method, we utilize the one presented in (19), where the utility of action a is defined as

$$V(a) = \sum_{G \in \mathcal{G}} \sum_{y \in \Upsilon_{G,a}} P(y|G,a,D)P(G|D)U(G,a,y,D) , \qquad (10)$$

where $\mathcal{G}$ is our set of possible DAGs and $\Upsilon_{G,\,a}$ denotes the set of possible observations that G can produce given that intervention a has been made. Note that action a can be an empty intervention (i.e., an observation) or it can consist of perturbing one or several nodes. The utility function is defined as $U(G, a, y, D) = \log P(G|a, y, D)$, which includes the assumption that every intervention costs the same. The most beneficial action is then selected from the set of all possible actions A as the one with maximal utility, i.e., $a^* = \text{argmax}_{a \in A} V(a)$.

Evaluating the whole Eq. 10 would require summation over the space of DAGs and possible observations, which is, again, impossible and stochastic sampling methods must be used instead. For sampling, it is possible to use, for example, importance sampling.

To demonstrate the use of an active learning method in a realistic situation and assessing its performance compared to a nonactive learner (i.e., picking interventions randomly), two competing schemes can be set up: (1) active learner, which makes at each step the measurement (intervention or observation) suggested by the active learning algorithm based on the sampled graphs, available measurements, sampled observations, and data collected so far, and (2) random learner, which makes a measurement randomly without relying on an active learning algorithm. In both cases, after each measurement, the MCMC chains are run for a between-measurement burn-in period, followed by taking new samples of DAGs from the chains.

Our test case consisted of taking from the Sachs dataset[2] a sample with 100 observational data points and 20 data points per

[2]Sachs dataset: consists of flow cytometry measurements from a signaling network with 11 nodes, of which five have been perturbed in some measurements (21). These interventions contain both inhibitions and activations of the nodes. The data was discretized into ternary values, and uniform Dirichlet priors for parameters and uniform structural priors were used.

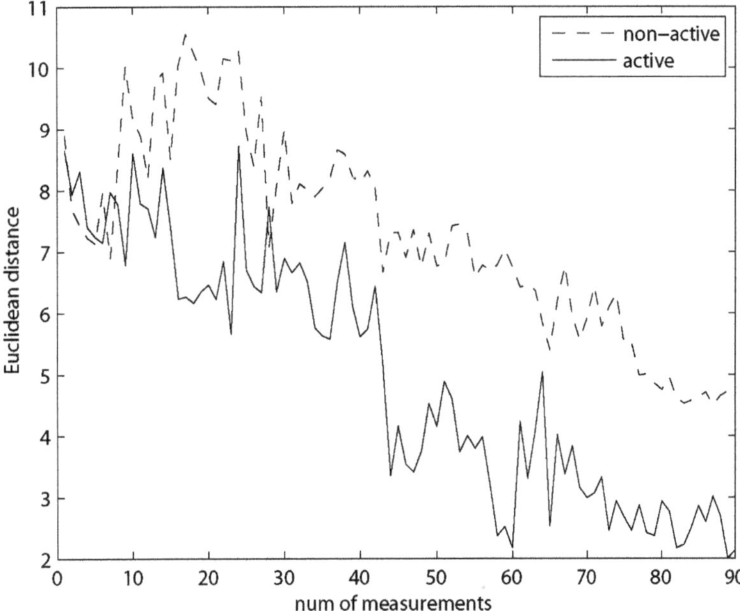

Fig. 3. Averaged results from four different runs showing Euclidean distance from edge posterior probabilities, calculated using samples from chains run with active and nonactive learning methods, to the "steady-state" posterior distribution. The system and data were the 11 node network from (21). Number of measurements shows the number of data points sampled after the initial 40 observational datapoints. For each run, the initial burn-in was $2 \cdot 10^5$, between-measurement burn-in was 5,000, graph sample size 5,000, and sampled observations 300.

intervention, totaling 220 measurements. Each test was initiated by taking a set of observations as initial data and running two MCMC chains in parallel for a long initial burn-in period. Following this, samples were taken from both chains, and the convergence of the chains was checked by comparing distributions of edge posterior probabilities calculated from both samples. If the distributions were similar (i.e., chains were converged), either sample was used as the initial sample for both active and non-active learners. Figure 3 shows the result when comparing both learners to the steady-state distribution obtained with a normal "batch" MCMC run.

The results show the clear benefit of active learning methods as guides to choosing experiments, since, as shown in this example, the same distance from the "true" structure can be obtained in an active learning setting with half the measurements needed for non-active learning.

Although the active learning method was demonstrated only for static BNs, it is also suitable for DBNs. We note that this does not hold for active learning based on breaking equivalence classes.

3. Notes

1. Bayes Net Toolbox is available (for Matlab) with extensive functionality (22).
2. For a list of other BN software, refer to (23).
3. If prior knowledge about the structure is available, it is sensible to use informative structure priors. Informative priors may be derived from expert knowledge or from multiple datasets, refer to for example (24). If such priors are not available, uninformative (uniform) priors (i.e., $P(G_i) = P(G_j)$, for all i and j) can be used or priors penalizing for growing complexity, such as (13): $P(G) \propto \prod_{i=1}^{N} \binom{N-1}{|Pa_G(X_i)|}^{-1}$.
4. Using MCMC over ordering of nodes instead of structures can considerably speed up computations by improving convergence of MCMC. However, the ordering introduces an inherent bias via structure priors, and trying to include additional prior information is intractable (for discussion on this topic, see, e.g., (12)). A large dataset can overweigh the priors, but for smaller datasets typically present in systems biology applications or when fusing different data types is required, it may be better to use structure MCMC.
5. The active learning method discussed in Subheading 2.2 (from (19)) is stochastic in nature, and, although in average performs better than random learning, it is possible that it give worse results at times. Thus, care must be taken when running the method and large enough samples should be used.
6. The discussed active learning method predicts for each iteration the intervention/observation most beneficial to make and the same intervention can be suggested recurringly, unlike in the equivalence class-based method (18), which outputs the sequence of interventions. Thus, the first method is more flexible, but because it is computationally demanding and since sampling can affect the precision of the method, using the method of (18) becomes more attractive after about $n > 12$.

Acknowledgements

This work was supported by the Academy of Finland (application numbers 135320 and 213462, Finnish Programme for Centres of Excellence in Research 2006–2011), and FP7 EU project SYBILLA.

References

1. Heckerman D (1998) A tutorial on learning with Bayesian networks. In: Jordan MI (ed) Learning in graphical models, pp 301–354. MIT Press, Cambridge
2. Husmeier D (2005) Introduction to learning Bayesian networks from data. In: Husmeier D, Dybowski R, Roberts S (eds) Probabilistic modeling in bioinformatics and medical informatics. Springer, Berlin, pp 17–57
3. Cooper G, Herskovits E (1992) A Bayesian method for the induction of probabilistic networks from data. Mach Learn 9:309–347
4. Geiger D, Heckerman D (1994) Learning Gaussian networks. Proceedings of tenth conference on uncertainty in artificial intelligence (UAI '94), Seattle, WA, pp 235–243
5. Heckerman D (1995) A Bayesian approach to learning causal networks. Proceedings of the eleventh conference annual conference on uncertainty in artificial intelligence (UAI '95), pp 285–295
6. Verma TS, Pearl J (1990) Equivalence and synthesis of causal models. Proceedings of the sixth annual conference on uncertainty in artifcial intelligence (UAI '90), pp 220–227
7. Heckerman D, Geiger D, Chickering D (1995) Learning Bayesian networks: the combination of knowledge and statistical data. Mach Learn 20:197–243
8. Cooper G, Yoo C (1999) Causal discovery from a mixture of experimental and observational data. Proceedings of the fifteenth annual conference on uncertainty in artificial intelligence (UAI '99), pp 116–125
9. Eaton D, Murphy K (2007) Exact Bayesian structure learning from uncertain interventions. Proceedings of the 23rd conference on uncertainty in artificial intelligence and statistics (AISTAT), pp 107–114
10. Madigan D, York J (1995) Bayesian graphical models for discrete data. Int Stat Rev 63:215–232
11. Castelo R, Kočka T (2003) On inclusion-driven learning of Bayesian networks. J Mach Learn Res 4:527–574
12. Grzegorczyk M, Husmeier D (2008) Improving the structure MCMC sampler for Bayesian networks by introducing a new edge reversal move. Mach Learn 71:265–305
13. Friedman N, Koller D (2003) Being Bayesian about network structure. A bayesian approach to structure discovery in Bayesian networks. Mach Learn 50:95–125
14. Kovisto M, Sood K (2004) Exact Bayesian structure discovery in Bayesian networks. J Mach Learn Res 5:549–573
15. Silander T, Myllymäki P (2006) A simple approach for finding the globally optimal Bayesian network structure. Proceedings of the twenty-second conference annual conference on uncertainty in artificial intelligence (UAI '06), pp 445–452
16. Eaton D, Murphy K (2007) Bayesian structure learning using dynamic programming and MCMC. Proceedings of the twenty-third conference on uncertainty in artificial intelligence (UAI '07) , pp 101–108
17. Lähdesmäki H, Shmulevich I (2008) Learning the structure of dynamic Bayesian networks from time series and steady state measurements. Mach Learn 71:185–217
18. Pournara I, Wernisch L (2004) Reconstruction of gene networks using Bayesian learning and manipulation experiments. Bioinformatics 20 (17):2934–2942
19. Murphy K (2001) Active learning of causal Bayes net structure. Technical Report, University of California, Berkeley, USA
20. Tong S, Koller D (2001) Active learning for structure in Bayesian networks. Proceedings of the seventeenth international joint conference on artifcial intelligence, Seattle, WA, USA, pp 863–869
21. Sachs K, Perez O, Peer DA, Lauffenburger DA, Nolan GP (2005) Protein-signaling networks derived from multiparameter single-cell data. Science 308:523–529
22. Bayes Net Toolbox for Matlab. http://code.google.com/p/bnt/ Cited 31 Dec 2010
23. Murphy K. Software packages for graphical models/Bayesian networks. http://www.cs.ubc.ca/~murphyk/Software/bnsoft.html Cited 31 Dec 2010
24. Bernard A, Hartemink A (2005) Informative structure priors: joint learning of dynamic regulatory networks from multiple types of data. Pacific symposium on biocomputing 2005 (PSB05), pp 459–470

Chapter 5

Supervised Inference of Gene Regulatory Networks from Positive and Unlabeled Examples

Fantine Mordelet and Jean-Philippe Vert

Abstract

Elucidating the structure of gene regulatory networks (GRN), i.e., identifying which genes are under control of which transcription factors, is an important challenge to gain insight on a cell's working mechanisms. We present SIRENE, a method to estimate a GRN from a collection of expression data. Contrary to most existing methods for GRN inference, SIRENE requires as input a list of known regulations, in addition to expression data, and implements a supervised machine-learning approach based on learning from positive and unlabeled examples to account for the lack of negative examples.

Key words: Gene regulatory network, Reverse engineering, Inference, Machine learning, Gene expression

1. Introduction

The regulation of gene expression in living cells is a complex process involving many actors. Among them, cis-regulation by transcription factors (TF) is of crucial importance to allow cells to adapt their behavior in response to external stimuli and to meet their needs depending on environmental conditions. Elucidating the structure of gene regulatory networks (GRN), i.e., identifying which genes are under control of which TF, is therefore an important challenge to gain insight on a cell's working mechanisms and may pave the way, e.g., to rational, predictive and personalized medicine (1).

The experimental characterization of transcriptional cis-regulation at a genome scale remains, however, a daunting task even for well-studied model organisms. *In silico* methods that attempt to reconstruct the GRN from prior biological knowledge and available genomic and post-genomic data therefore constitute an interesting

Hiroshi Mamitsuka et al. (eds.), *Data Mining for Systems Biology: Methods and Protocols*, Methods in Molecular Biology, vol. 939, DOI 10.1007/978-1-62703-107-3_5,

direction towards the elucidation of these networks. This task, often referred to as *reconstruction*, *inference*, or *reverse engineering* of GRN in the literature, is often based on the analysis of gene expression data across samples and experimental conditions (2). Indeed, large volumes of expression data are easily collected by DNA microarray and next-generation sequencing technologies, and expression levels are directly affected by changes in transcriptional regulation. A variety of approaches have been proposed to reverse engineer GRN from expression data, including clustering techniques to infer co-regulation (3), detection of relationship between a TF and target gene by correlation or mutual information (4–6), models of conditional probability distribution (7), differential equations (8–13), or Boolean relationships (14).

All aforementioned approaches infer a GRN from expression data, based on various assumptions on the influence of the GRN on expression. Since some GRN in model organisms have already been partially deciphered through decades of experimental biology, it is possible to assess the performance of these methods by comparing the GRN they predict to known regulations (6). However, it should be pointed out that none of these methods makes use of known regulations to guide the inference of new ones. In opposition to these approaches which perform de novo inference of GRN from expression data, we and others have proposed that using known regulations, in addition to expression data, in order to infer new ones may be a more promising paradigm to improve the accuracy of inference methods for GRN and other biological networks (15–19). We refer to this later family of methods as *supervised* methods, as opposed to de novo methods, because they follow the paradigm of supervised inference in statistics and machine learning where data known to have a property are used to guide the search of other data sharing the same property (20).

In this chapter, we describe the SIRENE method (Supervised Inference of REgulatory Networks), first proposed by (19), to infer a GRN from a set of expression data and a set of known regulations. SIRENE is a supervised method, intuitively based on the following inference principle: if a gene A has an expression profile "similar" to a gene B known to be regulated by a given transcription factor, then gene A is likely to be also regulated by the transcription factor. The fact that genes with similar expression profiles are likely to be co-regulated has been used for a long time in the construction of groups of genes by unsupervised clustering of expression profiles (3). The novelty in SIRENE is to use this principle in a supervised classification paradigm, where the notion of similarity is optimized to characterize the set of genes regulated by a given TF, based on known regulated genes. This inference paradigm has the advantage that no particular hypothesis is made regarding the relationship between the expression level of a TF and those of regulated genes. In fact, expression data for the TF are not even needed in this approach.

In practice, SIRENE treats each TF in turn, and learns so-called *local models* to characterize the genes regulated by each TF, an approach pioneered by (18) in the case of protein-protein interaction and metabolic networks. To estimate the local model of a given TF, we make use of genes known to be regulated by the TF to learn a scoring function able to predict the probability that other genes are also regulated by the TF. Since biological experiments are usually designed to prove the existence of an interaction—and not its absence—and since databases seldom report negative results, we rarely have in advance a list of genes that we know are *not* regulated by the TF. Calling a gene positive or negative depending on whether or not it is regulated by a TF of interest, this means that we have a set of positive examples from which we need to learn a model to predict the label of other genes. While learning a scoring function from positive examples only is a well-studied field in statistics and machine learning (21), it has been shown both theoretically and in practice that learning from positive *and* unlabeled examples (the so-called *PU learning* paradigm) is often more efficient than learning from positive examples only (22–31). In our setting, this amounts to learning a model from expression data (to represent all genes), a list of known regulated genes (positive examples), and the list of all other genes which are candidate targets for the TF (the unlabeled examples). SIRENE implements a supervised strategy to infer GRN from gene expression data and known regulations, based on the PU learning method proposed in (31), which we describe in detail below.

2. Materials

The method we describe is meant to infer GRN for any organism for which expression and regulation data are available. For the sake of clarity, we take the example of the bacteria *Escherichia coli* as a running example in this chapter.

2.1. Microarray Data

The first ingredient we need to collect in order to run this analysis is a collection of expression data. Typically, data obtained from DNA microarrays in different experimental collections can be gathered from public repositories or in-house databanks. Standard technology-dependent preprocessing must be carried out to assign to each gene of interest (typically, all known genes) a measure of expression in each experiments. The result should therefore be a $n \times p$ matrix of expression for n genes across p experimental conditions.

We used in our example the expression data collected by (6) for *E. coli*, which can be downloaded from http://gardnerlab.bu.edu/data/PLoS_2007/data_and_validation.html (see Note 1).

The expression data consist of a compendium of $p = 445$ *E. coli* Affymetrix Antisense2 microarray expression profiles for $n = 4,345$ genes. The microarrays were collected under different experimental conditions such as PH changes, growth phases, antibiotics, heat shock, different media, varying oxygen concentrations, and numerous genetic perturbations. We normalize the expression data for each gene to zero mean and unit standard deviation.

2.2. Regulation Data

The second ingredient we need is a list of TF, for which we will predict target genes and a list of known targets for these TF. In our case, we collect the list of 328 genes coding for known TF in *E. coli*, and the list of known targets for these TF from the RegulonDB (32) database. We take the data prepared by (6) who extracted 3, 293 experimentally confirmed regulation interactions between 154 known TF and 1, 211 target genes, including auto-regulations. These regulations are available at http://gardnerlab.bu.edu/data/PLoS_2007/reg.tar.gz. In addition, in the particular case of bacterial genomes, it is useful to collect regulon information since genes belonging to the same operon are regulated by the same TF. In our case, we downloaded the list of 899 known operons in *E. coli* from RegulonDB. Each operon contains one or several genes, and each gene belongs to at most one operon. Genes not present in any operon of the RegulonDB were considered to form an operon by themselves, resulting in a total of 3, 360 operons for the 4, 345 genes (see Note 2).

3. Methods

3.1. A Local Model Approach

We decompose the global GRN inference problem into a series of local problems, each dedicated to predicting new regulated genes for a single TF. Afterwards, the predictions of all local models are combined into a global network. We now explain how to make a single local model for a given TF of interest, which we refer to as $\mathcal{T}$ below. We denote by $\mathcal{P}$ the set of known targets of $\mathcal{T}$ (positive examples) and by $\mathcal{U}$ the set of genes among which we want to predict new targets. Usually, $\mathcal{U}$ is only made of all genes which are not in $\mathcal{P}$.

3.2. Learning from Positive and Unlabeled Examples

Given the sets of genes $\mathcal{P}$ and $\mathcal{U}$, where each gene is described by a vector of expression values across different experiments, we want to predict which genes in $\mathcal{U}$ have a positive label. While a frequent strategy is to score unlabeled examples based on their similarity to known positive examples (see Note 3), SIRENE follows the so-called PU learning paradigm where the scoring function is learned from both the positive and the unlabeled examples.

Algorithm 1 describes our PU learning method for identifying new regulated targets for TF A from $\mathcal{P}$ and $\mathcal{U}$. The general idea of the procedure is to estimate many scoring functions by discriminating $\mathcal{P}$ versus random subsamples of $\mathcal{U}$ and then average all scoring functions to assign a final score to each unlabeled example. The idea to use supervised machine learning procedures to discriminate $\mathcal{P}$ from $\mathcal{U}$ has been successfully used for PU learning, and our procedure builds on this idea (see Note 4). It is akin to bagging (33) to learn to discriminate $\mathcal{P}$ from $\mathcal{U}$ with two important specificities:

- First, only $\mathcal{U}$ is subsampled. This is to account for the fact that elements in $\mathcal{P}$ are known to be positive, and moreover that the number of positive examples is often limited (see Note 5).
- Second, the size of subsamples is a parameter K which is preferably smaller than the size of $\mathcal{U}$. Each time a random subsample $\mathcal{U}_t$ of $\mathcal{U}$ is generated, a classifier is trained to discriminate $\mathcal{P}$ from $\mathcal{U}_t$, and used to assign a predictive score to any element of $\mathcal{U} \setminus \mathcal{U}_t$.

Eventually, the score of any element $x \in \mathcal{U}$ is obtained by averaging the predictions of the classifiers trained on subsamples that did not contain x (see Note 6). As such, no point of $\mathcal{U}$ is used simultaneously to train a classifier and to test it. Note that the counter $n(x)$ in Algorithm 1 simply counts the number of scoring functions that were estimated without x as training example.

Algorithm 1 A PU learning algorithm for detecting target genes of a TF

INPUT : $\mathcal{P}, \mathcal{U}, K =$ size of bootstrap samples, $T =$ number of bootstraps
OUTPUT : a score $s : \mathcal{U} \to \mathbb{R}$
Initialize $\forall x \in \mathcal{U}, n(x) \leftarrow 0, f(x) \leftarrow 0$
for $t = 1$ to T **do**
  Draw a random sample $\mathcal{U}_t$ of size K from $\mathcal{U}$.
  Train a classifier f_t to discriminate $\mathcal{P}$ against $\mathcal{U}_t$.
  For any $x \in \mathcal{U} \setminus \mathcal{U}_t$, update:

$$f(x) \leftarrow f(x) + f_t(x)\,,$$
$$n(x) \leftarrow n(x) + 1\,.$$

end for
Return $s(x) = f(x)/n(x)$ for $x \in \mathcal{U}$

3.3. Choosing a Classification Method

At the heart of Algorithm 1, we need an algorithm to train a classifier to discriminate two sets of genes based on their expression data. In practice, any classification method can be used to train f_t, e.g., logistic regression or support vector machines (SVM). In SIRENE, we use a SVM, a popular algorithm to solve general supervised binary classification problems considered state of the art in many applications (34, 35). Many free and public implementations of SVM are available (see Note 7). An important ingredient needed to run an SVM is a kernel function $K(x, y)$ between any two genes x and y that can often be thought of as a measure of similarity between the genes. In our case, the similarity between genes is measured in terms of expression profiles. Given a set of n genes $x_1, \ldots, x_n$ that belong to two classes, denoted arbitrarily -1 and $+1$, an SVM estimates a scoring function for any new gene x of the form:

$$f(x) = \sum_{i=1}^{n} \alpha_i K(x_i, x) .$$

The weights α_i in this expression are optimized by the SVM to enforce as much as possible large positive scores for genes in the class $+1$ and large negative scores for genes in the class -1 in the training set. A parameter, often called C, allows to control the possible overfitting to the training set. The scoring function $f(x)$ can then be used to rank genes with unknown class by decreasing score, from the most likely to belong to class $+1$ to the most likely to belong to class -1.

The kernel $K(x, y)$ defines the similarity measure used by the SVM to build the scoring function. In our experiments, we want to infer regulations from gene expression data. Each collection of gene expression data is a vector, so we simply use the common Gaussian radial basis function kernel between vectors u and v:

$$K(u, v) = \exp\left(-\frac{||u - v||^2}{2\sigma^2} \right),$$

where $\sigma > 0$ is the bandwidth parameter of the kernel.

3.4. Parameter Selection

Each SVM has therefore two parameters, C and σ. To select the values of these parameters, we followed the following principles (see Note 6):

- Concerning the C parameter, it is often useful to check different values and choose the most appropriate one by cross-validation. This requires however a sufficient number of positive examples and might not be appropriate when the positive set is too small, like in our case. It addition, it considerably increases the computation time of the procedure. Alternatively, if time is a constraint or if there are not enough known positive examples to perform cross validation, one may fix an arbitrary value for C. A useful default choice is a large C value, e.g., $C = 1,000$, which corresponds approximately to learning a so-called hard-margin SVM (34). This choice is particularly relevant with a Gaussian kernel, where the bandwidth σ of the kernel allows to control overfitting (36).

- The σ parameter was simply set to unique value $\sigma = \sqrt{2p}/4$, where $p = 445$ is the number of expression data available for each gene. This choice was based on the observation that the expression profile for each gene was scaled to zero mean and unit standard deviation for each gene, meaning that each gene is represented by a vector of dimension p and of Euclidean norm $\sqrt{p}$. Two completely independent genes should be represented by roughly orthogonal vectors, implying that the Euclidean distance between two unrelated genes is of order of $\sqrt{2p}$. We expect that a bandwidth of the order of $\sigma = \sqrt{2p}/4$, which puts two orthogonal profiles at about 4σ from each other, is a safe default choice to make sure that unrelated genes have no effect on each other (see Note 8).

  In addition to SVM parameters, algorithm 1 also needs two additional input parameters:

- K, the size of subsamples. $K = N_P$ or up to $5 \times N_P$ seems a safe choice. To our knowledge, K does not have a significant influence on the performance (31). Note that the SVM weights should be balanced if K is considerably higher than NP, the size of the positive set.
- T, the number of bootstraps. In theory, the higher T, the better the performance but also the longer it takes for the algorithm to run. In practice, we observed that $T = 30$ led to good results, but we advocate to increase that value for smaller values of K if time is not an issue.

3.5. Results

In this section, we present a few results obtained from experiments on the *E. coli* dataset which we previously described. We restrict ourselves to 31 TFs with at least 8 known regulated genes and for each TF, and run a double three-fold cross validation with an internal loop on each training set to select parameter C of the SVM (or ν for the 1-class SVM). Following (19), we perform a particular cross validation scheme to ensure that operons are not split between folds. We test two variants of bagging SVM, setting K successively to N_P and $5 \times N_P$. These choices are denoted respectively by *bagging1 SVM* and *bagging5 SVM*. We compare them to the biased SVM, which is considerably longer to train, as well as to two methods that estimate the scoring function from positive examples only: the baseline method, which scores the unlabeled examples by their mean Euclidean distance to known positives, and the one-class SVM method (see Note 3).

Figure 1 shows the average precision/recall curves of all methods tested. Overall, we observe that all three PU learning methods give significantly better results than the two methods which use only positive examples (Wilcoxon paired sample test at 5% significance level). No significant difference was found between the three PU learning methods. This confirms again that for different values

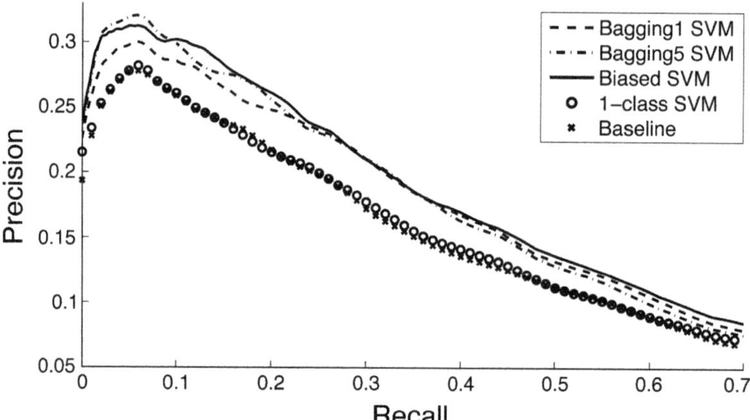

Fig. 1. Precision-recall curves to compare the performance between the bagging SVM, the bagging5 SVM, the biased SVM, the 1-class SVM and the baseline method.

of K, bagging SVM matches the performance of biased SVM. More details about the performance of SIRENE, in particular comparisons with de novo GRN inference methods, are presented and discussed in (19, 31, 37).

4. Notes

1. We used the RMA-normalized expression values for all probe sets that were annotated as genes and all conditions. More recent and complete datasets for *E. coli* are available from http://m3d.bu.edu/norm/.
2. More recent and complete datasets for operons can be downloaded from http://regulondb.ccg.unam.mx/data/OperonSet.txt.
3. A possible strategy to overcome the lack of negative examples is to disregard unlabeled examples during training and simply learn from the positive examples, e.g., by ranking the unlabeled examples by decreasing similarity to the mean positive example (38) or using more advanced learning methods such as 1-class SVM (21, 24, 36, 39).
4. Our starting point to learn a classifier in the PU learning setting is the observation that learning a scoring function to discriminate positive from unlabeled samples is a good proxy to our objective, which is to learn a scoring function to discriminate positive from negative samples (23, 27, 30). In practice, it is useful to train classifiers to discriminate $\mathcal{P}$ from $\mathcal{U}$ by penalizing more false-negative than false-positive errors, in order to account for the fact that positive examples are known to be

positive, while unlabeled examples are known to contain hidden positives. Using soft-margin SVM while giving high weights to false-negative errors and low weights to false positive errors leads to the biased SVM approach described by (27), while the same strategy using a logistic regression leads to the weighted logistic regression approach of (29). Both methods, tested on text categorization benchmarks, were shown to be very efficient in practice and in particular, outperformed all other existing approaches for PU learning. The PU learning method we propose in this chapter, based on bagging, is an extension of the biased SVM, which can lead to more accurate and computationally more efficient models (31):

- First, it can improve the performance of the model by limiting overfitting. In particular, assigning a score to an unlabeled example that has been used as negative training example, as does the biased SVM, can be problematic. Indeed, if the classifier fits too tightly the training data, a false-negative x_i can hardly be given a high training score when used as a negative. In a related situation in the context of semi-supervised learning, (40) showed, e.g., that unlabeled examples used as negative training examples tend to have underestimated scores when an SVM is trained with the classical hinge loss. More generally, most theoretical consistency properties of machine learning algorithms justify predictions on samples outside of the training set, raising questions on the use of all unlabeled samples as negative training samples at the same time.
- Second, it improves scalability and speed of biased SVM. Indeed, a typical machine-learning algorithm, such as an SVM, trained on N samples has time complexity proportional to N^α, with α between 2 and 3. Therefore, biased SVM which is trained to discriminate $\mathcal{P}$ from $\mathcal{U}$ has complexity proportional to $(N_P + N_U)^\alpha$, where N_P and N_U are respectively the sizes of $\mathcal{P}$ and $\mathcal{U}$. On the other hand, bagging SVM's complexity is proportional $T * (N_P + K)^\alpha$, where T is the number of times we randomly subsample unlabeled examples, and K is the size of each subsample. With the default choice $K = N_P$, the ratio of CPU time needed to train the biased SVM versus the bagging SVM is therefore $((N_P + N_U)/(2N_P))^\alpha/T$. Then, we conclude that bagging SVM is faster than biased SVM as soon as $N_U/N_P > 2T^{1/\alpha} - 1$. For example, taking $T = 35$ and $\alpha = 3$, bagging SVM is faster than biased SVM as soon as $N_U/N_P > 6$, a situation very often encountered in practice where the ratio N_U/N_P is more likely to be several orders of magnitude larger.

5. In this example, known regulations are manually curated and experimentally validated. Therefore, they are believed to be high-confidence interactions. However, if the positive training set is believed to be noisy, one might consider bootstrapping $\mathcal{P}$ too.
6. In practice, it is useful to ensure that all elements of $\mathcal{U}$ are not too often in $\mathcal{U}_t$, in order to average the predictions over a sufficient number of classifiers. This can be achieved, e.g., by repeatedly splitting $\mathcal{U}$ into N/K non overlapping groups of size K and training a model with each group taken in turn as negative examples. Besides, although we propose to aggregate the scores by a simple average, there exist alternative aggregation techniques (31).
7. A list of SVM implementations is available at http://www.support-vector-machines.org. We used the free and publicly available `libsvm` library (41), which can be called directly as a *C* program of via many languages such as R, MATLAB or PYTHON.
8. Correctly choosing parameters can be of paramount importance to obtain good performance with machine learning algorithms. The default choice we propose have the advantage to be easy and fast to implement, since they require no parameter optimization. However, it is always safe to check that these choices are not too bad for the problem to be solved. In our case, we performed preliminary experiments with different values of C and σ, which did not result in significant improvement or decrease of performance. This suggests that the behavior of our algorithm is robust to variations in its parameters around the default values we propose and that ignoring further parameter optimization is a safe decision. We strongly recommend to analyze the effect of parameters on a case-by-case basis.

References

1. Hood L, Heath JR, Phelps ME, Lin B (2004) Systems biology and new technologies enable predictive and preventative medicine. Science 306(5696):640–643
2. Bansal M, Belcastro V, Ambesi-Impiombato A, diBernardo D (2007) How to infer gene networks from expression profiles. Mol Syst Biol 3:78
3. Tavazoie S, Hughes JD, Campbell MJ, Cho RJ, Church GM (1999) Systematic determination of genetic network architecture. Nat Genet 22:281–285
4. Butte AJ, Tamayo P, Slonim D, Golub TR, Kohane IS (2000) Discovering functional relationships between RNA expression and chemotherapeutic susceptibility using relevance networks. Proc Natl Acad Sci USA 97 (22):12182–12186
5. Margolin AA, Nemenman I, Basso K, Wiggins C, Stolovitzky G, Dalla Favera R, Califano A (2006) ARACNE: an algorithm for the reconstruction of gene regulatory networks in a mammalian cellular contexts. BMC Bioinformatics 7 Suppl 1:S7
6. Faith JJ, Hayete B, Thaden JT, Mogno I, Wierzbowski J, Cottarel G, Kasif S, Collins JJ, Gardner TS (2007) Large-scale mapping and validation of Escherichia coli transcriptional regulation from a compendium of expression profiles. PLoS Biol 5(1):e8

7. Friedman N, Linial M, Nachman I, Pe'er D (2000) Using Bayesian networks to analyze expression data. J Comput Biol 7(3–4):601–620
8. Chen T, He HL, Church GM (1999) Modeling gene expression with differential equations. Pac Symp Biocomput 4:29–40
9. Tegner J, Yeung MKS, Hasty J, Collins JJ (2003) Reverse engineering gene networks: integrating genetic perturbations with dynamical modeling. Proc Natl Acad Sci USA 100 (10):5944–5949
10. Gardner TS, Bernardo D, Lorenz D, Collins JJ (2003) Inferring genetic networks and identifying compound mode of action via expression profiling. Science 301(5629):102–105
11. Chen K-C, Wang T-Y, Tseng H-H, Huang C-YF, Kao C-Y (2005) A stochastic differential equation model for quantifying transcriptional regulatory network in Saccharomyces cerevisiae. Bioinformatics 21(12):2883–2890
12. Bernardo D, Thompson MJ, Gardner TS, Chobot SE, Eastwood EL, Wojtovich AP, Elliott SJ, Schaus SE, Collins JJ (2005) Chemogenomic profiling on a genome-wide scale using reverse-engineered gene networks. Nat Biotechnol 23(3):377–383
13. Bansal M, Della Gatta G, Bernardo D (2006) Inference of gene regulatory networks and compound mode of action from time course gene expression profiles. Bioinformatics 22 (7):815–822
14. Akutsu T, Miyano S, Kuhara S (2000) Algorithms for identifying Boolean networks and related biological networks based on matrix multiplication and fingerprint function. J Comput Biol 7(3–4):331–343
15. Yamanishi Y, Vert J-P, Kanehisa M (2004) Protein network inference from multiple genomic data: a supervised approach. Bioinformatics 20: i363–i370
16. Vert J-P, Yamanishi Y (2005) Supervised graph inference. In: Saul LK, Weiss Y, Bottou L (eds) Advances in neural information processing systems, vol 17. MIT, Cambridge, MA, pp 1433–1440
17. Yamanishi Y, Vert J-P, Kanehisa M (2005) Supervised enzyme network inference from the integration of genomic data and chemical information. Bioinformatics 21:i468–i477
18. Bleakley K, Biau G, Vert J-P (2007) Supervised reconstruction of biological networks with local models. Bioinformatics 23(13):i57–i65
19. Mordelet F, Vert J-P (2008) SIRENE: Supervised inference of regulatory networks. Bioinformatics 24(16):i76–i82
20. Bishop C (2006) Pattern recognition and machine learning. Springer, Berlin
21. Schölkopf B, Platt JC, Shawe-Taylor J, Smola AJ, Williamson RC (2001) Estimating the Support of a High-Dimensional Distribution. Neural Comput 13:1443–1471
22. Denis F, Gilleron R, Letouzey F (2005) Learning from positive and unlabeled examples. Theoret Computer Sci 348(1):70–83
23. Scott C, Blanchard G (2009) Novelty detection: unlabeled data definitely help. In: van Dyk V, Welling M (ed) Proceedings of the twelfth international conference on artificial intelligence and statistics (AISTATS), vol 5. Clearwater Beach, Florida, pp 464–471
24. Manevitz LM, Yousef M (2001) One-class SVMs for document classification. J Mach Learn Res 2:139–154
25. Liu B, Lee WS, Yu PS, Li X (2002) Partially supervised classification of text documents. In: ICML '02: Proceedings of the Nineteenth International Conference on Machine Learning, San Francisco, CA, USA. Morgan Kaufmann Publishers, USA, pp 387–394
26. Li X, Liu B (2003) Learning to classify texts using positive and unlabeled data. In: IJCAI'03: Proceedings of the 18th international joint conference on Artificial intelligence San Francisco, CA. Morgan Kaufmann Publishers, USA, pp 587–592
27. Liu B, Dai Y, Li X, Lee WS, Yu PS (2003) Building text classifiers using positive and unlabeled examples. In: Proceedings of the third IEEE International Conference on Data Mining IEEE Computer Society, Washington DC, USA, pp 179–186
28. Yu H, Han J, Chang KC-C (2004) PEBL: Web page classification without negative examples. IEEE Trans Knowl Data Eng 16(1):70–81
29. Lee WS, Liu B (2003) Learning with positive and unlabeled examples using weighted logistic regression. In: Fawcett T, Mishra N (ed) Machine learning, proceedings of the twentieth international conference (ICML 2003). AAAI Press, USA, pp 448–455
30. Elkan C, Noto K (2008) Learning classifiers from only positive and unlabeled data. In: KDD '08: Proceeding of the 14th ACM SIGKDD international conference on knowledge discovery and data mining. ACM, New York, USA, pp 213–220
31. Mordelet F, Vert J-P (2010) A bagging SVM to learn from positive and unlabeled examples. Technical Report HAL:00523336
32. Salgado H, Gama-Castro S, Peralta-Gil M, Díaz-Peredo E, Sánchez-Solano F, Santos-Zavaleta A, Martínez-Flores I, Jiménez-Jacinto V, Bonavides-Martínez C, Segura-Salazar J, Martínez-Antonio A, Collado-Vides J (2006) RegulonDB

(version 5.0): Escherichia coli K-12 transcriptional regulatory network, operon organization, and growth conditions. Nucleic Acids Res 34 (Database issue):D394–D397

33. Breiman L (1996) Bagging predictors. Mach Learn 24(2):123–140
34. Vapnik VN (1998) Statistical learning theory. Wiley, New York
35. Schölkopf B, Tsuda K, Vert J-P (2004) Kernel methods in computational biology. MIT, Cambridge, MA
36. Vert R, Vert J-P (2006) Consistency and convergence rates of one-class SVMs and related algorithms. J Mach Learn Res 7:817–854
37. Mordelet F (2010) Learning from positive and unlabeled examples in biology. Ph.D. thesis, Mines ParisTech
38. Joachims T (1997) A probabilistic analysis of the Rocchio algorithm with TFIDF for text categorization. In: ICML '97: Proceedings of the fourteenth international conference on machine learning, Nashville, Tennessee. Morgan Kaufmann Publishers, USA, pp 143–151
39. De Bie T, Tranchevent L-C, vanOeffelen LMM, Moreau Y (2007) Kernel-based data fusion for gene prioritization. Bioinformatics 23(13):i125–i132
40. Zhang K, Tsang I, Kwok J (2009) Maximum margin clustering made practical. IEEE Trans Neural Network 20(4):583–596
41. Chang C-C, Lin C-J (2001) LIBSVM: a library for support vector machines. Software available at http://www.csie.ntu.edu.tw/~cjlin/libsvm

Chapter 6

Mining Regulatory Network Connections by Ranking Transcription Factor Target Genes Using Time Series Expression Data

Antti Honkela, Magnus Rattray, and Neil D. Lawrence

Abstract

Reverse engineering the gene regulatory network is challenging because the amount of available data is very limited compared to the complexity of the underlying network. We present a technique addressing this problem through focussing on a more limited problem: inferring direct targets of a transcription factor from short expression time series. The method is based on combining Gaussian process priors and ordinary differential equation models allowing inference on limited potentially unevenly sampled data. The method is implemented as an R/Bioconductor package, and it is demonstrated by ranking candidate targets of the p53 tumour suppressor.

Key words: Gaussian process, Reverse engineering, Gene regulatory network, Ordinary differential equation

1. Introduction

Understanding the function and regulation of all human genes is one of the most important challenges that needs to be solved to fully reap the benefits of the Human Genome Project and the genomic era.

There are several regulatory mechanisms affecting protein expression. We will focus on transcriptional regulation, which is mediated by transcription factors (TFs). They are proteins that bind the DNA to activate or repress the transcription of their target genes. The relationships between TFs and their target genes can be represented as a graph which is called the gene regulatory network. In reality, this network is context sensitive with many connections being, for instance, tissue specific.

Inferring the gene regulatory network from data is a very challenging problem. Even with high-throughput measurement

Hiroshi Mamitsuka et al. (eds.), *Data Mining for Systems Biology: Methods and Protocols*, Methods in Molecular Biology, vol. 939, DOI 10.1007/978-1-62703-107-3_6, © Springer Science+Business Media New York 2013

techniques providing data on a genome-wide scale, the amount of data is tiny compared to the potential complexity of the regulatory network. Taking a conservative estimate of 1,500 human TFs (1) regulating some 22,500 genes (2) yields an astronomical number of more than 10^{450} potential networks. Even with a more realistic assumption of at most 5 regulating TFs for each gene, there are more than 10^{18} potential networks or from another perspective more than $6.3 \cdot 10^{13}$ potential sets of regulators for each gene. Assuming the regulatory network can be captured by a differential equation model with as many parameters, roughly twice as many experiments would be needed to identify all the parameters (3). If a simplified model of regulation with fewer parameters (such as a linear differential equation) can be assumed, the number of experiments will drop accordingly (4), but naturally such a model cannot capture all possible modes of combinatorial regulation.

The difficulty of general network inference even for simpler organisms was recently demonstrated by the DREAM5 Network inference challenge (see Note 1), where one of the tasks was to infer the regulatory network for yeast using practically all available expression data. As a result, the best-performing team in this subtask achieved an area under the ROC curve of 0.539, which is only marginally better than the result 0.5 corresponding to random guessing.

1.1. Probabilistic Dynamical Models of Gene Regulation

To avoid these difficulties, we will focus on the specific task of identifying the targets of a TF in a time series experiment where the TF activity is changing (5, 6). The method works based on expression data alone. The TF activity can be estimated using information from known target genes or using the TF mRNA, if the TF is assumed to be primarily under transcriptional control (see Note 2). Given an estimate of TF activity and a model of transcription based on this activity, we can rank candidate targets based on how well they fit the model of regulation by this TF.

As our model of gene transcription regulated by a TF and optionally TF protein translation we use the following system of linear ordinary differential equations (ODEs):

$$\frac{dp(t)}{dt} = f(t) - \delta p(t) \tag{1}$$

$$\frac{dm_j(t)}{dt} = B_j + S_j p(t) - D_j m_j(t) \tag{2}$$

where $p(t)$ is the TF protein at time t, $m_j(t)$ is the jth target mRNA concentration and $f(t)$ is the TF mRNA. The parameters B_j, S_j, and D_j are the baseline transcription rate, sensitivity and decay rate, respectively, for the mRNA of the jth target as described in (7). The parameter δ is the decay rate of the TF protein (6).

In order to infer the protein activity $p(t)$, we need a prior for it. As the functions $f(t)$, $p(t)$ and $m_j(t)$ are deterministically linked by the ODEs, this can be accomplished by placing a Gaussian process prior on $f(t)$ (see Note 3). This leads to a joint Gaussian process over the three continuous-time functions $f(t)$, $p(t)$ and $m_j(t)$. Data are observations of the expression levels at arbitrary specific times (not necessarily evenly spaced), and we assume a Gaussian noise model: $\hat{m}_j(t_i) \sim N(m_j(t_i), \sigma^2_{i,m_j})$ and $\hat{f}(t_i) \sim N(f(t_i), \sigma^2_{i,f})$ with known (derived from *puma*, see below) or estimated gene-specific noise variance parameters. The parameters of the model as well as other parameters of the Gaussian process covariance are optimised by maximising the marginal likelihood.

1.2. Related Approaches

Other approaches for ranking TF targets using similar data include the rHVDM package (8) which implements a non-probabilistic variant of the same transcription ODE model (7). rHVDM does not support using the TF mRNA observations to infer TF activity. Another alternative that defines the model based only on TF expression data without a possibility of using information from the targets to infer TF activity is TSNI (9).

In general, the main difficulty in linking putative target genes with their regulators using expression data is the different degradation rates of mRNA molecules of different genes. If the degradation rates are available and can be compensated for, estimation of the regulator activation profiles is greatly simplified (10). Our method can easily use such information as well.

2. Materials

The presented target-ranking approach is implemented in the *tigre* package which is available in Bioconductor (11) for R (since Bioconductor 2.6 for R-2.11). *tigre* can make effective use of error estimates for expression levels provided by preprocessing methods. For Affymetrix arrays these are most easily available from the Bioconductor package *puma* (12); for Illumina arrays the corresponding information is available from the Bioconductor package *lumi* (13).

3. Methods

3.1. Data Collection and Experimental Design

The primary data used by the model are expression time series. They can be measured using any quantitative technique such as mRNA sequencing (RNA-seq) or microarrays.

The application of our ranking method requires time series data. Due to financial constraints, most biological time series are very short, which poses problems for data modelling. We have applied the ranking to data sets with as few as 6–7 time points (6, 14), although longer time series (>10 time points) are significantly more informative.

Given a fixed experimental budget, it seems in general preferable to have more experimental perturbations and more finely sampled time series rather than more replicates of the same measurements, except possibly for some single key time points. The models can use continuity in the time series to compensate for the noise that is usually detected using the replicates.

3.2. Preprocessing

The expression data should be preprocessed using the best tools for that particular platform. The *tigre* ranking method can make use of error or variance estimates from preprocessing. At the time of writing, the first such tools for RNA-seq data are only beginning to emerge. For microarrays, for example, the *puma* (12) Bioconductor package provides such estimates for Affymetrix GeneChips, while the *lumi* (13) package provides this for Illumina Bead Arrays.

In order to use the data with *tigre*, it must be imported using the function `processData` that normalises the output from *puma* and *lumi*, or `processRawData`, that handles plain expression data matrices. These functions require information about the experimental setup of the data, including observation times of different samples and which replicate time series they belong to. An example of the preprocessing using the p53 activation data from (7) is presented below. More details are available in the vignettes and other documentation of the respective packages.

```
library(ArrayExpress)

## Get data from ArrayExpress
p53.affybatch <- ArrayExpress('E-MEXP-549')
## Sort the arrays in a sensible order
mynames <- rownames(pData(p53.affybatch))
I <- order(strtrim(mynames, 5),
           pData(p53.affybatch)['Factor.Value..time.'])
p53.affybatch <- p53.affybatch[,I]

## Run mmgMOS preprocessing
library(puma)
p53.exprReslt <- mmgmos(p53.affybatch)

## Run tigre data normalisation
exps <- rep(1:3, each=7)    # replicates: 1,...,1,2,...,2,3,...,3
p53.timeseries <- processData(p53.exprReslt, experiments=exps)
```

3.3. Ranking

Ranking candidate targets of a TF requires inferring the TF activity profile. This can be done using TF mRNA expression levels if the TF can be assumed to be under transcriptional control, or using expression data of known targets, or combining both. Both of these cases are handled by the *tigre* function `GPRankTargets`.

If known targets are available, the function first fits a model using these known targets to infer the TF activity. All other genes are then screened by fitting additional models by adding them to the target set one at a time. If there are no known targets, all genes are screened by simply fitting them one at a time. In our p53 example, this is achieved using the code below.

```
## Known target probe-sets from Barenco et al. (2006)
## These are the most informative probes for genes
## 'DDB2', 'CDKN1A', 'PA26', 'BIK', 'TNFRSF10B'
knownprobes <- c('203409_at', '202284_s_at', '218346_s_at',
                 '205780_at', '209295_at')

## Run the ranking.  This may take several hours to run.
## Results are saved to file 'p53ranking.RData'.
scores <- GPRankTargets(p53.timeseries, knownTargets=knownprobes,
                        scoreSaveFile='p53ranking.RData')

## Sort the list according to likelihood
scores <- sort(scores, descending=TRUE)

## Find 10 top-ranking probe sets
genes(scores)[1:10]

## Find the corresponding gene symbols
library(annotate)
mget(unlist(genes(s)[1:10]),
     getAnnMap('SYMBOL', annotation(p53.timeseries)))
```

3.4. Visualisation and Analysis

The most effective way to explore the ranking results is by looking at visualisations of the models. Given that more than 10,000 models are produced in the above analysis, this can be a daunting task without proper tools. A very useful tool for browsing the models is the *tigreBrowser* tool (see Note 4). The scores and the corresponding model visualisations can be exported for *tigreBrowser* as below.

```
## Export the scores and model visualisations to tigreBrowser
export.scores(scores, datasetName='Barenco2006',
              experimentSet='GPSIM_5_known',
              database='p53_results.sqlite',
              preprocData=p53.timeseries)
```

The results can then be viewed most easily by running the `tigreServer.py` script and pointing it to the saved `p53_results.sqlite` file. A screen shot of the browser is shown in Fig. 1.

As there is no suitable genome-wide validation data available, we used a sequence-based method of validation. To do this, we looked for strong occurrences of the known p53 binding motif in the promoters of highly ranked targets. This was done by using the `p53scan` tool of Smeenk et al. (15) on the sequences of 5,000 bp upstream of annotated transcription starts of RefSeq genes with annotated 5' UTRs downloaded from hg19 assembly in the UCSC Genome Browser (16). Figure 2 shows the enrichment of the strongest 10% of the binding motif instances identified by `p53scan` as ranked by the score on the promoters of highly ranked genes. While these high-scoring motif instances are the most likely true binding sites, some of them may still be non-functional. On the other hand, it is likely that many other genes are regulated through a weaker site on the promoter or a more distant site which cannot be detected without additional information.

4. Notes

1. See http://wiki.c2b2.columbia.edu/dream/results/DREAM5/?c=4_1.
2. One critical question in applying *tigre* is whether to use the TF mRNA data. The extra information can potentially greatly help in identifying the TF activity profile and hence making better predictions of new targets, but it can also lead to incorrect predictions if it turns out incorrect. Based on our experience, the critical question is whether some outside influence, such as a external signal, is the rate-limiting factor in the production of the active TF. Simple post-translational modifications such as dimerisation do not appear to significantly hinder the use of TF mRNA data, as witnessed by the results in (6) for Drosophila TFs Mef2 and Twi which are known to function as dimers.
3. If the TF protein is under significant post-translational regulation, Eq. 1 may be omitted and the prior placed directly

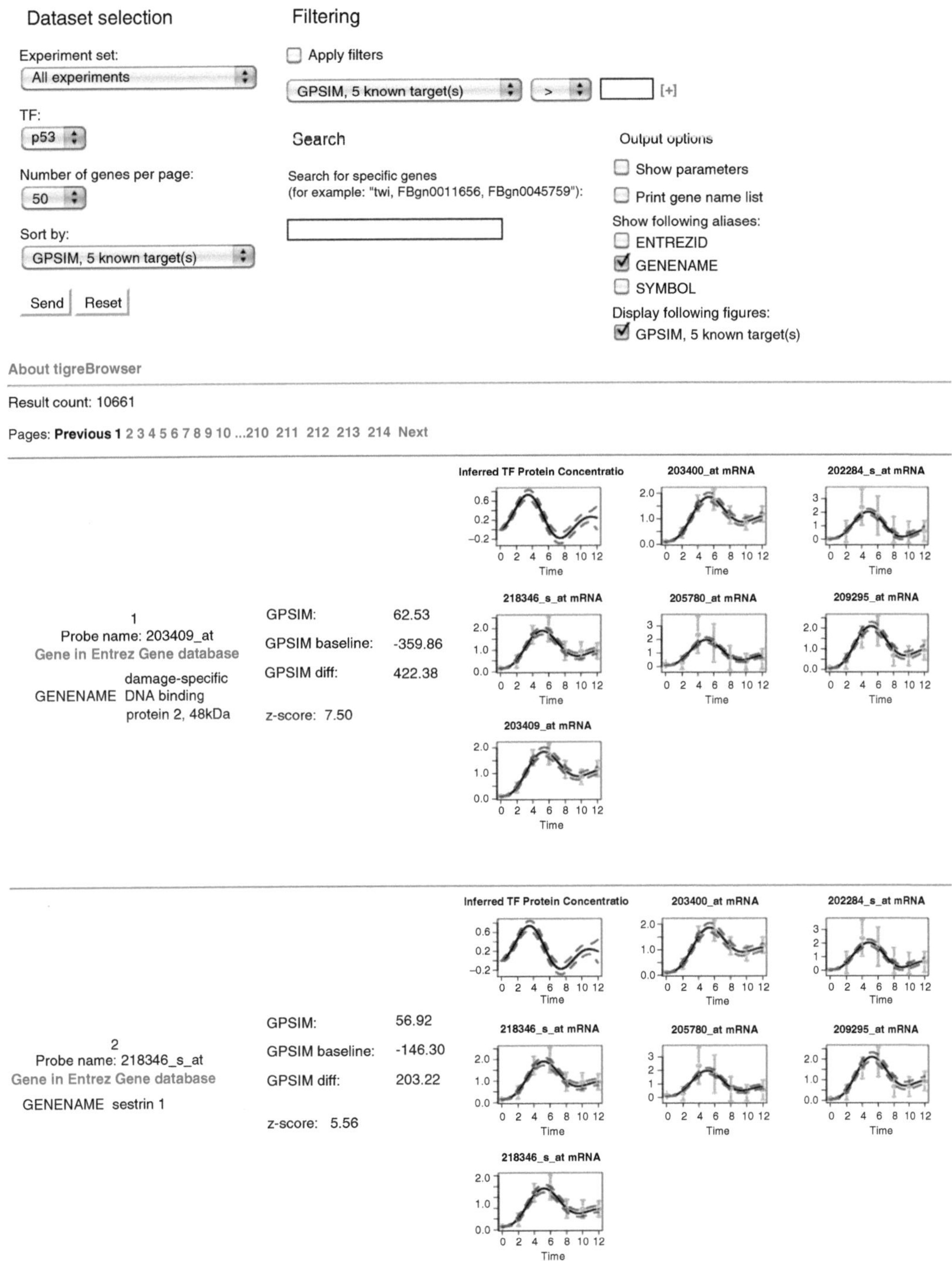

Fig. 1. Screen shot of the tigreBrowser showing the top results of the p53 experiment.

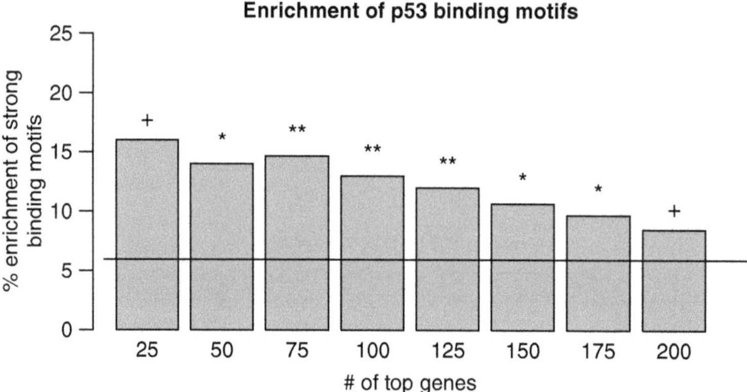

Fig. 2. Enrichment of strong p53 binding motifs in the promoters of selected number of top predicted targets. p-values of the enrichments are denoted by "***": $p < 0.001$, "**": $p < 0.01$, "*": $p < 0.05$, "+": $p < 0.1$ (tail probability in hypergeometric distribution).

on $p(t)$. In this case, multiple known targets are needed to reliably infer $p(t)$.

4. tigreBrowser is available for download at http://www.hiit.fi/u/ahonkela/tigre/.

References

1. Vaquerizas JM, Kummerfeld SK, Teichmann SA, Luscombe NM (2009) A census of human transcription factors: function, expression and evolution. Nat Rev Genet 10:252–263
2. Pertea M, Salzberg SL (2010) Between a chicken and a grape: estimating the number of human genes. Genome Biol 11:206
3. Sontag ED (2002) For differential equations with r parameters, 2r+1 experiments are enough for identification. J Nonlinear Sci 12:553–583
4. Stark J, Brewer D, Barenco M, Tomescu D, Callard R, Hubank M (2003) Reconstructing gene networks: what are the limits? Biochem Soc Trans 31:1519–1525
5. Gao P, Honkela A, Rattray M, Lawrence ND (2008) Gaussian process modelling of latent chemical species: applications to inferring transcription factor activities. Bioinformatics 24: i70–i75
6. Honkela A, Girardot C, Gustafson EH, Liu Y-H, Furlong EEM, Lawrence ND, Rattray M (2010) Model-based method for transcription factor target identification with limited data. Proc Natl Acad Sci USA 107:7793–7798
7. Barenco M, Tomescu D, Brewer D, Callard R, Stark J, Hubank M (2006) Ranked prediction of p53 targets using hidden variable dynamic modeling. Genome Biol 7:R25
8. Barenco M, Papouli E, Shah S, Brewer D, Miller CJ, Hubank M (2009) rHVDM: An R package to predict the activity and targets of a transcription factor. Bioinformatics 25:419–420
9. Della Gatta G, Bansal M, Ambesi-Impiombato A, Antonini D, Missero C, di Bernardo D (2008) Direct targets of the TRP63 transcription factor revealed by a combination of gene expression profiling and reverse engineering. Genome Res 18:939–948
10. Barenco M, Brewer D, Papouli E, Tomescu D, Callard R, Stark J, Hubank M (2009) Dissection of a complex transcriptional response using genome-wide transcriptional modelling. Mol Syst Biol 5:327
11. Gentleman RC et al (2004) Bioconductor: open software development for computational biology and bioinformatics. Genome Biol 5:R80
12. Pearson RD, Liu X, Sanguinetti G, Milo M, Lawrence ND, Rattray M (2009) puma: A bioconductor package for propagating uncertainty in microarray analysis. BMC Bioinformatics 10:211
13. Du P, Kibbe WA, Lin SM (2008) lumi: A pipeline for processing illumina microarray. Bioinformatics 24:1547–1548

14. Honkela A, Milo M, Holley M, Rattray M, Lawrence ND (2010) Ranking of gene regulators through differential equations and Gaussian processes. Proceedings of 2010 IEEE international workshop on machine learning for signal processing (MLSP 2010), Kittilä, Finland, pp 154–159
15. Smeenk L, van Heeringen SJ, Koeppel M, van Driel MA, Bartels SJJ, Akkers RC, Denissov S, Stunnenberg HG, Lohrum M (2008) Characterization of genome-wide p53-binding sites upon stress response. Nucleic Acids Res 36:3639–3654
16. Fujita PA, Rhead B, Zweig AS, Hinrichs AS, Karolchik D, Cline MS, Goldman M, Barber GP, Clawson H, Coelho A, Diekhans M, Dreszer TR, Giardine BM, Harte RA, Hillman-Jackson J, Hsu F, Kirkup V, Kuhn RM, Learned K, Li CH, Meyer LR, Pohl A, Raney BJ, Rosenbloom KR, Smith KE, Haussler D, Kent WJ (2011) The UCSC Genome Browser database: update 2011. Nucleic Acids Res 39:D876–D882

Chapter 7

Identifying Pathways of Coordinated Gene Expression

Timothy Hancock, Ichigaku Takigawa, and Hiroshi Mamitsuka

Abstract

Methods capable of identifying genetic pathways with coordinated expression signatures are critical to advance our understanding of the functions of biological networks. Currently, the most comprehensive and validated biological networks are metabolic networks. Complete metabolic networks are easily sourced from multiple online databases. These databases reveal metabolic networks to be large, highly complex structures. This complexity is sufficient to hide the specific details on which pathways are interacting to produce an observed network response. In this chapter we will outline a complete framework for identifying the metabolic pathways that relate to an observed phenomenon. To illuminate the functional metabolic pathways, we overlay microarray experiments on top of a complete metabolic network. We then extract the functional components within a metabolic network through a combination of novel pathway ranking, clustering, and classification algorithms. This chapter is designed as a simple tutorial which enables this framework to be applied to any metabolic network and microarray data.

Key words: Metabolic network, Gene pathway, Micorarray expression, Ranking, clustering, Classification

1. Introduction

Metabolic networks are maps of chemical reaction pathways that are known to occur within a cell. The functions of a metabolic network are controlled by the coordinated activation and interaction of specific genes. The metabolic network can therefore be viewed as a set of organized gene interactions which combine together to form well-defined pathways. For many organisms, the network structure and genetic dependencies of metabolism have been identified and are stored within online databases such as KEGG (1), MetaCyc (2), and Reactome (3, 4). These databases serve as maps which describe the known metabolic processes that occur within the cell. These maps can be analyzed at many different resolutions from large-scale views of interacting pathways and subnetworks through to the analysis of individual compounds and their genetic dependencies.

Hiroshi Mamitsuka et al. (eds.), *Data Mining for Systems Biology: Methods and Protocols*, Methods in Molecular Biology, vol. 939, DOI 10.1007/978-1-62703-107-3_7,

Pathways of coordinated gene expression determine which metabolic compounds can be synthesized and thus can define the function of metabolic networks. Additionally, microarray experiments have allowed researchers to measure gene expression under various experimental conditions. The widespread use of microarray expression has led to the creation of many bioinformatics methods, most notably gene set enrichment analysis (GSEA) (5, 6) and subgraph extraction methods (7–10). These methods combine microarray expression with metabolic networks in order to identify the pathways that drive an observed response. Although these methods all combine metabolic networks with gene expression they largely ignore the known pathway structure of metabolism.

In our original paper (11), we proposed our complete framework for the extraction and analysis of the most correlated metabolic pathways within gene expression data. Our framework is a combination of three complementary methods. Firstly, a pathway-ranking method is used to extract the most correlated metabolic pathways within gene expression data (12). Then to analyze the structure within the extracted pathway list, we developed Markov mixture models for both unsupervised (13) and supervised analyses (14, 15). In this chapter, we present a tutorial on our proposed framework, and we clearly discuss the major steps, assumptions, and limitations of our methodology.

In the *Materials* section (Subheading 2), we will discuss how to prepare a metabolic network for pathway analysis. This procedure involves three steps:

Creating a gene-interaction network: We describe how to convert the compound-reaction structure of metabolic networks into the gene-interaction network.

Defining the pathway targets: We discuss the specification of start and end nodes required for pathway mining and how this will affect the interpretation of the result.

Weighting the gene interactions: We show how to use microarray expression information to weight the edges of a gene-interaction network and highlight the metabolic network features which are related to specific observations.

In the *Methods* section (Subheading 3), we will discuss how to extract and analyze pathways of coordinated gene-expression from a metabolic network.

Pathway extraction: We describe how to efficiently identify the top K most important pathways from the weighted gene interaction network.

Pathway clustering and classification methods: We provide examples of pathway clustering and classification algorithms which allow for the analysis of large numbers of pathways.

A complete implementation of the methods described in this chapter is available in the `PathRanker` R package which can be downloaded from http://www.bic.kyoto-u.ac.jp/pathway/tim hancock/index.html.

2. Materials

Metabolic networks are collections of known biologically present pathways. Within online databases such as KEGG, known metabolic pathways are commonly represented as compound-reaction graphs. A compound-reaction graph focuses on how metabolic compounds are produced and used within the network by connecting compound nodes using reaction edges which define substrate and product compound dependencies. The entire metabolic network is then built by connecting all available pathway compound-reaction graphs. The resultant metabolic network is highly redundant where multiple paths exist between any two compounds. This high level of redundancy means that care must be taken to ensure that any identified paths through the network are faithful representations of the underlying biological processes. In this section, we describe a protocol for processing metabolic networks from the original compound-reaction graph stored by the online databases into a gene-interaction graph which captures the structure of the original network but allows for the extraction of paths comprising of pairwise gene interactions.

2.1. Creating a Gene Interaction Network

The top network in Fig. 1 presents an example compound-reaction graph in a form commonly found in online metabolic databases such as KEGG. This network contains five metabolic compounds [$C1$, $C2$, $C3$, $C4$, $C5$] connected by three reactions [$R1$, $R2$, $R3$], each of which are catalyzed by a subset of five genes [$G1$, $G2$, $G3$, $G4$, $G5$]. A compound reaction graph defines the nodes as metabolic compounds and labels the edges with reactions. Compound-reaction graphs focus on high-level reaction interactions and their compound dependencies. However, if we are to overlay microarray data onto a metabolic network and analyze gene interactions, a more detailed gene-interaction network must be created. Therefore, we must first convert this compound-reaction graph in Fig. 1 into a gene-interaction graph which defines the nodes as genes, and the edges are labeled by the compounds produced by the interaction between two genes. The compound-reaction graph in Fig. 1 has some interesting features that are common within metabolic networks but which highlight the assumptions we must make when we define a gene-interaction network. These features are:

1. There are two reactions, $R1$ and $R3$, which produce two product compounds: $C1 \xrightarrow{R1} (C2, C3)$ and $C3 \xrightarrow{R3} (C4, C5)$.

Start Compounds: [C1]
End Compounds: [C4, C4]
Metabolic Compounds: [C1, C2, C3, C4, C5]
Genes: [G1, G2, G3, G4, G5]
Reactions: [R1 = [G1,G2,G3], R2 = [G4], R3 = [G2,G5]]

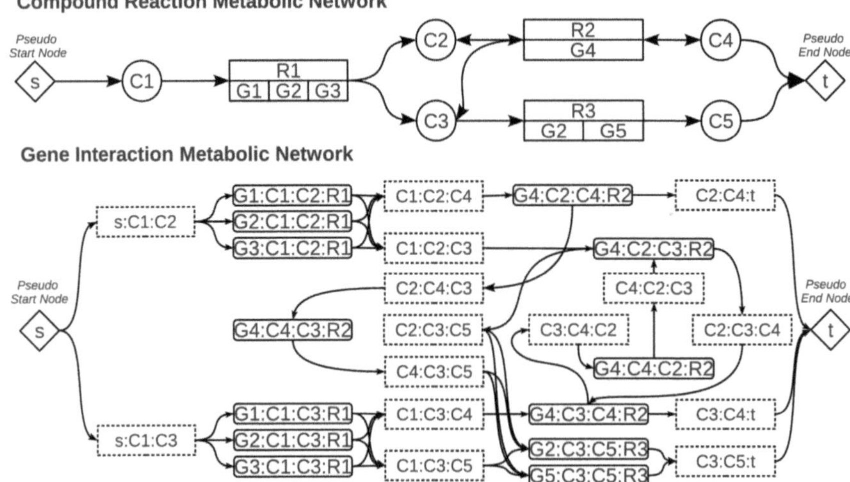

Fig. 1. Example conversion from a metabolic network to a gene interaction network.

2. Reaction $R2$ requires both $C2$ and $C3$ to proceed: $(C2, C3) \xrightarrow{R2} C4$.
3. Reaction $R2$ is reversible meaning that $C4 \xrightarrow{R2} (C2, C3)$ is also a valid pathway.
4. Gene $G2$ occurs in two connected reactions $R1$ and $R3$.

To convert a compound-reaction graph into a gene-interaction graph requires the definition of node annotations which include function and location information of each gene. To achieve this, we define each node by the gene name, substrate compound, product compound, reaction, and pathway within which it is observed. In the gene-interaction graph at the base of Fig. 1, each rounded rectangle is a resulting gene node where each annotation is a string delimited by a colon, for example, annotation *G4:C2:C4:R2* corresponds to the gene $G4$ observed in reaction $R2$ using substrate compound $C2$ and producing compound $C4$. For clarity of representation within Fig. 1, we removed the pathway annotations.

Edge annotations are also added to the gene-interaction graph in Fig. 1 and are displayed as dashed rectangles. These edge annotations allow for easy visual mapping between the compound-reaction and gene-interaction graphs. The edge annotations simply summarize the interaction between two genes by the substrate compound, middle compound, and final product compound produced. For example, edge label *C1:C2:C4* defines a transition from

C1 → *C2* and then from *C2* → *C4*. The edge annotates help to reduce the visual complexity of the network by allowing for grouping of edges with the same compound transitions. This can help as each reaction can be catalyzed by multiple genes but still only related to a single compound transition.

Using this annotation scheme, the following compound-reaction network properties are retained:

1. Reaction direction information is maintained by the creation of separate gene nodes for each direction. For example, *G4:C4:C2:R2* is the reverse direction of *G4:C2:C4:R2*. These different nodes each have different connectivity within the graph and therefore will be used in paths.
2. The different reaction memberships of *G2* are also simply converted into different nodes, *G2:C1:C2:R1*, *G2:C1:C3:R1* and *G2:C3:C5:R3*. However, *G2* is now directly connected to itself (with different annotations) which will cause trouble when we considered weighting the edges using microarray data as both nodes will contain the same expression information. We deal with this issue in Subheading 2.3.

The disadvantages of this annotation scheme are:

1. Reactions that produce multiple products or require multiple substrates are broken up into single pathways. For example, the multiple substrate reaction $(C2, C3) \xrightarrow{R2}$ is treated as two separate pathways leading to *C4*: $C2 \xrightarrow{R2} C4$ and $C3 \xrightarrow{R2} C4$. This separation of reaction function may not always completely represent the underlying biology but does allow for a view of the network in terms of pairwise gene interactions.
2. The gene-interaction network appears much more complex than the original compound-reaction graph. However, this disadvantage motivates our development of pathway-mining tools to extract the important structure from such complex networks.

2.2. Defining Pathway Targets

A pathway in biological terms has multiple definitions. We define a pathway to be a connected sequence of connected gene interactions between prespecified start and end points (see Note 1). These start and end points are defined to be sets of compounds within the metabolic network. In Fig. 1, we specified the start compound to be *C1* and the end compound to be either *C4* or *C5*. These start and end compound sets are stored as pseudo nodes in the network from which we add edges to each relevant compound. For example, in Fig. 1, the start pseudo node **s** is connected to *C1*, and the end pseudo node **t** connected to both *C4* and *C5*. This connection is done before the conversion to the gene-interaction network:

- In the `PathRanker` R package, the specification of the start and end compounds and the conversion to a gene-interaction network can be done using a combination of the `getKEGG` and `computeGeneNetwork` functions.

The specification of the start and end compounds is also specifying the type of research question. In (12) and (14, 15) (see Note 3), we specified single start and end compounds. Our goal in these analyses was to identify pathway branches from otherwise conserved paths which explain a specific set of experimental observations. A separate approach taken in (11) defined the set of start compounds to be all source compounds and the set of end compounds to be all sink compounds within the entire KEGG metabolic network. In this analysis our goal was to extract paths with common structure that could be clustered together to identify larger subnetworks that corresponded to specific experimental conditions.

2.3. Weighting the Gene Interactions

The specification of pseudo start and end nodes means that we are seeking the most important paths of coordinated gene expression between $\mathbf{s} = C1$ and either $C4$ or $C5$, $\mathbf{t} = [C1, C5]$. There are many potential paths which fulfill the criteria, and two potential paths could be:

1. $\mathbf{s} \to s : C1 : C2 \to G1 : C1 : C2 : R1 \to C1 : C2 : C4$
   $\to G4 : C2 : C4 : R2 \to C2 : C4 : t \to \mathbf{t}.$
2. $\mathbf{s} \to s : C1 : C3 \to G2 : C1 : C3 : R1 \to C1 : C3 : C4 \to G4 :$
   $C3 : C4 : R2 \to C3 : C4 : C2 \to G4 :$
   $C4 : C2 : R2 \to C4 : C2 : C3 \to G4 :$
   $C2 : C3 : R2 \to C2 : C3 : C5 \to G5 :$
   $C3 : C5 : R3 \to C3 : C5 : t \to \mathbf{t}.$

These two paths finish at different end compounds and have a remarkably different structure. In this section we describe how to use microarray data to weight the edges of the network. These edge weights will then be combined together to compute a pathway probability score which will be used to all rank pathways in order of importance.

Microarray data is presented in a matrix where the expression levels of each gene are defined as the columns and the rows are the experiments. In Subheading 2.1 we construct a gene iteration network where each gene is a network node. A correlation coefficient of the expression values between two connected genes can now be used to measure the importance of each edge within the gene-interaction network. In (11, 12), we use the correlation between two neighboring gene expression vectors to define a set of edge weights for a metabolic network:

- It is common for each experiment to be labeled by an experimental condition. In this case separate edge weights can be computed from each experimental condition separately.

To illustrate the results and highlight some pitfalls of this weighted gene-interaction network we follow these steps:

1. Download the KEGG *saccharomyces cerevisiae* (budding yeast) metabolic network and process it into a gene-interaction network as in Fig. 1 (see Note 2).
2. Download the benchmark Gasch microarray data (16) from the `gaschYHS` Bioconductor package (17, 18). The Gasch microarray data measure the gene expression response of yeast under multiple stress conditions.
3. Extract the heat shock experiments as our target experimental condition and group together all other stress conditions into one group called other.
4. For both heat shock and other microarray observations, compute the Pearson correlation coefficient for all neighboring genes within the gene-interaction network:
   (a) For the correlation coefficient computation to reduce the effect of outlying observations we take the median of 100 bootstrapped samples.
   (b) The edge weights between the pseudo start s and end t nodes and the genes they are connected to are set to 1.

   The results of this procedure are presented in Fig. 2:

   - In the `PathRanker` R package, the bootstrapped edge correlations can be computed using the `assignEdgeWeights` function.

In Fig. 2, the left-most column of histograms shows the distribution of all neighboring gene correlation coefficients for each stress condition. Note here that both histograms, specifically for the other group, appear have a peak correlation of $r = 1$. This is due to the effect of the same gene being connected to itself. This situation is

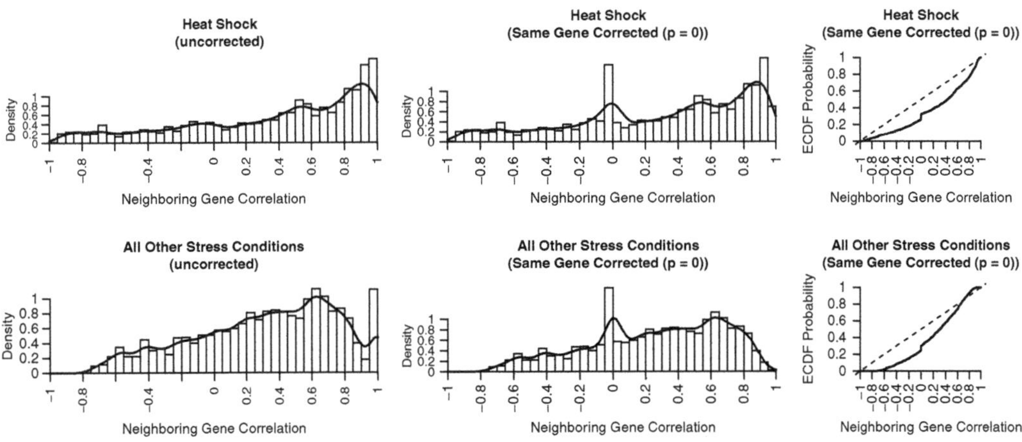

Fig. 2. Distribution of gene-interaction (edge weight) correlations.

shown in the example network in Fig. 2 with *G2* being connected to itself (*G2:C1:C3:R1* $\rightarrow$ *G2:C3:C5:R3*). The same microarray expression must be used for all three nodes related to *G2* in the gene-interaction network, and the correlation between all of these nodes will be $r = 1$. We call this situation a same gene edge.

It is common in metabolic networks for the same gene to catalyze a chain of reactions, see, for example, the fatty acid metabolism pathways of KEGG. Therefore, same gene edges must be very common in the gene-interaction network. However, as most analyses are interested in the interaction between different genes, these same edges may become somewhat artificial and uninteresting. One way to reduce the impact of the same gene edges within any pathway analysis is to manually set the correlation of these edges to a prespecified value as done in (11). We perform this correction to the edge correlations in Fig. 2 and set all same gene edges to have a correlation of ($p = 0$) and present the results in the middle column of histograms in Fig. 2. The same gene-corrected histograms in Fig. 2 clearly show the artificial perfect ($r = 1$) correlations induced by same gene edges create a skew within the edge correlation histograms:

1. Assigning the same gene edge correlations to a value of $p = 0$ has the effect of emphasizing positively correlated edges. Positively corrected connected genes are assumed to support each other and therefore create a path. Negative correlations are assumed to create a block and not allow the path to proceed. Setting the same gene edge correlations to a value of 0 will have the effect of a path-ranking algorithm to select a same gene edge rather than a negatively correlated edge.
2. The setting of same gene correlations to a low value may have the effect of blocking some pathways, such as fatty acid metabolism.

While it is reasonable to use the correlation value as an edge weight, the procedure can be generalized by using the probability of each edge weight as computed by an empirical cumulative distribution function (ecdf probability) over all edge weights (12). The ecdf probability computes the probability of each edge weight given all observed edge correlations. There are two advantages of using the ecdf as the edge weight:

1. It allows for the computation of a nonparametric p value for any path within the network (12).
2. It can be computed for any edge weight, not simply correlation but also allows any other metrics of interest to be specified.

The ecdf probabilities for the edge correlations are presented in the right most column of plots in Fig. 2. Once the edges have been weighted with microarray expression information we can now consider methods of extracting and analyzing pathway through the weighted gene-interaction network.

3. Methods

3.1. Pathway Extraction

Once a weighted gene-interaction network has been created by following the instructions in Subheading 2 we use the procedure as defined in (12) for ranking and extracting the top K pathways. The pathway-ranking procedure is a greedy algorithm to find the shortest loopless paths through the gene interaction network between any pair of start, **s**, and end, **t**, nodes. In practice, this is solved efficiently in polynomial time with the Yen–Lawler algorithm (19, 20). In this chapter, we do not discuss the implementation details of Yen–Lawler algorithm but instead discuss the practical use of the path-ranking algorithm.

To highlight the potential problems that may be encountered when extracting the top K pathways, we have to consider the pathway scoring metric. Consider a path of length L containing genes $[g_1, \ldots, g_L]$, then the path score metric used in (12) is defined to be

$$P_{score}(g_1, \ldots, g_L) = \min\left\{ -\sum_{l=2}^{L} \log\left[ecdf(g_{l-1}, g_l)\right] \right\}, \quad (1)$$

where ecdf is the empirical cumulative distribution probability of the correlation between g_{l-1} and g_l. There are several important implications of this scoring system that must be considered:

1. The ecdf value is the probability of each edge correlation observed within the data. Therefore, an ecdf value of 1 corresponds to the most highly correlated gene pair.
2. The closer the ecdf value is to 1, the closer its log value will be to 0.
3. Therefore, to minimize, we simply take the shortest most highly correlated gene path between any start and end vertex.

Depending on how the **s** and **t** nodes are specified, minimizing P_{score} can be biased to short paths. For example, in Fig. 3, we show a simple example network where naively minimizing P_{score} may lead to problems. In Fig. 3, we have predefined start nodes [$S1$, $S2$, $S3$] and end node [$T1$, $T2$] which are connected through a gene network. The ecdf values of each edge are also displayed. From Fig. 3 we can see the following features:

1. The top-ranked path will clearly be from $S1$ to $T1$ by $G1$ and $G3$. The P_{score} of this path is 0.0305, and it contains one gene interaction so the path length is 1.
2. The longest path is from $S3$ to $T2$ by genes $G6$, $G8$, $G9$, $G7$ and $G5$. The of this path is 0.1202, and it contains four gene interactions, so its path length is 4.

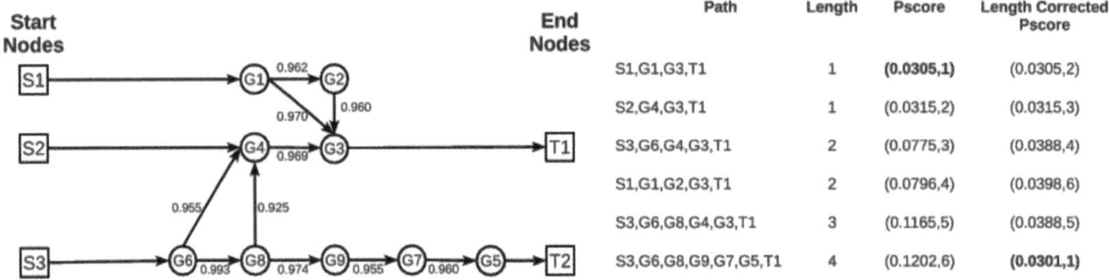

Path	Length	Pscore	Length Corrected Pscore
S1,G1,G3,T1	1	**(0.0305,1)**	(0.0305,2)
S2,G4,G3,T1	1	(0.0315,2)	(0.0315,3)
S3,G6,G4,G3,T1	2	(0.0775,3)	(0.0388,4)
S1,G1,G2,G3,T1	2	(0.0796,4)	(0.0398,6)
S3,G6,G8,G4,G3,T1	3	(0.1165,5)	(0.0388,5)
S3,G6,G8,G9,G7,G5,T1	4	(0.1202,6)	**(0.0301,1)**

Fig. 3. P_{score} pathway ranking versus *p* value ranking.

If we use the P_{score} metric, the longest paths are ranked last, and the shortest path are ranked first. The table of ranked paths is displayed on the right-hand side of Fig. 3. However, the initial ranking highlights the strong bias of the P_{score} metric to path length. If we correct by path length (by simply dividing by the path length), we see a complete reversal in the pathway rankings. The longest path now has the minimum corrected score of 0.0301 and is ranked first. Therefore, naively using P_{score} to rank pathways can lead to a collection of overly short and potentially biologically uninteresting pathways:

1. It is not possible to efficiently extract paths of minimum average score as this is a known NP hard problem (12).
2. The path length bias problem is significantly reduced if specific start and end nodes are set. However, the problem is not completely removed, and it is recommended that some path length correction is performed on all analyses.

To efficiently overcome the path length bias of the metric, we propose two heuristic methods:

Set a minimum path length and only extract paths which are greater than this value: We used path length thresholding in (11) and found that searching for the optimal minimum path length can yield insights into the complexity of specific metabolic phenomena. However, this approach requires an external method to evaluate the quality of the extracted paths.

Perform a post-hoc correction for path length: In (12), we proposed a method to convert the P_{score} into a P_{value} and correct for path length bias. However, for this procedure to be successful, a large number of paths must be extracted, and there is no guarantee that the best path in terms of *p* value will be found.

To create a pathway dataset for further analysis, we extract the top 2,000 paths for both heat shock and other with a minimum path length of five gene interactions. A visual display of this dataset is presented in Fig. 4. In Fig. 4, the top plot is an image of all paths where the rows correspond to an extracted path and the columns to a gene. A color indicates that a gene has been selected within a path. The paths are grouped in terms of their experimental condition and sorted in order of P_{score}. The P_{score} value for each path is plotted in the right-hand side line plot. The bar plot at the base of the image displays the probability of each gene over all paths and experimental conditions:

- In the `PathRanker` R package pathway extraction can be done using the `pathRanker` function.

An immediate striking feature of Fig. 4 is that there seems to be clusters of similar paths within each experimental condition. This clustering effect is due to path-ranking method getting caught in highly correlated subnetworks of genes which are related to specific experimental conditions. However, there are too many genes, over 300 selected in all paths, and therefore, analyzing the structure of

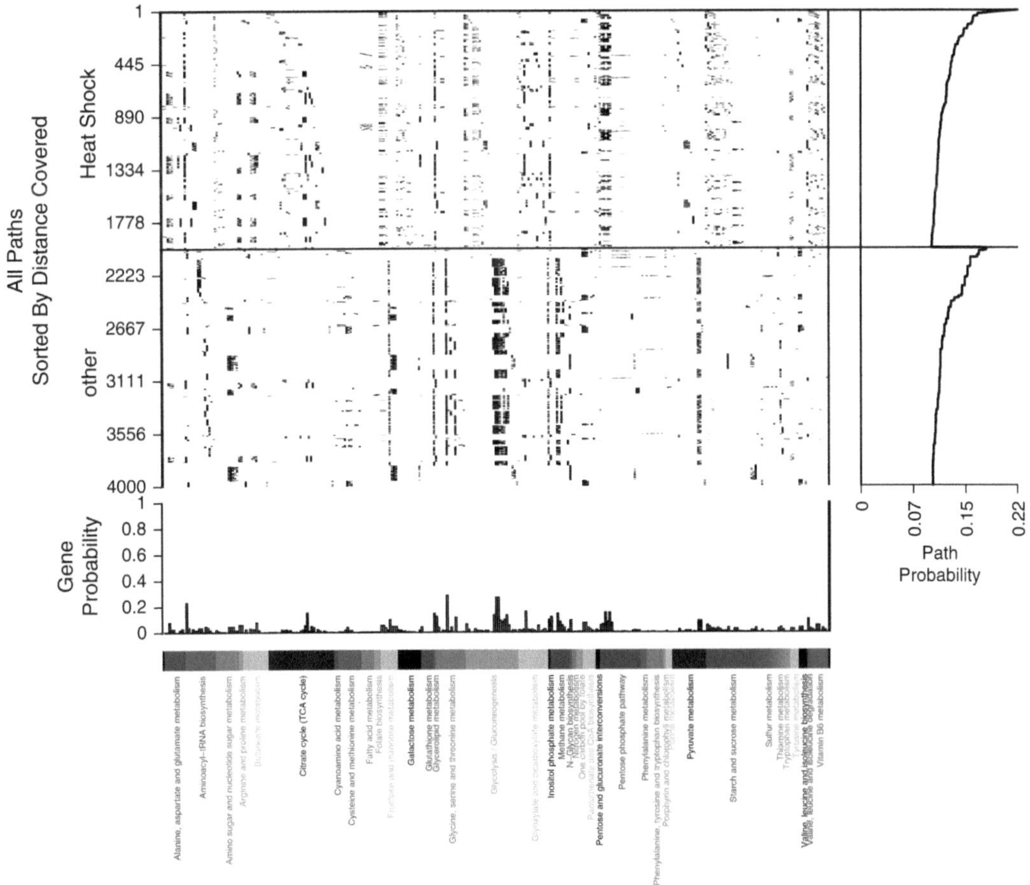

Fig. 4. Top 2,000 extracted paths for "heat shock" and all "other" stress conditions using the Gasch microarray and the KEGG yeast metabolic network.

all paths simply by looking at Fig. 4 is quite difficult. Therefore, to make this task possible, we now describe pathway-specific clustering and classification algorithms.

3.2. Pathway Clustering and Classification Methods

In (13–15), we proposed pathway-specific probabilistic models for clustering with 3M and classification with HME3M respectively. These models have the same probabilistic assumption that a pathway is a first-order Markov chain of gene interactions. In this chapter, we do not describe the derivation of these models but only present the model definitions and describe their practical use for analyzing a real pathway dataset.

To begin an analysis of the extracted pathways with 3M or HME3M the following steps are taken:

1. Both 3M and HME3M construct a binary pathway matrix X which is of the structure in Fig. 4. The pathway X defines each row as a pathway and each gene as a column. If gene j occurs within a pathway i, then $x_{ij} = 1$ otherwise $x_{ij} = 0$.
2. Once the pathway matrix X is constructed, the model structure of 3M and HME3M are,

   (a) **3M**:

$$p(x) = \sum_{m=1}^{M} \pi_m \prod_{k=2}^{K} p(g_k | g_{k-1}; \theta_{km})$$

   (b) **HME3M**:

$$p(y|x) = \sum_{m=1}^{M} \pi_m p(y|X, \beta_m) \prod_{k=2}^{K} p(g_k | g_{k-1}; \theta_{km}),$$

   where

   - M is the number of pathway clusters to be found.
   - π_m is the probability of each pathway cluster.
   - $p(g_k | g_{k-1}; \theta_{km})$ is the conditional probability of g_k being within each path given gene g_{k-1}. This conditional probability is estimated by the parameters θ_{km}.
   - K is the length of each path.
   - y is a binary response variable indicating the target experimental condition to be classified.
   - $p(y|X, \beta_m)$ is a classification model trained within each pathway cluster and estimates the probability of classifying the response variable y with parameters β_m.
3. The 3M or HME3M model parameters can be estimated using the expectation maximization (EM) algorithm (21).

The difference between 3M and HME3M is that HME3M is a supervised model and includes experimental condition information through the classification probabilities $p(y|X, \beta_m)$. In contrast the 3M model is an unsupervised clustering model. The choice of 3M or HME3M for any given situation depends on the problem:

1. The unsupervised nature of 3M means it is better suited to identifying pathways of similar structure regardless of observed experimental condition.
2. The supervised nature of HME3M means it is suited to identifying pathways which define a specific experimental condition.

The HME3M classification probabilities, $p(y|X, \beta_m)$, are estimated within each pathway cluster M by separate weighted penalized logistic regression (PLR) models (22). This penalization is an L2 or ridge penalization, and its purpose is to stabilize each estimate of the parameter vector β_m to cope with a large number of genes and the sparse nature of the pathway matrix X:

1. The PLR model contains a penalization parameter λ to control the degree of penalization. In (11, 14, 15), we set this $\lambda = 1$ and did not optimize this parameter (see Note 3). In our experience, the optimization of λ in the context of HME3M does not dramatically impact the classification performance but does affect the convergence of the EM optimization algorithm.
2. The parameters β_m for each component are the log odds that each gene can correctly classify the response the label in component m. The β_m values may be positive or negative which indicates the presence or absence of a gene within each pathway component m respectively.
3. The optimization of each pathway cluster PLR model is not done as a separate step after training an initial 3M model but at the same time as the pathway clustering through the mixture of expert framework (23). The mixture of expert framework uses an EM algorithm to share information between HME3M's pathway clustering and classification steps to ensure that each pathway cluster found is optimized for classification performance.

The optimization of 3M or HME3M involves estimating the of number pathway clusters M to be extracted. This can be done in two ways:

1. In (14, 15), we searched for the optimum M through a greedy search between $M = 1$ to 10. This procedure assumes that there are multiple pathway clusters within each experimental condition, and the number of subclusters can be greater than the number of experimental conditions. For example, in (14, 15), we found 4–5 pathway subclusters within two experimental conditions

2. In (11) we set M to be the known number of groups and then search for the optimum minimum path length in the pathway-ranking algorithm defined in Subheading 3.1. This approach assumes that there is one pathway cluster for each experimental condition, but the size and complexity of this structure are unknown. By increasing the minimum path length that can be extracted by the path-ranking method, we are increasing the complexity of each path. As we have shown in (11), increasing the complexity of these paths can be used to optimize the performance of 3M and HME3M.
3. For any given parameter settings, the performance of either 3M or HME3M should be evaluated by a cross validation regime:
   - In the `PathRanker` R package, 3M pathway clustering can be performed using the `pathCluster` function, and HME3M pathway classification can be performed using the `pathClassifier` function.

To show the differences in the results obtained by 3M and HME3M, we run both models on the dataset presented in Fig. 5. To produce the images in Fig. 5, perform the following steps:

1. Optimize the number of components by performing five repeats of fivefold cross validation for each model for every size $M = 2$ to $M = 10$. The results from the performance optimization are presented in Fig. 5a. In Fig. 5a we compare the performance of each model using a normalized mutual information (NMI). The optimal value of the NMI which indicates perfect agreement between the response label and predicted label is an $NMI = 1$ and the minimum value is 0 (see Note 4).

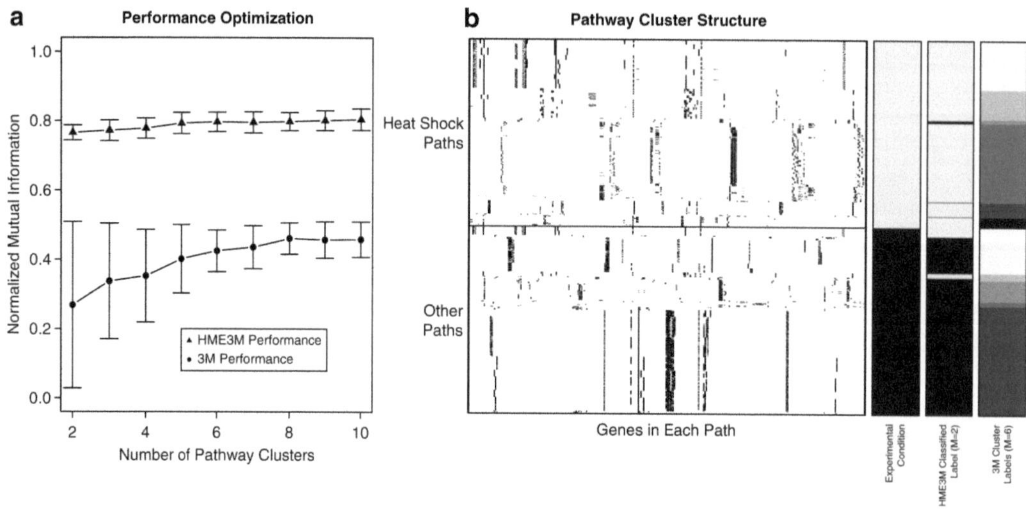

Fig. 5. Pathway clustering and classification results.

2. Identify the optimal 3M and HME3M model size and plot the optimal pathway clusters by running each model once on the entire pathway dataset. The optimal pathway clusters are plotted in Fig. 5b.

The performance optimization results (Fig. 5a) clearly show the improvement in classification accuracy that can be gained by using the supervised model HME3M as opposed to the unsupervised approach of 3M:

1. The additional response information included within the HME3M clearly increases the NMI between the known response labels and predicted model labels.
2. The HME3M model also has reduced variance of the predictions which leads to a simpler optimal model.
3. To pick the optimal model size, we take a conservative approach for our model selection and pick the HME3M model size to be $M = 2$ and the 3M model to be $M = 6$.

To show the difference between the 3M and HME3M model, we produce an image of the identified clusters in Fig. 5b:

1. Figure 5b displays the same pathway dataset in Fig. 4; however, we have sorted the pathways in order of the known experimental conditions then by 3M cluster label and finally by HME3M predicted label.
2. The three color bars to the right of the pathway image in Fig. 5b show the known experimental conditions, the HME3M predicted label, and 3M cluster labels respectively. Each color represents a different cluster label.

From observation of the pathway clusters for each model in Fig. 5b, we clearly see the difference between the goals of HME3M and 3M:

1. The HME3M-assigned labels are clearly seen to agree very well with the known experimental conditions, with a few expected assignment errors:
   (a) This good classification performance however is not extracting the cluster structure of the pathways. To observe the cluster structure of a HME3M model, you have to analyze each component individually and take into account the logistic regression parameters β_m.
   (b) This result highlights the upside of HME3M which is improved predictive accuracy, but the downside is the additional analysis complexity introduced by the model parameters β_m.

2. The 3M model label assignments appear to agree with the pathway cluster structure within the dataset. However, these labels do not necessarily agree with the response labels:
   (a) This result highlights the upside of 3M to clearly identify the cluster structure within a pathway dataset, but this structure may not be representative of the known experimental conditions.

4. Notes

We have described a complete framework for the extraction and analysis of co-ordinated gene expression pathways within metabolic networks. All methods described in this chapter are freely available as an R package which can be downloaded from the website http://www.bic.kyoto-u.ac.jp/pathway/timhancock/index.html.

Although we have discussed implementation issues throughout the chapter, we now briefly address some broader issues related our framework:

1. Care should be taken when analyzing metabolic pathways of gene expression as expression levels are only indicative of gene RNA levels. However, it is not the RNA levels that perform the enzymatic function but their proteins, and it is not clear to what extent increasing RNA levels leads to increased protein production.
2. In this work, we used the KEGG metabolic networks; however, other networks exist which may contain more or different annotations and information. However, as different qualities of evidence exist to assign genes to metabolic reactions, we recommend care when constructing metabolic networks.
3. In our initial HME3M (14, 15) work, we used a different pathway extraction technique to the one described in this chapter. This initial HME3M pathway technique cannot be applied for the analysis of large networks as described here.
4. The paths are extracted summaries of the structure within the original microarray dataset, and model performances based on these paths do not necessarily reflect the actual performance on the original dataset. These performances cannot be compared to models which directly classify using the original microarray dataset.

References

1. Kanehisa M, Goto S (2000) KEGG: kyoto encyclopedia of genes and genomes. Nucleic Acids Res 28:27–30
2. Caspi R, Altman T, Dale JM, Dreher K, Fulcher CA, Gilham F, Kaipa P, Karthikeyan AS, Kothari A, Krummenacker M, Latendresse M,

Mueller LA, Paley S, Popescu L, Pujar A, Shearer AG, Zhang P, Karp PD (2010) The MetaCyc database of metabolic pathways and enzymes and the BioCyc collection of pathway/genome databases. Nucleic Acids Res 38 (Database issue):D473–9

3. Matthews L, Gopinath G, Gillespie M, Caudy M, Croft D, de Bono B, Garapati P, Hemish J, Hermjakob H, Jassal B, Kanapin A, Lewis S, Mahajan S, May B, Schmidt E, Vastrik I, Wu G, Birney E, Stein L, D'Eustachio P (2009) Reactome knowledgebase of human biological pathways and processes. Nucleic Acids Res 37: D619–22
4. Vastrik I, D'Eustachio P, Schmidt E, Gopinath G, Croft D, de Bono B, Gillespie M, Jassal B, Lewis S, Matthews L, Wu G, Birney E, Stein L (2007) Reactome: a knowledge base of biologic pathways and processes. Genome Biol 8(3):R39
5. Subramanian A, Tamayo P, Mootha VK, Mukherjee S, Ebert BL, Gillette MA, Paulovich A, Pomeroy SL, Golub TR, Lander ES, Mesirov JP (2005) Gene set enrichment analysis: a knowledge-based approach for interpreting genome-wide expression profiles. PNAS 102(43):15545–15550
6. Liu Q, Dinu I, Adewale AJ, Potter JD, Yasui Y (2007) Comparative evaluation of gene-set analysis methods. BMC Bioinformatics 8:431
7. Sanguinetti G, Noirel J, Wright PC (2008) MMG: a probabilistic tool to identify submodules of metabolic pathways. Bioinformatics 24 (8):1078–84
8. Hanisch D, Zien A, Zimmer R, Lengauer T (2002) Co-clustering of biological networks and gene expression data. Bioinformatics 18: S145–54
9. Ulitsky I, Shamir R (2007) Identification of functional modules using network topology and high-throughput data. BMC Syst Biol 1:8
10. Ideker T, Ozier O, Schwikowski B, Siegel AF (2002) Discovering regulatory and signalling circuits in molecular interaction networks. Bioinformatics 18:S233–40
11. Hancock T, Takigawa I, Mamitsuka H (2010) Mining metabolic pathways through gene expression. Bioinformatics 26(17):2128–2135
12. Takigawa I, Mamitsuka H (2008) Probabilistic path ranking based on adjacent pairwise coexpression for metabolic transcripts analysis. Bioinformatics 24(2):250–257
13. Mamitsuka H, Okuno Y, Yamaguchi A (2003) Mining biologically active patterns in metabolic pathways using microarray expression profiles. SIGKDD Explorations 5(2):113–121
14. Hancock T. and Mamitsuka H (2009) A Markov Classification Model for Metabolic Pathways, Algorithms in Bioinformatics, 9th International Workshop, (WABI), Philadelphia, PA, USA. Proceedings, WABI, volume 5724, Springer, Eds: Salzberg, Steven, and Warnow, Tandy
15. Hancock T, Mamitsuka H (2010) A Markov classification model for metabolic pathways. Algorithms Mol Biol 5(1):10
16. Gasch AP, Spellman PT, Kao CM, Carmel-Harel O, Eisen MB, Storz G, Botstein D, Brown PO (2000) Genomic expression programs in the response of yeast cells to environmental changes. Mol Biol Cell 11(12):4241–57
17. Gentleman R, Carey V, Huber W, Irizarray R, Dudoit S (2005) Bioinformatics and Computational Biology Solutions Using R and Bioconductor. Springer, Berlin
18. Gasch A, Carey V (2010) gaschyhs: Expressionset for response of yeast to heat shock and other environmental stresses. www.bioconductor.org
19. Yen J (1971) Finding the k-shortest loopless paths in a network. Manage Sci 17:712–716
20. Lawler E (1972) A procedure for computing the k best solutions to discrete optimization problems and its application to the shortest path problem. Manage Sci 18:401–405
21. Dempster AP, Laird NM, Rubin DB (1977) Maximum likelihood from incomplete data via the EM algorithm, J Roy Stat Soc B, 39(1), 1–38.
22. Park MY, Hastie T (2008) Penalized logistic regression for detecting gene interactions. Biostatistics 9(1):30–50
23. Jordan M, Jacobs R (1994) Hierarchical mixtures of experts and the EM algorithm. Neural Comput 6(2):181–214

Chapter 8

Mining Frequent Subtrees in Glycan Data Using the Rings Glycan Miner Tool

Kiyoko Flora Aoki-Kinoshita

Abstract

This chapter describes the Glycan Miner Tool, which is available as a part of the Resource for INformatics of Glycomes at Soka Web site. It implements the α-closed frequent subtree algorithm to find significant subtrees from within a data set of glycan structures, or carbohydrate sugar chains. The results are returned in order of *p*-value, which is computed based on the probability of the reproducibility of the returned structures. There is also a user-friendly manual that allows users to apply glycan array data from the Consortium for Functional Glycomics. Thus, glycobiologists can take the glycan structures that bind to a particular glycan-binding protein, for example, to retrieve the glycan subtrees that are deemed to be important for the binding to occur.

Key words: Glycans, Glycobiology, Complex carbohydrates, Data mining, α-Closed frequent subtrees

1. Introduction

The Web-based tool called the Glycan Miner Tool introduced in this chapter implements the data mining of glycan structures based on the concept of α-closed frequent subtree mining (1). Frequent subtree mining in general has been applied to various research areas, including computer networks, Web mining, and bioinformatics (2). These consisted of finding the most frequently occurring subtrees within a forest of data. For example, tree mining was applied to molecular databases, where tree-shaped molecular fragments frequently found in active molecules, but also infrequent in inactive molecules, were extracted. These frequently occurring fragments could be thereby used to analyze Quantitative Structure-Activity Relationships (QSAR) (3).

Because of the large number of potential subtrees that can be produced by mining frequent subtrees, new methods for mining

Hiroshi Mamitsuka et al. (eds.), *Data Mining for Systems Biology: Methods and Protocols*, Methods in Molecular Biology, vol. 939, DOI 10.1007/978-1-62703-107-3_8, © Springer Science+Business Media New York 2013

trees were developed. One of these is the mining of *maximal frequent subtrees* (4), whereby the most frequent supertrees in a data set could be found. In other words, only those frequent subtrees that are not subtrees of other frequent subtrees are returned. Another mines *closed frequent subtrees* (5), whereby for any support value, only the most frequent supertrees are returned. As a result, one can see that the set of maximal frequent subtrees is a subset of closed frequent subtrees, which is itself a subset of frequent subtrees in general.

Mining maximal and closed frequent subtrees produced results that could be considered sufficient for such applications as drug discovery and others (6). However, in order to allow more flexibility in mining frequent subtrees, the idea of an additional parameter α was introduced. By specifying a threshold by which to mine frequent subtrees, users could control the range of results that could be obtained. Thus, the concept of α-closed frequent subtree was proposed (1), which was used as the basis of the Glycan Miner Tool.

2. Materials

2.1. Glycobiology

Glycobiology is the study of complex carbohydrates, or glycans, which are chains of monosaccharides (five- or six-carbon rings) that oftentimes take branched configurations (7). Thus, glycans can be represented as tree structures, whereby the nodes represent monosaccharides and edges represent glycosidic bonds. Because glycosidic bonds can be formed between any of several hydroxyl groups, or carbon atoms, on monosaccharides, a single monosaccharide may be bonded (via its first or second carbon) with several other monosaccharides via different carbon atoms, which are usually carbon numbers 2, 3, 4, or 6. Figure 1 is an illustration of a disaccharide structure; Galactose is linked to Glucose via a glycosidic linkage in a beta 1–4 configuration. In this case, the "parent" would be considered the glucose, and the "child" would be the galactose, because glycan structures are usually drawn from right to left.

2.2. α-Closed Frequent Subtree Mining

In order to explain the usage of the Glycan Miner Tool, we first describe some notations to introduce the methodology behind this tool called α-closed frequent subtree mining. A *graph* is a data object that contains nodes and edges, where an edge connects two nodes. A tree, usually denoted as T, is a graph without any cycles. A *rooted* tree is a tree that contains a root node from which all other nodes emanate. A *labeled* tree is a tree whose nodes (and at times edges) contain labels. A node n on a unique path from the root node to a node c is an *ancestor* of node c, and conversely, c is a

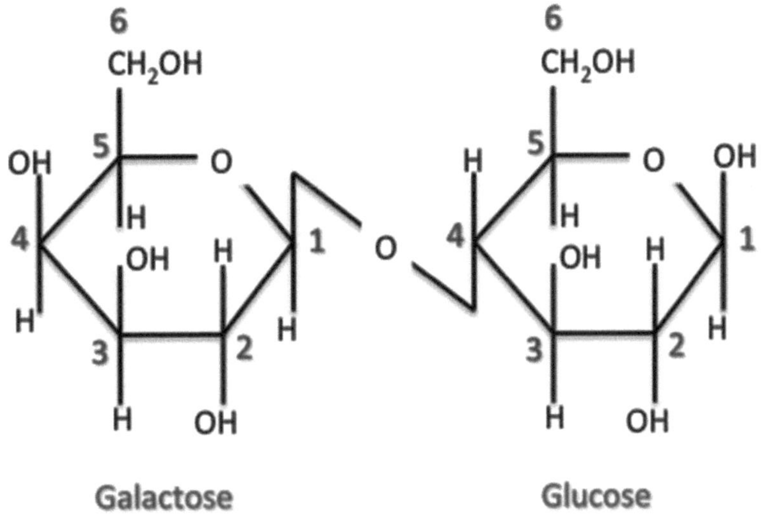

Fig. 1. An illustration of a disaccharide structure, where Galactose is linked to Glucose via a glycosidic linkage in a beta 1–4 configuration. In this case, the "parent" would be considered the glucose, and the "child" would be the galactose.

descendant of *n*. A *parent* of a node *n* is defined as the closest ancestor of *n*, and a *child* is the closest descendant. A *leaf* is a node that does not have any children. Internal nodes are nodes that are neither leaves nor the root. An *ordered tree* is a tree containing nodes whose children have an ordering.

In the context of glycobiology, glycans are regarded as rooted, labeled, ordered trees, which are hereafter simply called trees. A tree *S* is a called a *subtree* of tree T if *S* consists of nodes and edges which form a connected (and rooted) subset of the nodes and edges of *T*. Conversely, *T* can then be called a *supertree* of *S*. The size of a tree T refers to the number of edges in *T*, denoted by $|T|$. An *immediate supertree* T of S satisfies $|T| = |S| + 1$.

2.3. Gycan Miner Tool

In order to use the Glycan Miner Tool, we next describe the text format called KCF, which is used to describe glycan structures. This format is currently required to specify the input glycan data.

KEGG Chemical Function format, or KCF, was first described for use in the KEGG GLYCAN database (8). It uses a graph concept to describe glycan structures, where nodes represent monosaccharides and edges represent glycosidic bonds. An example of the tri-mannose *N,N'*-diacetyl chitobiose core structure of *N*-glycans, usually depicted as in Fig. 2, is described in KCF format as follows:

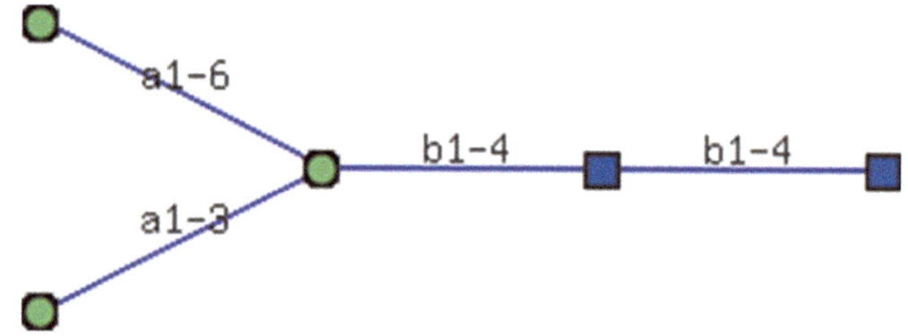

Fig. 2. The tri-mannose *N,N*-diacetyl chitobiose core structure of *N*-glycans.

```
ENTRY     G00311                   Glycan
NODE      5
          1   GlcNAc   15   0
          2   GlcNAc    5   0
          3   Man      -5   0
          4   Man     -15   5
          5   Man     -15  -5
EDGE      4
          1   2:b1    1:4
          2   3:b1    2:4
          3   4:a1    3:6
          4   5:a1    3:3
///
```

The KCF file format must contain the NODE and EDGE sections as a minimum. The ENTRY line is used to denote the name or ID of the represented structure, if any. In this example, the glycan ID is G00311. The NODE section starts with a number indicating the number of residues being represented. The same number of rows follows this line, each containing the following information: residue number, residue name, and x- and y-coordinates for drawing the structure. The EDGE section follows similarly with a number indicating the number of glycosidic bonds in the structure, which will usually be one less than the number of residues. The same number of rows follows, with each row containing the following information: bond number, node numbers of residues being bound, anomeric configuration, and hydroxyl groups in the bond. For example, the following row

```
3 4:a1 3:6
```

represents the third bond, which indicates that node number 4 (Man at position −15,5) is bound to node number 3 (Man at position −5,0) in an α1–6 configuration.

The Resource for INformatics of Glycomes at Soka (RINGS) Web site (Akune 2010) provides a convenient drawing tool whereby users can draw a glycan structure and immediately obtain its KCF format. RINGS also has several format conversion utilities such that data that may already be in a particular format can be easily converted to KCF format. Conversely, there are tools that can translate from KCF to other formats, such as LinearCode® (9) and as image files.

2.4. Glycan Databases

2.4.1. KEGG GLYCAN Database

The KEGG GLYCAN database is available at http://www.genome.jp/kegg/glycan/ and is a database of glycan structures, accumulated from the original CarbBank database (8). It has since been refined and updated with structures from the literature. All the data is freely available from the Web and can also be accessed via an application programming interface (API), which is described at http://www.genome.jp/kegg/docs/keggapi_manual.html.

2.4.2. Consortium for Functional Glycomics

The Consortium for Functional Glycomics (CFG) glycan structure database was originally developed as a part of the bioinformatics core of the CFG. In addition to their own database of glycan-binding proteins (GBP) and glycosyltransferases, they had initially accumulated N- and O-linked glycans from the CarbBank database as well as the glycan structure data from GlycoMinds, Ltd. Since then, they have added their own synthesized glycans from their glycan array library and characterized glycans from their tissue and cell profiling data.

All of the glycan profiling data, glycan array data, and knockout mouse data generated by the CFG are also available as data resources over the Web. Glycan profiling data consists of the mass spectral data for various human and mouse tissue samples, which have been annotated using Cartoonist. The glycan array data consist of binding affinity information of various glycans for different GBP, viruses, bacteria, etc. Each data set focuses on a particular GBP or other binder and lists the binding affinity for each glycan structure on the array. Glycan arrays have developed over the years, and the latest version contains over 600 glycan structures (10).

2.4.3. GlycomeDB

In addition to KEGG GLYCAN and CFG, there are several glycan structure databases available around the world. GlycomeDB was developed to serve as a portal to all of these structure databases via a single interface. It was a major project that entailed with integration of all the structures in the various databases, which was complicated mostly due to the inconsistent naming conventions of monosaccharides. As a result, this database was able to integrate seven different glycan databases and includes over 30,000 distinct glycan structures. The URL is http://www.glycome-db.org/ (11).

3. Methods

3.1. α-Closed Frequent Subtree Mining

Given a set of trees D, the *support* of a subtree S is the number of trees in D that contain S as a subtree, which we denote by *support (S)*. A *frequent* subtree is thus the subtree whose support is larger than or equal to a given threshold value denoted by *minsup*. A *maximal* frequent subtree is a frequent subtree, none of

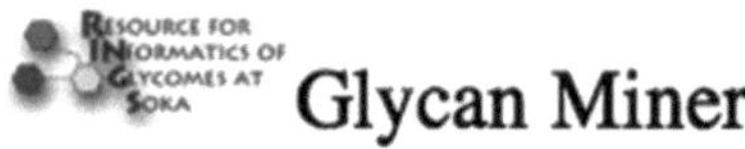

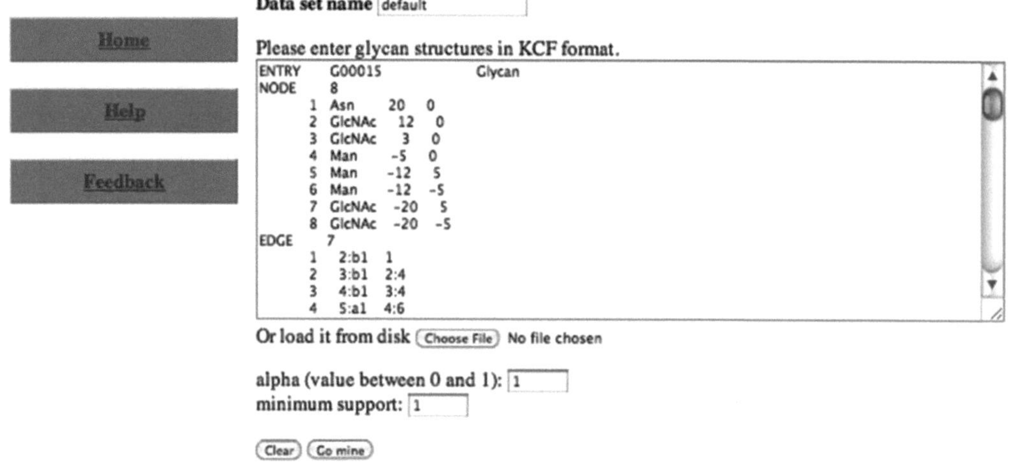

Fig. 3. The Glycan Miner Tool where default values have been supplied.

whose proper supertrees are frequent. A subtree is *closed* if none of its proper supertrees has the same support as it has. With the concept of maximal and closed frequent subtrees defined, we can then formally define an α-closed frequent subtree as a frequent subtree *S*, none of whose frequent supertrees has support greater than or equal to a fraction α of *support(S)*.

Given a data set of trees, the mining of α-closed frequent subtrees from within this data set entails the enumeration of all possible subtrees and then determining their support values. This is in fact a difficult problem because the enumeration of all possible subtrees can grow exponentially. However, by considering the support of supertrees as subtrees are enumerated, potential subtrees can be pruned so that frequent subtrees can be found efficiently. Because the details of this method are beyond the scope of this chapter, interested readers are referred to (1).

3.2. Gycan Miner Tool

The Glycan Miner Tool is available in the RINGS resource at http://www.rings.t.soka.ac.jp/ (12). By registering as a user, any uploaded data is automatically saved in the user data space, which is password protected. Thus all executions of any program on RINGS can be recorded and managed by the user.

The input to the Glycan Miner Tool is a list of glycan structures in KCF format. There are also two parameters, minsup and alpha, which take on a value between zero and one. A screenshot of the Glycan Miner Tool where default values have been supplied is illustrated in Fig. 3. Given these inputs, the tool then computes

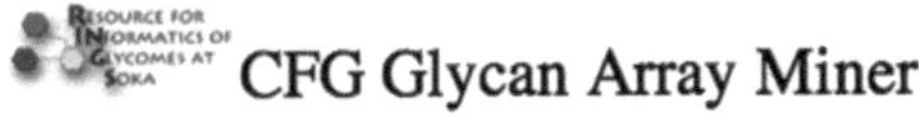

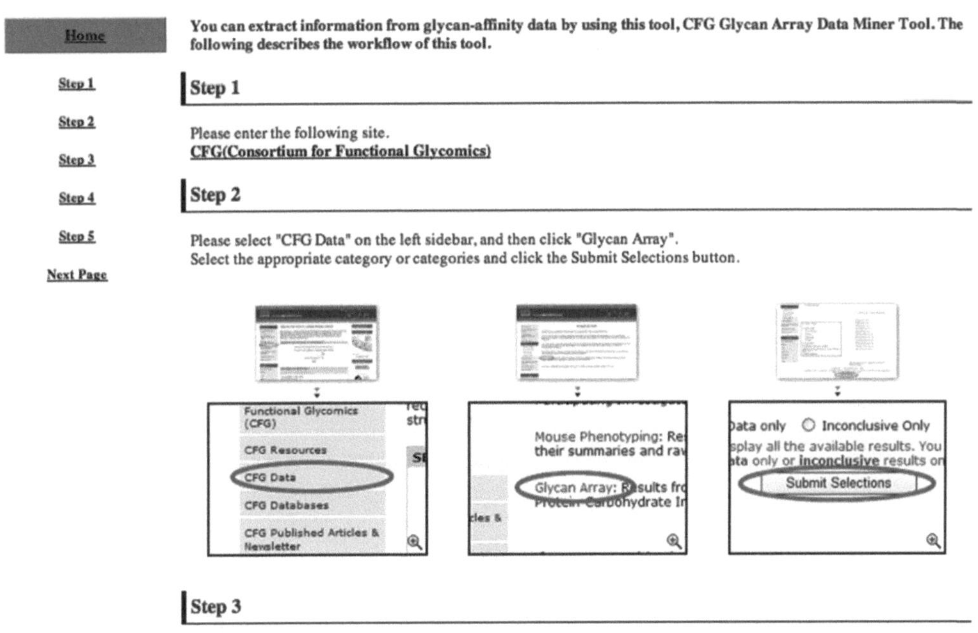

Fig. 4. A screenshot of part of the step-by-step manual for using the Glycan Miner Tool with glycan array data from the CFG. Each step is listed in a menu on the left.

the α-closed frequent subtrees from among the input data, and returns the results in order of p-value, which is computed as the following. First, the list of α-closed frequent subtrees are broken down into parent–child pairs of linked monosaccharides. Then taking the shape of each of the resulting structures, the probability of regenerating the same structure based on the distribution of original parent–child pairs is computed as the p-value for each structure. The final results are then returned in order of increasing p-value.

As an example of the usage of the Glycan Miner Tool, a step-by-step manual was developed for glycobiologists such that the glycan array data of the CFG, for example, can be applied directly to this tool. The link to the manual is provided at the top of the input screen of this tool (Fig. 3). The top part of this manual is listed in Fig. 4, where each step is listed in a menu on the left. First, users can download the spreadsheet files for each GBP that was analyzed on the glycan array. These files include the binding affinity for each glycan on the array, so the user must select the strongly binding glycan structures for analysis. Conversely, it may be possible to

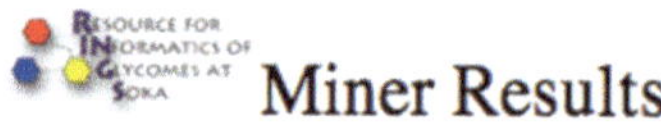

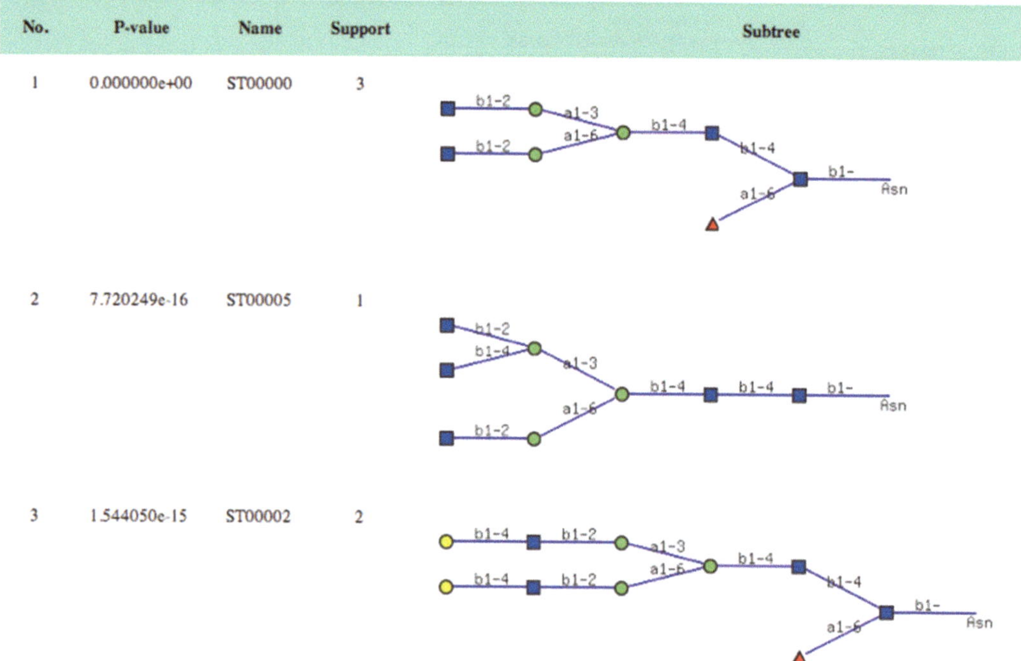

Input size: 6
alpha: 1.0
Minsup: 1
Subtrees: 7

No.	P-value	Name	Support	Subtree
1	0.000000e+00	ST00000	3	
2	7.720249e-16	ST00005	1	
3	1.544050e-15	ST00002	2	

Fig. 5. A screenshot of the results of the default input for the Glycan Miner Tool.

select the weakly binding ones to compare those subtrees that must NOT be in a glycan for binding to occur. In any case, the glycan structures are depicted in IUPAC format, so the data needs to be translated into KCF format for use in the Glycan Miner Tool. RINGS provides a conversion utility for this purpose, so the selected structures from the spreadsheet can be copy-and-pasted into the converter. The resulting KCF data can then be used as input to the Glycan Miner Tool. Once the minsup and alpha values are specified, the α-closed frequent subtrees will be returned in order of increasing *p*-value. Figure 5 is a screenshot of the results of the default input.

4. Notes

1. The results of the Glycan Miner Tool may return with no structures. When this occurs, the parameters of minsup and alpha should be adjusted. Minsup should be at most the

number of structures that were input; alpha should be between 0 and 1. The simplest input would be a minsup value of one (1) and alpha of one (1). These are the default values of the tool.

References

1. Hashimoto K, Takigawa I, Shiga M, Kanehisa M, Mamitsuka H (2008) Mining significant tree patterns in carbohydrate sugar chains. Bioinformatics 24:i167–i173
2. Zaki MJ (2002) Efficiently mining frequent trees in a forest. In: W. Wang and J. Yang (eds.). IEEE Transaction on Knowledge and Data Engineering, Special Issue on Mining Biological Data. Knowledge discovery and data mining. Edmonton, pp. 71–80
3. Ruckert U, Kramer S (2004) Frequent free tree discovery in graph data. In: H. Jamil and R. Meo (eds.). Proceedings of the 2004 ACM symposium on applied computing. ACM, Nicosia, pp. 564–570
4. Xiao Y, Yao J.-F (2003) "Efficient data mining for maximal frequent subtrees," ICDM 2003. Third IEEE International Conference on Data Mining, on November 19–22, pp. 379–386
5. Chi Y, Muntz R, Nijssen S, Kok J (2005) Frequent subtree mining: an overview. Fundamenta Informaticae 66:161–198
6. Chi Y, Xia Y, Yang Y, Muntz RR (2005) Mining closed and maximal frequent subtrees from databases of labeled rooted trees. IEEE Trans Knowl Data Eng 17:190–202
7. Varki A (2009) Essentials of glycobiology. Cold Spring Harbor Laboratory Press, Cold Spring Harbor
8. Hashimoto K, Goto S, Kawano S, Aoki-Kinoshita K, Ueda N, Hamajima M, Kawasaki T, Kanehisa M (2006) KEGG as a glycome informatics resource. Glycobiology 16:63R–70R
9. Banin E, Neuberger Y, Altshuler Y, Halevi A, Inbar O, Nir D, Dukler A (2002) A novel Linear Code((R)) nomenclature for complex carbohydrates. Trends in Glycoscience and Glycotechnology. 14:127–137
10. Raman R, Venkataraman M, Ramakrishnan S, Lang W, Raguram S, Sasisekharan R (2006) Advancing glycomics: implementation strategies at the consortium for functional glycomics. Glycobiology 16:82R–90R
11. Ranzinger R, Herget S, von der Lieth C-W, Frank M (2011) GlycomeDB—a unified database for carbohydrate structures. Nucleic Acids Res 39:D373–D376
12. Akune Y, Hosoda M, Kaiya S, Shinmachi D, Aoki-Kinoshita KF (2010) The RINGS resource for glycome informatics analysis and data mining on the Web. OMICS 14:475–486

Chapter 9

Chemogenomic Approaches to Infer Drug–Target Interaction Networks

Yoshihiro Yamanishi

Abstract

The identification of drug–target interactions from heterogeneous biological data is critical in the drug development. In this chapter, we review recently developed in silico chemogenomic approaches to infer unknown drug–target interactions from chemical information of drugs and genomic information of target proteins. We review several kernel-based statistical methods from two different viewpoints: binary classification and dimension reduction. In the results, we demonstrate the usefulness of the methods on the prediction of drug–target interactions from chemical structure data and genomic sequence data. We also discuss the characteristics of each method, and show some perspectives toward future research direction.

Key words: Drug–target interactions, Compound–protein interactions, Chemical genomics, Genomic drug discovery, Bipartite graph, Supervised network inference

1. Introduction

The completion of the human genome sequencing project has made it possible for us to analyze the "genomic space" consisting of possible proteins coded in the human genome, and various postgenomic approaches are being undertaken to utilize the genome information, such as for discovery of new therapeutic protein targets and personalized medicine. At the same time, many efforts have also been devoted to the constitution of molecular databanks to explore the entire "chemical space" of possible compounds. These public or private databanks contain synthesized molecules or natural molecules extracted from animal, plants, or microorganisms. They are available as physical and virtual databanks, to be screened in various biological assays or virtual screens. However, there is little knowledge about the relationship between the chemical and genomic spaces. For example,

Hiroshi Mamitsuka et al. (eds.), *Data Mining for Systems Biology: Methods and Protocols*, Methods in Molecular Biology, vol. 939, DOI 10.1007/978-1-62703-107-3_9, © Springer Science+Business Media New York 2013

the US PubChem database at NCBI (National Center for Biotechnology Information) stores more than sixty million compounds, but the number of compounds with information on their target proteins is very limited (1).

Most drugs are small compounds which interact with their target proteins and inhibit or activate the biological behavior of the proteins. Therefore, the identification of drug–target interactions, which are defined as interactions between drugs (or drug candidate compounds) and target proteins (target candidate proteins), is an important part of genomic drug discovery. Although high-throughput screening (HTS) is becoming available, experimental determination of drug–target interactions remains challenging and very expensive even nowadays. Therefore, there is a strong incentive to develop new methods capable of predicting potential drug–target interactions, in order to reduce the experimental work to be done.

So far, a variety of computational approaches have been developed to analyze and predict drug–target interactions or compound–protein interactions. Traditional computational approaches are categorized into ligand-based approach and docking approach. Ligand-based approach like quantitative structure activity relationship (QSAR) compares a candidate ligand to the known ligands of a target protein to predict its binding using machine learning methods (2, 3). However, the performance of the ligand-based approach is poor when the number of known ligands for a target protein of interest decreases. The docking is a powerful approach, but the docking cannot be applied to proteins whose 3D structures are unknown (4). This limitation is serious for membrane proteins such ion channels and G protein-coupled receptors (GPCRs), so it is difficult to use the docking on a genome-wide scale. Recently, a classification of target proteins based on their ligand structures has been performed (5) and an analysis of the drug–target network has revealed characteristic features of its network topology (6). However, neither protein sequence information nor chemical structure information was taken into consideration simultaneously. Another unique approach is the text mining techniques which are usually based on keyword searching in a huge number of literatures (7), but it suffers from an inability to detect new biological findings and the problem of redundancy in the compound names and protein names in the literature.

In that domain, the importance of chemogenomics research has recently grown fast to investigate the relationship between the chemical space and the genomic space (8–10). A key issue in chemogenomics is computational prediction of drug–target interactions or compound–protein interactions on a genome-wide scale. Recently, a variety of in silico chemogenomic approaches have been developed to predict drug–target interactions or compound–protein interactions. The underlying idea is that similar ligands are likely to interact with similar proteins, and the prediction is performed based on compound chemical structures, protein sequences, and the currently known drug–target interactions. A straightforward statistical approach for predicting drug–target interactions is to use binary classification methods where they take drug–target pairs as an input for machine learning classifiers such as neural network (11) and support vector machine (SVM) (12–15). The combination of many target-specific and drug-specific local classifiers was also proposed to detect missing interactions between known drugs and known target proteins (16). Another statistical approach for predicting drug–target interactions is the dimension reduction that map drugs and target proteins into a unified feature space in which known interacting drugs and target proteins are close to each other, then to infer potentially new drug–target interactions between other pairs of drugs and target proteins that were newly mapped close to each other in the unified space (17, 18).

Another promising approach for predicting drug–target interactions is to use pharmacological information of drugs. The use of side-effect similarity has been recently proposed, which is based on the assumption that drugs with similar side effects are likely to interact with similar target proteins (19). However, the method requires drug package inserts that describe the detailed side-effect information, so it is applicable only to marketed drugs for which side-effect information is available. Therefore, it is not possible to predict potential interactions between new drug candidate compounds and target proteins. To overcome this limitation, a method for predicting unknown pharmacological information of any compounds from their chemical structures has been proposed (20, 21), which enables us to predict drug–target interactions on a large scale.

In this chapter, we review recently developed in silico chemogenomic approaches to predict drug–target interactions from chemical data of drugs and genomic data of target proteins. Especially, we introduce several kernel-based statistical methods from two different viewpoints: binary classification (12–16) and the dimension reduction (17, 18). In the results, we show the usefulness of the methods on the predictions of drug–target interactions from chemical structure data and genomic sequence data. We also discuss the characteristics of each method and show some perspectives toward future research direction.

2. Materials

2.1. Drug–Target Interactions

The information about drug–target interactions was obtained from the KEGG BRITE (22), SuperTarget (23) and DrugBank databases (24). In this study, we focus on drug–target interactions involving four pharmaceutically useful protein classes: enzymes, ion channels, GPCRs, and nuclear receptors. We constructed a set of drug–target interactions, where the number of known interactions involving enzymes, ion channels, GPCRs, and nuclear receptors is 2,926, 1,476, 635, and 90, respectively. The number of known drugs targeting enzymes, ion channels, GPCRs, and nuclear receptors are 445, 210, 223, and 54, respectively, and the number of target proteins in these classes is 664, 204, 95, and 26, respectively. This is the same data used in (17). These data sets are used as gold standard data to evaluate the prediction performance.

2.2. Chemical Structures

Chemical structures of the drugs were obtained from the KEGG DRUG database (22). We computed the kernel similarity value of chemical structures between drugs using the SIMCOMP algorithm (25), where the similarity value between two drugs is computed by Tanimoto coefficient defined as the ratio of common substructures between two drugs based on the chemical graph alignment. Applying this operation to all drug pairs, we construct a similarity matrix.

2.3. Target Protein Sequences

Amino acid sequences of the human proteins were obtained from the KEGG GENES database (22). We computed the sequence similarities between the target proteins using Smith-Waterman scores based on the local alignment between two amino acid sequences (26). Applying this operation to all target protein pairs, we construct a similarity matrix.

2.4. Computation of Kernel Similarity Matrices

In this study, we used the above similarity measures as kernel functions, because these measures are very popular, efficient, and widely used in the field of chemistry and genomics. However, the graph-based Tanimoto coefficients and the Smith–Waterman scores are not always positive definite, so we added an appropriate identity matrix such that the corresponding kernel Gram matrix is positive definite. All the kernel matrices are normalized such that all diagonals are ones. Note that other kernel functions can be used in the same framework in the methods introduced in this chapter (see note 1).

3. Methods

The drug–target interaction network can be regarded as a bipartite graph with drugs (or drug candidate compounds) and target proteins (or target candidate proteins) as heterogeneous nodes and their interactions as edges, which is mathematically represented by a bipartite graph $G = (U + V, E)$, where $U = \{x_1, \ldots, x_{n_x}\}$ is a set of drug nodes, $V = \{y_1, \ldots, y_{n_y}\}$ is a set of target protein nodes and

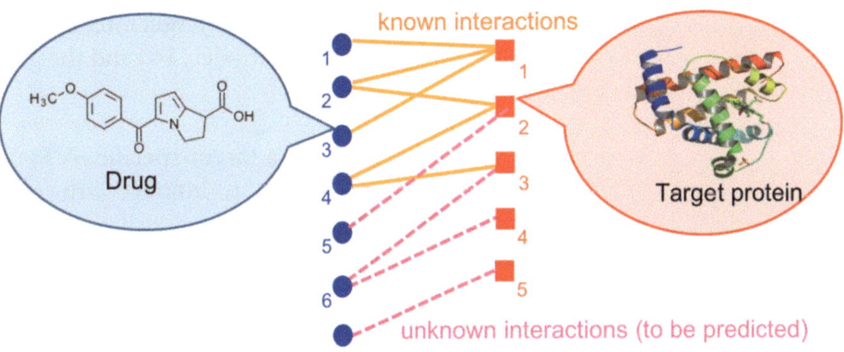

Fig. 1. An illustration of the problem of drug–target interaction prediction.

$E \subset (U \times V)$ is a set of drug–target interaction edges. From the viewpoint of statistics and machine learning, the prediction of drug–target interactions can be formulated as the problem of supervised bipartite graph inference. The question is to predict the presence or absence of edges between heterogeneous objects known to form the nodes of the bipartite graph, based on the observed data about the heterogeneous objects. Figure 1 shows an illustration of this problem, where solid lines indicate known interactions and dot lines indicate unknown interactions to be predicted. In the following section, we assume that we have a set of drugs $\{x_i\}_{i=1}^{n_x}$ and a set of target proteins $\{y_j\}_{j=1}^{n_y}$ and their interaction information. We consider the situation where we want to predict unknown interactions involving any given drug candidate compound x' and any given target candidate protein y'.

3.1. Binary Classification Approach

A straightforward approach for drug–target interaction prediction is to use a binary classification method. Among many binary classification algorithms, the SVM is recently gaining popularity in bioinformatics (27) and in chemoinformatics (28) because of its high-performance classification ability and applicability to structured data. Therefore, we focus on the use of SVM in this chapter. An SVM basically learns how to classify an object z' into two classes $\{-1, +1\}$ from a set of labeled objects $\{z_1, z_2, \ldots, z_n\}$. The resulting classifier is formulated as

$$f(z') = \sum_{i=1}^{n} \tau_i k(z_i, z'), \tag{1}$$

where z' is any new object to be classified, n is the number of training objects, $k(\cdot,\cdot)$ is a positive definite kernel, that is, a symmetric function $k : \mathcal{Z} \times \mathcal{Z} \rightarrow \mathbf{R}$ satisfying $\Sigma_{i,\, j=1}^{n} a_i a_j k(z_i, z_j) \geq 0$ for any a_i, $a_j \in \mathbf{N}$, and $\{\tau_1, \tau_2, \ldots, \tau_n\}$ are the parameters learned. If $f(z')$ is positive, z' is classified into class + 1. On the contrary, if $f(z')$ is negative, z' is classified into class − 1.

All SVM-based methods in the drug–target interaction prediction problem are classified into the local model (16) and the global model (12–15).

3.1.1. Local SVM

1. A simple way is to construct a target-specific SVM classifier in order to predict a given drug x' to interact with target protein y_j or not, as follows:

$$f_{y_j}(x') = \sum_{i=1}^{n_x} \alpha_i k_x(x_i, x') \quad (j = 1, 2, \ldots, n_y), \tag{2}$$

where $k_x(\cdot,\cdot)$ is a kernel function for drugs and $\{\alpha_i\}_{i=1}^{n_x}$ are the parameters learned. If $f_{y_j}(x')$ is positive, drug x' and target protein y_j are predicted to interact with each other. On the contrary, if $f_{y_j}(x')$ is negative, drug x' and target protein y_j are predicted not to interact. We repeat the process for all n_y target proteins. The concept of constructing a classifier for a specific target protein is similar to traditional ligand-based virtual screenings such QSAR(2, 3).

2. Likewise, we can construct a drug-specific SVM classifier in order to predict a given target protein y' to interact with drug x_i or not, as follows:

$$f_{x_i}(y') = \sum_{j=1}^{n_y} \beta_j k_y(y_j, y') \quad (i = 1, 2, \ldots, n_x), \tag{3}$$

where $k_y(\cdot,\cdot)$ is a kernel function for target proteins and $\{\beta_j\}_{j=1}^{n_y}$ are the parameters learned. If $f_{x_i}(y')$ is positive, drug x_i and target protein y' are predicted to interact with each other. On the contrary, if $f_{x_i}(y')$ is negative, drug x_i and target protein y' are predicted not to interact. We repeat the process for all n_x drugs.

3.1.2. Pairwise SVM with Pairwise Kernels

1. Another approach is to construct a global SVM classifier by regarding each drug–target pair as an object (12–15). In this case, we construct an SVM classifier to classify a given drug–target pair (x', y') into two classes $\{-1, +1\}$ from a set of labeled drug–target pairs $\{x_i, y_j\}$ $(i = 1, \ldots, n_x, j = 1, \ldots, n_y)$. The resulting classifier is formulated as

$$f(x', y') = \sum_{i=1}^{n_x} \sum_{j=1}^{n_y} \tau_{ij} k_{pair}((x_i, y_j), (x', y')), \tag{4}$$

where (x', y') is any new drug–target pair to be classified, $k_{\text{pair}}(\cdot,\cdot)$ is a positive definite kernel drug–target pairs, and τ_{ij} are the parameters learned. If $f(x', y')$ is positive, drug x' and target protein y' are predicted to interact with each other. On the contrary, if $f(x', y')$ is negative, drug x' and target protein y' are predicted not to interact. Therefore, the essential question here is how to design the kernel function for drug–target pairs.

2. We consider a vector representation of a drug–target pair (x, y). Suppose that a drug x is represented by a vector $\Phi_x(x) \in \mathbf{R}^{d_x}$, which corresponds to physico-chemical molecular descriptors or substructure fingerprint (29). Likewise, suppose that a target protein y is represented by a vector $\Phi_y(y) \in \mathbf{R}^{d_y}$, which corresponds to the features related with amino acid composition or functional motif profiles, for example. We then consider representing a drug–target pair (x, y) by a vector $\Phi(x, y)$. A simple vector representation is to concatenate $\Phi_x(x)$ and $\Phi_y(y)$ as $\Phi(x, y) = (\Phi_x(x)^T, \Phi_y(y)^T)^T$ (12, 13). Note that the size of the vector is $(d_x + d_y)$ in this case.
3. Another vector representation approach is to use the set of all possible products of features of x and y by the tensor product as follows (14, 15):

$$\Phi(x, y) = \Phi_x(x) \otimes \Phi_y(y). \tag{5}$$

Note that the tensor product is a vector of size $(d_x \times d_y)$, so it requires prohibitive computational burden. To avoid such a computational problem, an efficient technique has been proposed by using a property of tensor product (14, 15). The use of a classical property of tensor products enables us to compute the inner product between tensor products can be computed by

$$k_{pair}((x, y), (x', y')) = k_x(x, x') \times k_y(y, y'), \tag{6}$$

where $k_x(x, x') = \Phi_x(x)^T \Phi_x(x')$ and $k_y(y, y') = \Phi_y(y)^T \Phi_y(y')$. This implies that, as soon as we obtain the drug kernel $k_x(x, x')$ and the target protein kernel $k_y(y, y')$, we can compute the kernel for the corresponding drug–target pair.
4. Finally, the pairwise kernel is used as an input in the SVM classifier in order to predict whether drug–target pairs are likely to interact or not. Note that pairwise SVM (P-SVM) requires considerable computational burden (see note 2).

3.2. Dimension Reduction Approach

Here, we introduce two dimension reduction approaches based on kernel regression model (KRM) (17) and kernel distance learning (KDL) (18). Both methods consist of the following two steps:

- Learn two mappings f and g in order to embed drugs and target proteins into a unified Euclidean space representing the network topology, where interacting drugs and target proteins are close to each other.
- Apply the mappings f and g to any drugs and target proteins, respectively, and predict new interactions between drugs and target proteins if the distance between mapped drugs and target proteins is smaller than a threshold.

The difference between the two methods is that KRM *explicitly* embeds drugs and target proteins in a unified feature space, while KDL *implicitly* embeds drugs and target proteins in a unified feature space.

3.2.1. Kernel Regression Model

1. We represent the bipartite graph structure of a drug–target interaction network by an Euclidian space such that both drugs and target proteins are represented by sets of d-dimensional feature vectors $\{\mathbf{u}_{x_i}\}_{i=1}^{n_x}$ and $\{\mathbf{u}_{y_j}\}_{j=1}^{n_y}$, respectively. To do so, based on the shortest distance between drugs and target proteins on the bipartite graph, we first construct a graph-based similarity matrix $L = \begin{pmatrix} L_{cc} L_{cp} \\ L_{cp}^T K_{pp} \end{pmatrix}$, where the elements of L_{cc}, L_{pp}, and L_{cp} are computed by using Gaussian functions as follows: $(L_{cc})_{ij} = \exp(-s(x_i, x_j)^2/h^2)$ for $i, j = 1, \ldots, n_x$, $(L_{pp})_{ij} = \exp(-s(y_i, y_j)^2/h^2)$ for $i, j = 1, \ldots, n_y$, and $(L_{cp})_{ij} = \exp(-s(x_i, y_j)^2/h^2)$ for $i = 1, \ldots, n_x$, for $j = 1, \ldots, n_y$, where $s(\cdot, \cdot)$ is the shortest distance between all possible objects (including drugs and target proteins) on the interaction network with a bipartite graph structure, h is a width parameter, and the distance between unreachable object pairs is treated as infinity. Note that the size of the resulting matrix L is $(n_x + n_y) \times (n_x + n_y)$. The matrix L is not always positive definite, so an appropriate identity matrix is added to the L such that the matrix L meets the positive definite property. Another possibility of constructing the graph-based similarity matrix L is to use the diffusion kernel (30).
2. Similarly as kernel principal component analysis (31), we apply the eigen-value decomposition of L as $L = \Gamma\Lambda^{1/2}\Lambda^{1/2}\Gamma^T$, where the diagonal elements of matrix Λ are eigenvalues and columns of matrix Γ are eigenvectors, and we construct a $(n_x + n_y) \times d$ feature matrix U as $U = \Gamma_d\Lambda_d^{1/2}$ by using the d largest eigenvalues and associated eigenvectors. Then, we represent all drugs and target proteins by using the row vectors of the feature matrix $U = (\mathbf{u}^{x_1}, \ldots, \mathbf{u}_{x_{nx}}, \mathbf{u}^{y_1}, \ldots, \mathbf{u}_{y_{ny}})^T$. The space spanned by features $\mathbf{u}_x$ and $\mathbf{u}_y$ is referred to as "interaction feature space".
3. We consider a model representing the correlation between the data similarity space and the interaction feature space. To do so, we propose to apply a variant of the KRM as follows:

$$\mathbf{u} = f(z) = \sum_{i=1}^{n} k(z, z_i)\mathbf{w}_i + \epsilon, \tag{7}$$

where z is an object, n is the number of training samples, f is the projection from a similarity space to an Euclidean space,

$k(\cdot,\cdot)$ is a kernel similarity function, $\mathbf{w}_i$ is a weight vector of size d, and ϵ is a noise vector. For simplicity, we assume that all the feature values are centered. The model fitting can be done by finding $\mathbf{w}_i$ which minimizes $\|U - KW\|_F^2$, where K is an $n \times n$ similarity matrix, $W = (\mathbf{w}_1, \ldots, \mathbf{w}_n)^T$, and $\|\cdot\|_F$ is Frobenius norm. In this study, we learn two models: f_x for the correlation between the chemical structure similarity space and the interaction feature space with respect to drugs, and f_y for the correlation between the genomic sequence similarity space and the interaction feature space with respect to target proteins, respectively.

4. Given a new drug x', we apply the model f_x and map the new drug x' onto the interaction feature space as

$$\mathbf{u}_{x'} = f_x(x') = \sum_{i=1}^{n_x} k_x(x', x_i)\mathbf{w}_{x_i}, \tag{8}$$

where $\mathbf{w}_{x_i}$ is a weight vector and $k_x(\cdot,\cdot)$ is a chemical structure kernel function.

5. Given a new target protein y', we apply the model f_g and map the new target protein y' onto the interaction feature space as

$$\mathbf{u}_{y'} = f_y(y') = \sum_{j=1}^{n_y} k_y(y', y_j)\mathbf{w}_{y_j}, \tag{9}$$

where $\mathbf{w}_{y_j}$ is a weight vector and $k_y(\cdot,\cdot)$ is a sequence kernel function.

6. Finally, we compute the closeness between drugs and target proteins by the inner product of the features between the corresponding drugs and target proteins in the interaction feature space. Then, drug–target pairs whose closeness is larger than a threshold are predicted to interact with each other.

3.2.2. Kernel Distance Learning

1. We consider two functions $f: U \to \mathbf{R}$ and $g: V \to \mathbf{R}$ to map drugs and target proteins in a unified feature space, where interacting drugs and target proteins are close to each other. A possible criterion to assess whether interacting drug–target pairs are mapped onto close points in $\mathbf{R}$ is the following:

$$R(f, g) = \frac{\sum_{(x_i, y_j) \in E} (f(x_i) - g(y_j))^2}{\sum_{(x_i, y_j) \in U \times V} (f(x_i) - g(y_j))^2}, \tag{10}$$

where $E \subset (U \times V)$ is a set of drug–target interaction edges on the graph. A small value of $R(f, g)$ ensures that connected drug–target pairs tend to be closer than all possible drug–target pairs in the sense of quadratic error.

2. To avoid the over-fitting problem and obtain meaningful solutions, we regularize the criterion Eq. 10 by a smoothness functional on f and g based on a classical approach in statistical learning (32, 33). We assume that f and g belong to the reproducing kernel Hilbert space (r.k.h.s.) $\mathcal{H}_U$ and $\mathcal{H}_V$ defined by the kernels k_x for drugs and k_y for target proteins and to use the norms of f and g as regularization operators. Let us define by $||f||$ and $||g||$ the norms of f and g in $\mathcal{H}_U$ and $\mathcal{H}_V$. Then, the regularized criterion to be minimized becomes:

$$R(f,g) = \frac{\sum_{(x_i,y_j)\in E}(f(x_i) - g(y_j))^2 + \lambda_1||f||^2 + \lambda_2||g||^2}{\sum_{(x_i,y_j)\in U\times V}(f(x_i) - g(y_j))^2}, \quad (11)$$

where λ_1 and λ_2 are regularization parameters which control the trade-off between minimizing the original criterion Eq. 10 and smoothing the functions, and the norms are set as $||f|| = ||g|| = 1$.

3. In order to obtain a d-dimensional feature representation of the objects, we iterate the minimization of the regularized criterion Eq. 11 under orthogonality constraints in the r.k.h.s., that is, we recursively define the l-th features f_l and g_l for $l = 1, \ldots, d$ as follows:

$$(f_l,g_l) = \arg\min \frac{\sum_{(x_i,y_j)\in E}(f(x_i) - g(y_j))^2 + \lambda_1||f||^2 + \lambda_2||g||^2}{\sum_{(x_i,y_j)\in U\times V}(f(x_i) - g(y_j))^2}, \quad (12)$$

under the orthogonality constraints: $f \perp f_1, \ldots, f_{l-1}$, and $g \perp g_1, \ldots, g_{l-1}$.

4. According to the representer theorem (34), for any $l = 1, \ldots, d$, the solution to Eq. 12 has the following expansions: $f_l(x) = \sum_{j=1}^{n_x}\alpha_{l,j}k_x(x_j, x)$ and $g_l(y) = \sum_{j=1}^{n_y}\beta_{l,j}k_y(y_j, y)$ for some vectors $\alpha_l = (\alpha_{l,1}, \ldots, \alpha_{l,n_x})^T \in \mathbf{R}^{n_x}$ and $\beta_l = (\beta_{l,1}, \ldots, \beta_{l,n_y})^T \in \mathbf{R}^{n_y}$. The optimization problem can be reduced to the generalized eigenvalue problem with respect to $\boldsymbol{\alpha}$ and $\boldsymbol{\beta}$. More details of the algorithm can be found in the original paper (18). Note that if the denominator and numerator in Eq. 11 are interchanged, the above minimization problem can be thought of as the maximization problem. In the implementation, the solution of the maximization problem is more stable than that of the minimization problem in solving the corresponding generalized eigenvalue problem from a numerical viewpoint.

5. Finally, we compute the closeness between drugs and target proteins by the cosine correlation coefficient of the features between

the corresponding drugs and target proteins in the interaction feature space. Then, drug–target pairs whose closeness is larger than a threshold are predicted to interact with each other.

3.3. Bipartite Local Model Approach with Score Aggregation

The combination of many local SVM (L-SVM) classifiers to predict potential drug–target interactions was proposed (16). This method is the bipartite graph version of the original local model approach which was proposed in the context of protein-protein interaction prediction (35).

3.3.1. Bipartite Local Model

In their approach, the presence or absence of interactions between drug x_i and target protein y_j are predicted in the following way.

1. Excluding target y_j, we make a list of all other known targets of drug x_i in the gold standard network, as well as a separate list of the targets not known to be targeted by drug x_i. The known targets are given a label + 1 and the others a label − 1.
2. We look for a classification rule that tries to discriminate the + 1 labeled data from the − 1 labeled data using the available genomic sequence data for the targets.
3. We take this rule and use it to predict the label of target y_j and hence an edge or nonedge between drug x_i and target y_j.
4. We fix the same target y_j, then, excluding drug x_i, we make a list of all other known drugs targeting y_j in the gold standard network, as well as a list of drugs not known to target y_j. Similarly to before, drugs known to target y_j are given the label + 1 and the others the label − 1.
5. We look for a classification rule that tries to discriminate the + 1 labeled data from the − 1 labeled data, using the available chemical structure data for the drugs.
6. We take this rule and use it to predict the label of drug x_i and hence an edge or nonedge between drug x and target y.

Part of the originality of the approach is the steps 4–6, where the goal is to make a second independent prediction of the same edge, whenever possible. Even though we are attempting to predict exactly the same edge in both cases, we are doing it with a different data set in each case and potentially a different classification rule (or class of rules). This gives us two independent predictions for the same edge, though with one caveat.

3.3.2. Score Aggregation

In practice, either the drug may have no known targets or the target may have no known targeting drug. Results in this chapter are therefore presented to give a clear idea of prediction accuracy in each of the following three cases:

Case 1. The drug has no known target, and the target has at least one known targeting drug.

Case 2. The target has no known targeting drug, and the drug has at least one known target.

Case 3. The drug has at least one known target, and the target has at least one known targeting drug.

The first two cases reflect the situation where we want to predict unknown interactions involving newly arriving drug candidate compounds or target candidate proteins outside of the training data set. The third case represents a kind of double application of the algorithm, treating each edge of the bipartite network as two directed edges pointing in opposite directions. In this case, we end up with two independent predictions for the same edge. Essentially, we then define a function $m(\cdot,\cdot)$ that aggregates the two (or even more) prediction scores for the same edge into a global score. A simple heuristic is the choice $m(a, b) = \max(a, b)$, which was used in the original work (16). Note that the score aggregation procedure can be used not only for the L-SVM method but also for other methods in the third case.

3.4. Performance Evaluation

3.4.1. Baseline Method

As a baseline method, we used the nearest neighbor method (NN), because this idea has been used in traditional molecular screening so far. Given a new drug candidate compound, we find a known drug (in the training set) sharing the highest structure similarity with the new compound, and predict the new compound to interact with target proteins known to interact with the nearest drug. Likewise, given a new target candidate protein, we find a known target protein (in the training set) sharing the highest sequence similarity with the new protein and predict the new protein to interact with drugs known to interact with the nearest target protein. Newly predicted drug–target interaction pairs are assigned prediction scores with the highest structure or sequence similarity values involving new compounds or new proteins.

3.4.2. Implementation

We tested the five different methods: NN, L-SVM, P-SVM, KRM, and KDL on their abilities to predict the drug–target interactions.

1. In the application of L-SVM, we used the LIBSVM (v.2.88) SVM implementation freely available for the MATLAB environment, and we fixed the *C* regularization parameter at 1.
2. In the application of P-SVM, it is impossible to apply standard SVM implementations such as LIBSVM and SVM^{light} (36), because the size of the kernel matrix for all possible drug–target pairs is too huge to construct explicitly in the memory. In fact, the space complexity is $O(n_x^2 \times n_y^2)$ which is just for storing the kernel matrix, where n_x and n_y are the numbers of drugs and target proteins, respectively. Therefore, we trained the models of the P-SVM by using an online learning algorithm which

processes one training example at each training step, so it is computationally and spatially efficient. In this study, we used PUMMA (37) whose solutions asymptotically converge to those by the SVM with the squared hinge loss. In the experiment, the hyperparameter for regularization in P-SVM is set to $C = 1$, and all of the training data were processed one time in the training phase.

3. In the application of KRM, the width parameter in the Gaussian function is set to $h = 2$
4. In the application of KDL, the regularization parameter is set to $\lambda = 2$ and the number of features is set to $d = 100$, but $d = 20$ is set for the nuclear receptor classes because of the data size.

3.4.3. Experiment for Predicting Interactions Involving New Drugs and NewTargets

1. We evaluated the performance for predicting unknown interactions involving new drug candidate compounds and target proteins by performing the following fivefold cross-validation in a block-wise manner: Both drugs and target proteins in the gold standard set was split into five subsets of roughly equal size, both each drug subset and each target subset were then taken in turn as test sets, and the training is performed on the remaining four sets. We draw the ROC curve, the plot of true positives as a function of false positives based on various thresholds, where true positives are correctly predicted interactions and false positives are predicted interactions that are not present in the gold standard interactions. The ROC curves are drawn for different sets of predictions depending on whether the drugs and/or the target proteins were in the initial training set or not. Drugs and target proteins in the training set are called "training," whereas those not in the training set are called "test." Three different classes are then possible: (1) test drugs vs training target proteins, (2) training drugs vs test target proteins, (3) test drugs vs test target proteins. The performance was evaluated by AUC (area under the ROC curve) score.
2. Table 1 shows the resulting AUC scores for four classes of drug–target interaction data. Note that NN and L-SVM cannot predict the interactions between test drugs and test target proteins, while this is possible with P-SVM, KRM, and KDL. One explanation of the worst performance of NN is that raw drug structure or protein sequence similarities do not always reflect the tendency of interaction partners in true drug–target interactions, which demonstrates the usefulness of the supervised learning of the other four methods: L-SVM, P-SVM, KRM, and KDL. Comparing the four different methods, KRM seems to have the best performance for all the four protein families, and consistently outperform the other methods.

Table 1
AUC (area under ROC curve) in cross-validation for predicting unknown interactions involving new drug candidate compounds and target proteins

Data	Method	AUC (i) Test drugs vs training targets	AUC (ii) Training drugs vs test targets	AUC (iii) Test drugs vs test targets
Enzyme	NN	0.667	0.898	–
	L-SVM	0.867	0.901	–
	P-SVM	0.869	0.903	0.799
	KRM	0.861	0.962	0.822
	KDL	0.840	0.935	0.800
Ion channel	NN	0.634	0.882	–
	L-SVM	0.738	0.938	–
	P-SVM	0.740	0.943	0.710
	KRM	0.746	0.951	0.681
	KDL	0.760	0.947	0.753
GPCR	NN	0.691	0.801	–
	L-SVM	0.886	0.903	–
	P-SVM	0.891	0.901	0.829
	KRM	0.886	0.918	0.838
	KDL	0.814	0.918	0.802
Nuclear receptor	NN	0.742	0.672	–
	L-SVM	0.817	0.792	–
	P-SVM	0.829	0.796	0.728
	KRM	0.849	0.851	0.740
	KDL	0.770	0. 807	0.739

3.4.4. Experiment for Detecting Missing Interactions Between Known Drugs and Known Targets

1. We evaluated the performance of the bipartite local model (BLM) approach with score aggregation for detecting missing interactions between known drugs and known targets. The score aggregation technique was proposed for L-SVM in the previous work (16), but it is possible to apply for the other methods. So we also tested the effect of the score aggregation for NN and KRM as well.
2. Table 2 shows comprehensive results of performing the leave-one-out cross-validation. The first column shows the AUC scores when performing leave-one-out on potential drugs, which reflects the situation where the drug has no known target and the target has at least one known targeting drug. The second column shows results when performing leave-one-out on potential target proteins, which reflects the situation where the target has no known targeting drug and the drug has at least one known target. The third column shows results when

Table 2
AUC (area under ROC curve) in cross-validation for detecting missing interactions between known drugs and known targets

		AUC		
Data	**Method**	**(i) Drug score**	**(ii) Target score**	**(iii) Score aggregation**
Enzyme	NN	0.682	0.899	0.930
	L-SVM	0.831	0.942	0.973
	KRM	0.828	0.929	0.967
Ion channel	NN	0.647	0.887	0.917
	L-SVM	0.745	0.935	0.970
	KRM	0.745	0.917	0.969
GPCR	NN	0.695	0.812	0.885
	L-SVM	0.823	0.872	0.953
	KRM	0.828	0.873	0.947
Nuclear receptor	NN	0.733	0.687	0.851
	L-SVM	0.812	0.536	0.858
	KRM	0.836	0.523	0.867

combining two or four leave-one-out predictions for the same edge, which reflects the situation where the drug has at least one known target and the target has at least one known targeting drug. The third column simulates the prediction of missing drug–target interactions in the known network. Here, we see the significant improvement in AUC score that can be achieved by aggregating the set of prediction scores for the same drug–target interaction (edge) into a global prediction score in all the methods NN, L-SVM, and KRM. This result suggests that the approach is useful for detecting missing interactions between known drugs and known targets in practical applications.

4. Notes

1. All the methods introduced in this chapter belong to a class of kernel methods (27), so the performance could be improved by using more sophisticated kernel similarity functions designed for drugs and target proteins. Considering the ligand–protein interaction mechanism, the incorporation of prior information about pharmacophores (38) and biding pockets (39) into the design of drug similarity and target protein similarity is an

important work (40). Another promising approach is to use pharmacological information such as side effect and drug efficacy in the design of drug–drug similarity function (19, 20).

2. One serious problem of the P-SVM is that the complexity of the "training" phase scales with the *square* of the "number of training drugs *times* the number of training target proteins," leading to computational difficulties for large-scale problems and requiring prohibitive computational cost.

References

1. Wang Y, Xiao J, Suzek T, Zhang J, Wang J, Bryant S (2009) PubChem: a public information system for analyzing bioactivities of small molecules. Nucleic Acids Res 37:D623–D633
2. Butina D, Segall M, Frankcombe K (2002) Predicting ADME properties in silico: methods and models. Drug Discov Today 7:S83–S88
3. Byvatov E, Fechner U, Sadowski J, Schneider G (2003) Comparison of support vector machine and articial neural network systems for drug/nondrug classification. J Chem Inf Comput Sci 43:1882–1889
4. Rarey M, Kramer B, Lengauer T, Klebe G (1996) A fast flexible docking method using an incremental construction algorithm. J Mol Biol 261:470–489
5. Keiser M, Roth B, Armbruster B, Ernsberger P, Irwin J, Shoichet B (2007) Relating protein pharmacology by ligand chemistry. Nat Biotechnol 25:197–206
6. Yildirim M, Goh K, Cusick M, Barabasi A, Vidal M (2007) Drug-target network. Nat Biotechnol 25:1119–1126
7. Zhu S, Okuno Y, Tsujimoto G, Mamitsuka H (2005) A probabilistic model for mining implicit "chemical compound-gene" relations from literature. Bioinformatics 21(Suppl 2): ii245–251
8. Kanehisa M, Goto S, Hattori M, Aoki-Kinoshita K, Itoh M, Kawashima S, Katayama T, Araki M, Hirakawa M (2006) From genomics to chemical genomics: new developments in KEGG. Nucleic Acids Res 34:D354–357
9. Stockwell B (2000) Chemical genetics: ligand-based discovery of gene function. Nat Rev Genet 1:116–125
10. Dobson C (2004) Chemical space and biology. Nature 432:824–828
11. Bock JR, Gough DA (2005) Virtual screen for ligands of orphan G protein-coupled receptors. J Chem Inf Model 45:1402–1414
12. Erhan D, Lheureux P-J, Yue SY, Bengio Y (2006) Collaborative ltering on a family of biological targets. J Chem Inf Model 46:626–635
13. Nagamine N, Sakakibara Y (2007) Statistical prediction of proteinchemical interactions based on chemical structure and mass spectrometry data. Bioinformatics 23:2004–2012
14. Faulon J, Misra M, Martin S, Sale K, Sapra R (2008) Genome scale enzyme-metabolite and drug-target interaction predictions using the signature molecular descriptor. Bioinformatics 24:225–233
15. Jacob L, Vert J-P (2008) Protein-ligand interaction prediction: an improved chemogenomics approach. Bioinformatics 24:2149–2156
16. Bleakley K, Yamanishi Y (2009) Supervised prediction of drug–target interactions using bipartite local models. Bioinformatics 25:2397–2403
17. Yamanishi Y, Araki M, Gutteridge A, Honda W, Kanehisa M (2008) Prediction of drug–target interaction networks from the integration of chemical and genomic spaces. Bioinformatics 24:i232–i240
18. Yamanishi Y (2009) Supervised bipartite graph inference. In Koller D, Schuurmans D, Bengio Y, Bottou L (eds) Advances in Neural Information Processing Systems, vol 21. MIT, Cambridge, MA, pp 1841–1848
19. Campillos M, Kuhn M, Gavin A, Jensen L, Bork P (2008) Drug target identification using side-effect similarity. Science 321 (5886):263–266
20. Yamanishi Y, Kotera M, Kanehisa M, Goto S (2010) Drug-target interaction prediction from chemical, genomic and pharmacological data in an integrated framework. Bioinformatics 26:i246–i254
21. Atias N, Sharan R (2011) An algorithmic framework for predicting side-effects of drugs. Journal of Computational Biology, 18, 207–218

22. Kanehisa M, Araki M, Goto S, Hattori M, Hirakawa M, Itoh M, Katayama T, Kawashima S, Okuda S, Tokimatsu T, Yamanishi Y (2008) KEGG for linking genomes to life and the environment. Nucleic Acids Res 36(Database issue):D480–D485
23. Gunther S, Guenther S, Kuhn M, Dunkel M et al (2008) SuperTarget and Matador: resources for exploring drug–target relationships. Nucleic Acids Res 36(Database issue): D919–D922
24. Wishart D, Knox C, Guo A, Cheng D, Shrivastava S, Tzur D, Gautam B, Hassanali M (2008) DrugBank: a knowledgebase for drugs, drug actions and drug targets. Nucleic Acids Res 36(Database issue): D901–D906
25. Hattori M, Okuno Y, Goto S, Kanehisa M (2003) Development of a chemical structure comparison method for integrated analysis of chemical and genomic information in the metabolic pathways. J Am Chem Soc 125:11853–11865
26. Smith T, Waterman M (1981) Identification of common molecular subsequences. J Mol Biol 147:195–197
27. Schölkopf B, Tsuda K, Vert J (2004) Kernel methods in computational biology. MIT, Cambridge, MA
28. Lodhi H, Yamanishi Y (2010) Chemoinformatics and advanced machine learning perspectives: complex computational methods and collaborative techniques. IGI Global, USA
29. Todeschini R, Consonni V (2002) Handbook of molecular descriptors. Wiley, New York
30. Kondor R, Lafferty J (2002) Diffusion kernels on graphs and other discrete input spaces. In: Faucett T, Mishra N (eds) Proceedings of the twentieth international conference on machine learning. AAAI Press, USA, pp 321–328
31. Scholkopf B, Smola A, Muller K-R (1998) Nonlinear component analysis as a kernel eigenvalue problem. Neural Comput 10:1299–1319
32. Wahba G (1990) Splines models for observational data: series in applied mathematics. SIAM, Philadelphia
33. Girosi F, Jones M, Poggio T (1995) Regularization theory and neural networks architectures. Neural Computation 7:219–269
34. Shawe-Taylor J, Cristianini N (2004) Kernel methods for pattern analysis. Cambridge University Press, Cambridge
35. Bleakley K, Biau G, Vert J-P (2007) Supervised reconstruction of biological networks with local models. Bioinformatics 23:i57–i65
36. Joachims T (2003) Learning to classify text using support vector machines: methods, theory and algorithms. Kluwer Academic, Dordrecht
37. Ishibashi K, Hatano K, Takeda M (2008) Online learning of approximate maximum p-norm margin classiers with biases. Proceedings of the 21st annual conference on learning theory (COLT2008), 69–80
38. Mahe P, Ralaivola L, Stoven V, Vert J (2006) The pharmacophore kernel for virtual screening with support vector machines. J Chem Inf Model 46:2003–2014
39. Kratochwil N, Malherbe P, Lindemann L, Ebeling M, Hoener M, Muhlemann A, Porter R, Stahl M, Gerber P (2005) An automated system for the analysis of G protein-coupled receptor transmembrane binding pockets: alignment, receptor-based pharmacophores, and their application. J Chem Inf Model 45:1324–1336
40. Jacob L, Hoffmann B, Stoven V, Vert J-P (2008) Virtual screening of gpcrs: an in silico chemogenomics approach. BMC Bioinformatics 9:363

Chapter 10

Localization Prediction and Structure-Based In Silico Analysis of Bacterial Proteins: With Emphasis on Outer Membrane Proteins

Kenichiro Imai*, Sikander Hayat*, Noriyuki Sakiyama*, Naoya Fujita*, Kentaro Tomii*, Arne Elofsson*, and Paul Horton*

Abstract

In this chapter, we first discuss protein localization in bacteria and evaluate some localization prediction tools on an independent dataset. Next, we focus on β-barrel outer membrane proteins (BOMPs), describing and evaluating new tools for BOMP detection and topology prediction. Finally, we apply general protein structure prediction methods on these proteins to show that the structure of most BOMPs in *E. coli* can be modeled reliably.

Key words: Protein localization, Topologyprediction, β-barrelmembrane proteins, SVM, HMM

Abbreviations

BAM	ß -barrel assembly machine complex
BBOMP	bacterial ß -barrel outer membrane protein
HMM	hidden markov model
SVM	support vector machine
ioM-profile	inside-outside-membrane profile

1. Introduction

Continuing rapid advances in sequencing technology have created a situation in which genomic and inferred amino acid sequence data is inundating our community. We now have complete sequence data

*contributed equally.correspondence regarding BOCTOPUS.

Hiroshi Mamitsuka et al. (eds.), *Data Mining for Systems Biology: Methods and Protocols*, Methods in Molecular Biology, vol. 939, DOI 10.1007/978-1-62703-107-3_10, © Springer Science+Business Media New York 2013

for organisms that otherwise have not been well characterized—to the extent that with metagenomic studies, we even have sequence data for organisms that we do not know how to cultivate. Often, these organisms are bacteria, which in some cases are disease causing or of interest for environmental monitoring, biofuel development, etc. Thus, it is important to try to infer as much as we can about such organisms based on sequence data alone.

The localization of a protein within a bacteria provides an important hint towards the function and potential interaction partners of that protein. Gram-negative bacteria, in particular, possess both inner and outer membranes, with outer membrane proteins (OMPs) often playing an important role in drug resistance. Fortunately, localization prediction from amino acid sequence is a well-studied problem that can be performed relatively reliably in many cases.

Structural analysis, such as structure-based remote homolog search and homology modeling, is another application of sequence analysis which can provide important clues regarding the function of a protein. With the steady accumulation of soluble (and recently membrane) protein structures for use as templates, along with improvements in computational methods, structure-based sequence analysis now enjoys increased coverage and accuracy.

In this chapter we describe some prediction software which can be used to predict the localization and in many cases some aspects of the structure of bacterial proteins based on amino acid sequence alone. In the second part of this chapter, we focus on methods aimed at OMPs, describing existing methods as well as our ongoing predictor projects. Finally, we apply state-of-the-art structural modeling methods to all proteins predicted to localize in the outer membrane of *E.coli*.

2. General Prediction of Subcellular Localization in Bacteria

2.1. Subcellular Localization of Proteins in Bacteria

Proteins must be transported into their appropriate subcellular compartment to maintain eukaryotic cellular function. Subcellular localization of proteins is also important for bacteria, even though they lack organelles, because they are surrounded by one or two lipid bilayer membranes, through which proteins cannot freely diffuse. Indeed, the presence or absence of an outer membrane, which can be visualized by Gram staining (bacteria with outer membranes are Gram-negative), serves as the first level of bacteria classification. Gram-positive bacteria have four localization sites: cytoplasm, cytoplasmic membrane, cell wall, and extracellular, while Gram negative bacteria have five: cytoplasm, inner membrane, periplasm, outer membrane, and extracellular.

2.1.1. Gram-Positive Bacteria

Since Gram-positive bacteria only posses a single membrane, the main localization events are secretion and integration of proteins into the cytoplasmic membrane. Both of these events can take place in a Sec-dependent manner. Many secreted and membrane proteins are synthesized with a cleavable N-terminal signal peptide which targets them, in an unfolded state, to the SecYEG translocon located in the cytoplasmic membrane. This N-terminal signal peptide consists of a positively charged N-terminus (N-region), followed by a hydrophobic core region (H-region) and the signal peptidase cleavage site (C-region) (1). Accurate prediction tools (2) have been developed for this relatively well-characterized sorting signal.

Gram-positive bacteria also use another pathway, termed the twin-arginine translocation (Tat) pathway, to transport proteins across the cytoplasmic membrane (3). The Tat system comprises of TatAC translocases in the cytoplasmic membrane, which transport proteins in a fully folded state. Proteins using the Tat pathway have a cleavable N-terminal signal peptide containing an almost invariant twin-arginine sequence motif. Signal peptides of the Tat pathway are less hydrophobic and tend to be longer than those of the Sec pathway, with a proline often appearing at the -6 position from the cleavage site (3). Recently, an artificial neural network-based prediction method for Tat signal peptides has been developed (4).

2.1.2. Gram-Negative Bacteria

Gram-negative bacteria have five localization sites, and thus, protein sorting is a bit more complicated (particularly, the secretion pathway). Like Gram-positive bacteria, Gram-negative bacteria use the Sec and Tat pathways to transport proteins across the inner membrane. Although the signal peptides of Gram-negative bacteria tend to be shorter and have less positively charged residues in their N-region than those of Gram-positive bacteria, integral membrane proteins targeted for the inner membrane are also integrated via the Sec system.

Almost all integral OMPs are β-barrel proteins, which are assembled by the outer membrane complex BAM. Details regarding the localization mechanism of β-barrel outer membrane proteins (BOMPs) are given in Subheading 3.2.

Secretion is complicated in Gram-negative bacteria. They have evolved at least six types (types I–VI), some of which are not well characterized and remain difficult to predict from sequence. However, they can be roughly classified as Sec-dependent and Sec-independent. The type II (T2SS) and type V (T5SS) secretion systems are Sec-dependent, while types I (T1SS), III (T3SS), IV (T4SS), and VI (T5SS) are Sec-independent (Fig. 1).

Sec-Dependent Pathway

Secretion by T2SS is a two-step process.Cargo proteins are transported to the periplasmvia either the Sec or Tat pathway and then through the outer membrane by the secretion, a protein complex

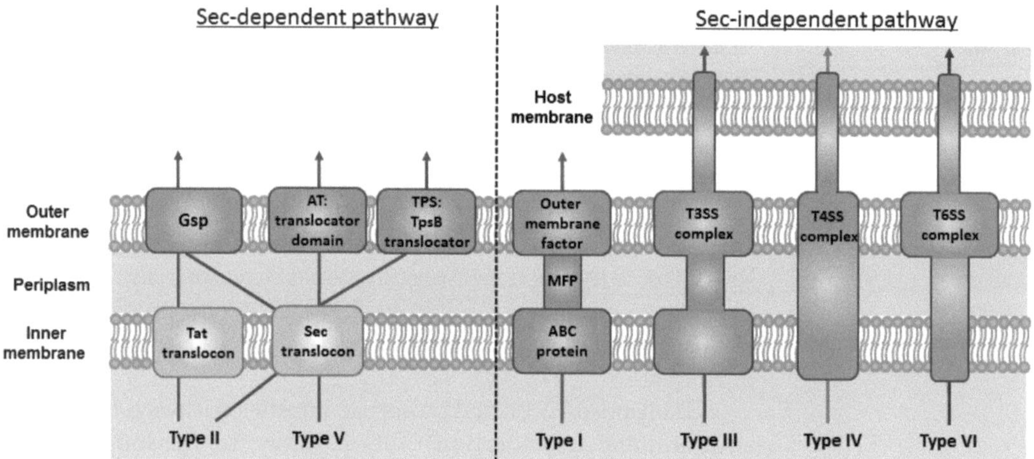

Fig. 1. Secretion System. Gram-negative bacteria secrete proteins through a variety of pathways.

spanning both the inner and the outer membrane(5). For example, the two proteins constituting cholera toxin (CT) aresynthesized with a cleavable N-terminal signal peptide by which they aretransported into the periplasm (6) and subsequently transported across theouter membrane via the type II secretion complex as a hetero-hexamercomplex.

Another Sec-dependent pathway, T5SS, involves autotransporters (ATs) and two-partner secretion (TPS) (7–9). ATs code for both a passenger domain to be transported out of the cell and a β-barrel-forming translocator domain—all in the same amino acid chain! TPS is similar to , but with the passenger and translocator domains found on separate proteins. In both AT and TPS, the β-barrel translocator domains are assembled by the β-barrel biogenesis pathway (see Subheading 3.2).

Sec-Independent Pathway

Proteins secreted by T1SS, T3SS, T4SS, and T6SS do not carry N-terminal signal peptides, but substrates of T1SS have a noncleavable secretion signal in their C-terminus (10). The T1SS machinery is a large complex, comprised of three proteins that span both the inner and outer membrane: an ATP-driven transporter (ABC protein), a membrane fusion protein (MFP), and an OMP belonging to the TolC family.

The T3SS is a needle-like protein complex of about 30 proteins with structural similarity to bacterial flagella. This system is not only able to secrete proteins out of the bacteria but can even inject them into the cytosol of eukaryotic cells. This mechanism is used by some pathogenic bacteria, in which case the injected proteins are called "effector proteins."

Structural proteins forming the T3SS apparatus and chaperones are also secreted by this system. The secretion signals of YopE (Yersinia outer protein E) and YopH are located at the N-terminus but share no similarity with the Sec-dependent pathway signal peptide (11). Interestingly, it has been shown that the secondary structure of Yop messenger RNA is important for secretion of their protein products (12). Recently, some prediction tools have been developed to predict effector proteins based on amino acid composition, G+C content, and physicochemical properties (13, 14).

T4SS also directs the translocation of virulence proteins into eukaryotic host cell, but it differs from TS33 in using a pilus-based system rather than a flagella-like tube. Conjugative transfer of DNA into bacteria also is depended on T4SS. Secretion signals of some effector proteins are located in their C-terminal region (15, 16). However, in the case of the CagA protein, the C-terminal translocation signal is not sufficient, because the deletion of the N-terminus also prevents its translocation (17).

Recently, a novel secretion system, designated T6SS, has been discovered in two bacterial pathogens. T6SS is a phage-like machinery with roles in virulence, symbiosis, inter-bacterial interactions, and anti-pathogenesis (18). However, the details of this secretion system are not yet well characterized.

2.2. Prediction Tools for Bacterial Subcellular Localization

2.2.1. Prediction Tools

Prediction of protein localization can provide important clues to function and can help to identify drug and vaccine targets in pathogenic bacteria. In principle, the localization site of proteins is determined by signals in their amino acid sequence, which in some cases can be recognized without structural information. Thus, many methods have been developed to predict localization from amino acid sequence. In this section, we summarize some general prediction methods and in later sections discuss methods specialized for the prediction of BOMPs (Table 1).

PSORTb is a predictive method which combines several sources of information: homology as detected by `blastp` search, a support vector machine (SVM) trained on sequence signatures associated with each localization site, PROSITE motifs, a transmembrane α-helix predictor, and a signal peptide predictor (19). Recently, PSORTb updated to version 3.0 (20), which can make predictions for all prokaryotes, including archaea and Gram-positive and Gram-negative bacteria. PSORTb is available as both a web server and an open-source stand-alone program. Several new subcategories (host associated, type III secretion, fimbrial, flagellar, and spore) can be predicted by PSORTb 3.0.

CELLO combines SVM-based prediction and homology search to predict localization for both Gram-positive and Gram-negative bacteria (also eukaryotes) (21). The SVM part uses a modified version of n-gram features (e.g., the frequency of dipeptides) which groups amino acid residues by physicochemical properties. CELLO is only available as a web server, but it does accept multiple sequences.

Table 1
Prediction tools for protein localization and β-barrel outer membrane proteins

Predictor	URL
General protein localization	
SOSUI-GramN	http://bp.nuap.nagoya-u.ac.jp/sosui/sosuigramn/sosuigramn_submit.html
Gpos-mPLoc	http://www.csbio.sjtu.edu.cn/bioinf/Gpos-multi/
Gneg-mPLoc	http://www.csbio.sjtu.edu.cn/bioinf/Gneg-multi/
PSORTb *	http://www.psort.org/psortb
CELLO *	http://cello.life.nctu.edu.tw/
β-barrel outer membrane (BBOMP)	
BOCTOPUS	http://boctopus.cbr.su.se/
BOMP	http://services.cbu.uib.no/tools/bomp
TMB-hunt	http://bmbpcu36.leeds.ac.uk/~andy/betaBarrel/AACompPred/aaTMB_Hunt.cgi
PROFtmb	http://www.predictprotein.org/
HHomp	http://toolkit.tuebingen.mpg.de/hhomp

Prediction tools and their public server website addresses are shown. PSORTb and CELLO are marked with an asterisk to indicate that we included them in our BBOMP detection comparison as well as our general localization prediction comparison

SOSUI-GramN is based on a canonical discriminant analysis using the physicochemical properties of the N- and C-terminal region, including targeting signals (22). The prediction procedure consists of three layers of filters: the first layer is prediction of inner membrane proteins by SOSUI (23); the second layer discriminates between Sec-dependent and Sec-independent pathway cargo based on N-terminal signals; and the third layer consists of several prediction modules for subclassification of extracellular proteins, OMPs, periplasmic proteins, and cytoplasmic proteins, based on physicochemical properties of the N- and C-termini. Like CELLO, the SOSUI-GramN system is only available as a web server, but it also accepts multiple sequences.

Gpos-mPLoc and Gneg-mPLoc were recently developed for predicting protein subcellular localization of Gram-positive and Gram-negative bacteria, respectively (24, 25). They use a pipeline consisting of Gene Ontology term search, domain search, and evolutionary information as detected by PSI-BLAST (26). Both tools are freely available at their web site, but only single-sequence queries are accepted, which makes this server difficult to use when analyzing large numbers of sequences.

2.2.2. Performance Evaluation

As mentioned above, many tools for localization prediction of bacterial proteins have been developed. However, direct comparison based on published accuracies is difficult, because the methods have been trained and evaluated on different data sets. Therefore, we report the results of a direct comparison here.

When preparing our dataset for comparison, we set aside proteins used to train {PSORTb, Gpos-mPLOC} and {PSORTb, SOSUI-GramN, Gneg-mPLoc}, for Gram-positive and Gram-negative bacteria respectively, as "trained on". We did not include the CELLO v2.5 training set because it was not available for download. As a test dataset, we extracted all Swiss-Prot (release 2011_07) records of bacterial proteins with firm localization annotation, excluding records marked as "by similarity," etc. We further excluded any proteins with significant sequence similarity (E-value $< 0.\ 001$, as computed by SSEARCH) to any of the trained on proteins. Finally, we used the CD-HIT program (27) with a cutoff of 30% identity to reduce the presence of pairs of similar sequences within the test set. Unfortunately, this left only one OMP, so we added two bacterial β-barrel outer membrane proteins (BBOMPs) with less than 25% sequence identity from PDBTM (28) (see also Subheading 3.3). In this way, we obtained 48 sequences as a test set for Gram-negative bacteria: 15 cytoplasmic, 12 periplasmic, 11 extracellular, 7 inner membrane, and 3 outer membrane proteins; and 55 sequences for Gram-positive bacteria: 21 cytoplasmic, 15 extracellular, 11 cytoplasmic membrane, and eight cell wall.

We performed prediction on the test proteins using the following web servers: PSORTb 3.0, CELLO v2.5, SOSUI-GramN, Gneg-mPLoc, and Gpos-mPLoc. For each localization site, we computed standard binary classification performance measures defined in terms of the number of true positives (TP), true negatives (TN), false positives (FP), and false negatives (FN), namely, recall, TP/(TP+FN); specificity, TN/(TN+FP); and Matthews correlation coefficient (MCC) (29):

$$MCC = \frac{TP \cdot TN - FP \cdot FN}{\sqrt{(TP + FP)(TP + FN)(TN + FP)(TN + FN)}}.$$

Two details slightly complicate this picture; for some proteins, some predictors output no sites ("unknown") or multiple sites. We treated no site prediction as predicting negative for all sites, so this tends to increase specificity at the cost of lower recall. For methods, such as CELLO, which give numerical scores, we did not allow multiple predictions but instead converted them to single-site predictions by taking the site with the maximum score. However, for methods, such as Gneg-mPLoc and Gpos-mPLoc, which do not give scores, we did allow multiple positive predictions, which tend to increase recall at the cost of specificity. For example, a prediction of both periplasm and extracellular for a protein annotated as extracellular was scored as a FP for periplasm and a TP for extracellular.

Table 2
Performance comparison for Gram-positive and Gram-negative bacteria protein localization prediction tools

	PSORTb 3.0			CELLO v2.5			Gpos-mPLoc		
Gram positive	**Rec**	**Spe**	**MCC**	**Rec**	**Spe**	**MCC**	**Rec**	**Spe**	**MCC**
Cytoplasmic	0.76	0.76	0.52	0.95	0.76	0.70	1.00	0.85	0.83
Cytoplasmic membrane	0.73	0.91	0.62	0.82	0.95	0.77	0.82	0.86	0.61
Cell wall	0.00	1.00	0.00	0.00	1.00	0.00	0.13	0.98	0.20
Extracellular	0.07	0.95	0.03	0.67	0.85	0.51	0.80	0.73	0.47
Average	0.39	0.91	0.29	0.61	0.89	0.49	0.69	0.86	0.53
Overall accuracy	80%			85%			84%		

	PSORTb 3.0			CELLO v2.5			Gneg-mPLoc			SOSUI-GramN		
Gram negative	**Rec**	**Spe**	**MCC**	**Rec**	**Spe**	**MCC**	**Rec**	**Spe**	**MCC**	**Rec**	**Spe**	**MCC**
Cytoplasmic	0.67	0.79	0.44	0.87	0.76	0.58	0.53	0.73	0.25	0.87	0.73	0.55
Inner membrane	0.29	0.95	0.30	0.29	0.98	0.38	0.86	0.46	0.23	0.71	0.95	0.67
Periplasmic	0.00	1.00	0.00	0.75	0.89	0.62	0.17	0.97	0.25	0.42	1.00	0.59
Outer membrane	0.33	1.00	0.56	1.00	0.93	0.68	0.33	0.96	0.29	0.67	0.93	0.48
Extracellular	0.00	1.00	0.00	0.27	0.95	0.30	0.45	0.92	0.42	0.18	0.89	0.09
Average	0.26	0.95	0.26	0.64	0.90	0.51	0.47	0.81	0.29	0.57	0.90	0.48
Overall accuracy	82%			85%			74%			84%		

Recall, specificity, and MCC are shown for each combination of predictor andlocalization site. The last row shows the overall accuracy

Table 2 shows our comparison results. CELLO v2.5 performed better than other predictors for both Gram-positive and Gram-negative bacteria, with an average MCC of 0.49 and 0.51, respectively, and also edged out Gpos-mPLoc and SOSUI-GramN by one point in overall accuracy—but recall that our test data may include some of the CELLO training data. For Gram-negative bacteria, SOSUI-GramN attained the 2nd highest average MCC (0.48) and the highest for detection of inner membrane proteins. We also computed overall accuracy (number of correct predictions divided by total number of predictions), with similar results: Unfortunately, no prediction tools could successfully predict cell wall proteins. However, in many cases, cell wall proteins were predicted as extracellular. If we consider cell wall proteins together with extracellular proteins, the MCC for this extended extracellular class improved to 0.29, 0.49, and 0.53 for PSORTb 3.0, CELLO v2.5, and

Gpos-mPLoc, respectively. Nevertheless, for all predictors, the prediction of extracellular proteins in Gram-negative bacteria was poor compared to other localization sites. As mentioned above, secretion in Gram-negative bacteria is complicated and involves a variety of secretion signals. We hope that continued efforts to incorporate new knowledge, such as the discovery of T6SS, into prediction methods will lead to improved accuracy.

3. Prediction of β-Barrel Outer Membrane Proteins

3.1. What Are β-BarrelOuter Membrane Proteins?

Gram-negative bacteria are enveloped by both an inner and outer membrane. The outer membrane, homologous to the outer membrane of mitochondria and plastids, shows distinct properties including asymmetrical distributions of phospholipids and lipopolysaccharides. Proteins crossing the inner membrane adopt an α-helical structure in their membrane-spanning regions, but almost all proteins in the outer membrane adopt β-barrel structures (see (30, 31) for two exceptions).

Due to its physical properties, the outer membrane functions as a selective barrier that prevents the entry of many toxic molecules into the cell, a property that is crucial for bacterial survival in many environments. In such an impermeable membrane, BBOMPs not only work as channels for ion transport, nutrient intake, enzymes, and signaling but also play an important role in virulence and multidrug resistance (32–35).

3.2. Biogenesis of β-BarrelOuter Membrane Proteins

The mechanism of BBOMP biogenesis is not yet completely clear; however, a network of proteins related to this process has been discovered in the past 10 years (36, 37). The process of sorting and assembly of BBOMPs to the outer membrane has multiple stages (Fig. 2). BBOMP precursors are synthesized with N-terminal signal peptides, which lead them to the Sec (secretion) translocon in the inner membrane (38). The cytoplasmic chaperon SecB plays an important role in this process by preventing the BBOMPs from folding in the cytosol.

After cleavage of their signal peptide by signal peptidase, nascent BBOMPs cross the periplasm in an unfolded state. The periplasmic chaperones (SurA, Skp, and DegP) interact with BBOMPs to prevent them from aggregation or misfolding and target them from the Sec translocon to the BAM (β-barrel assemble machinery) complex in the outer membrane. Periplasmic transport is thought to be handled by two pathways: the SurA pathway and the Skp-DegP pathway (39). SurA was reported to bind unfolded BBOMPs and aromatic residue-rich peptides such as Ar-X-Ar-X-P or Ar-X-Ar (40, 41) (where Ar indicates F,Y, or W).

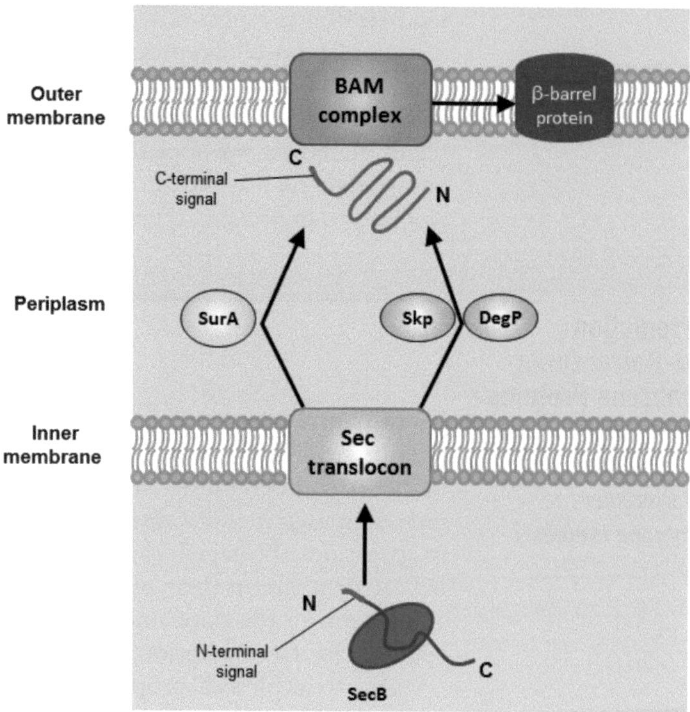

Fig. 2. Sorting and integration of BBOMPs. BBOMP precursors interact with the Sec translocon and cytoplasmic chaperones to mediate translocation across the inner membrane. Subsequently, nascent BBOMPs then are transported to the outer membrane by periplasmic chaperones in the SurA or Skp/DegP pathways. The BAM complex integrates BBOMPs into outer membrane and folds them.

These alternating aromatic residue patterns are much more common in BBOMPs than in soluble or inner membrane proteins (40, 41). Skp is a general chaperone which binds to denatured BBOMPs, and DegP also exhibits chaperone activity in a temperature-dependent manner (42, 43). Knockdown experiments suggest that Skp and DegP function in the same pathway, but SurA works in a separate parallel pathway (39). Several hypotheses regarding the Skp-DegP pathway have been suggested, but its precise role in BBOMP biogenesis remains unclear (36, 37).

The BAM complex assembles BBOMPs in the outer membrane when they reach the outer membrane. In *E. coli*, this complex is composed of the integral membrane β-barrel protein BamA (omp85/YaeT) and four lipoproteins: BamB(YfgL), BamC (NlpB), BamD(YfiO), and BamE(SmpA). BamA and BamD are essential genes, while the deletion of the other three lipoproteins also affects BBOMP biogenesis (37). BamA, the central component of this complex, is conserved in all Gram-negative bacteria and has homologs in both mitochondria (Sam50) and plastids (Toc75).

BamA has a soluble domain usually containing 3–7 polypeptide transport-associated (POTRA) repeat units on its periplasmic side (44). The POTRA domain enables BAM complex assembly by binding to other component lipoproteins (45). BamA is also essential for the recognition of BBOMP substrates. It has been reported that it may recognize the C-terminal Phe (or Trp) present in a large fraction of BBOMPs and the dyadic periodicity of hydrophobic residues common in membrane-spanning β-strands (46, 47).

3.3. Structural Features of β-Barrel Outer Membrane Proteins

In BBOMPs, cylindrical β-barrel structures are formed by hydrogen bonds between adjacent β-strands. So far, more than 40 nonredundant BBOMP structures have been solved (28). All of these BBOMPs contain between 4 and 24 transmembrane β-strands (28, 48, 49). Of these, the BBOMPs with the smallest number of β-strands form a trimeric 12-stranded β-barrel structure, to which each monomer contributes four β-strands, as exemplified by TolC and Hia (50, 51). In the case of monomeric β-barrels, the smallest number of β-strands is eight. At the other extreme, the recently solved structure of FimD (52) forms a 24-stranded β-barrel translocon channel.

Several rules can be inferred from these bacteria β-barrel structures (49):

- All β-strands are antiparallel and connected to their nearest primary sequence neighbors.
- The number of β-strands is even, and the N- and C-termini are found on the periplasmic side.
- The periplasmic side interstrand loops tend to be shorter than the external loops.
- In membrane-spanning β-strands, alternating residues tend to be less hydrophobic (they face the aqueous interior of the barrel).
- Tyrosine is often observed in β-barrel structures with its aromatic ring located near the membrane interface (53, 54).

Note however that the mitochondrial outer membrane protein VDAC breaks the first two rules by forming a 19-transmembrane β-strand structure (55–57).

3.4. Computational Detection Methods for β-Barrel Outer Membrane Proteins

3.4.1. Prediction Methods

The membrane-spanning β-strands of BBOMPs are shorter and less hydrophobic than their α-helical counterparts, making them somewhat more difficult to predict. As mentioned above, membrane-spanning β-strands often exhibit hydrophobic residues at alternating positions and a characteristic amino acid composition, both of which are commonly used as features for prediction. TMB-Hunt (58) is a *k*-nearest neighbor predictor using an amino acid composition-derived function as a protein–protein distance measure. BOMP (59) uses a pipeline requiring candidate BBOMPs to match a regular expression pattern at their 10 C-terminal residues, contain sequence segments fitting the statistical characteristics of membrane-spanning β-barrel strands (60), and have an appropriate overall amino acid composition. PROFtmb (61) utilizes a kind of hidden Markov model (HMM) with a topology designed to capture the different environments of residues relative to the membrane, with states representing periplasmic and extracellular loops and various positions in the transmembrane-spanning strands. States in their HMM emit PSI-BLAST (26)-type profiles instead of single characters, enabling the use of some evolutionary information. The HHomp (62) approach rests on the belief that all β-strand pairs (ββ-hairpins) in BBOMPs have a common origin (63). It uses a remote homology detection method based on transitive sequence similarity search, in which similarity is measured by pairwise HMM profile comparison.

In addition to these BBOMP-specific methods, general prediction methods (described in Subheading 2) for protein subcellular localization, such as CELLO (64) and PSORTb (20), can predict BBOMPs reasonably well, by treating all predicted OMPs as BBOMPs.

Prediction accuracy has improved year by year, with recent methods reporting over 90% accuracy. However, as in general localization prediction, direct comparison of published accuracies is somewhat problematic due to different datasets and different levels of redundancy. Therefore, we conducted a simple comparison on an independent test data.

3.4.2. Performance Evaluation

Using two databases, the Gram-negative BBOMPs database OMPdb (65) and the membrane protein structure database PDBTM (28), we gathered BBOMP sequences with no sequence similarity to the BBOMPs used to train existing BBOMP predictors (E-value by SSEARCH ≥ 0.001). As mentioned in Subheading 2.2.2, it is quite difficult to make an independent BBOMP dataset because almost all experimental verified BBOMPs have already been used in the training of existing BBOMP predictors. Since we found only 3 experimentally verified independent BBOMPs (Uniprot ID: Q72JD8, Q8ZPT3, and Q9Z6M6), we added 52 outer membrane proteins in OMPdb without known structures but with annotated β-barrel domains by InterPro and Pfam (66, 67). However, we must warn the reader that these 52 sequences cannot be considered as absolutely confirmed BBOMPs. In any case, we obtained 55 test sequences sharing less than 25% sequence similarity with each other. As mentioned above, these sequences do not show a significant SSEARCH hit against BBOMPs used to train existing prediction, although they probably are distant homologs as reflected in their classification by InterPro and Pfam. For negative data, we used the 45 non-BBOMP proteins described in Subheading 2.2.2.

Table 3
Performance comparison between representative BBOMP prediction methods

Predictor	TP	FP	FN	TN	Precision	Recall	Specificity	MCC
BOMP	39	0	16	45	**1.00**	0.71	**1.00**	0.72
TMB-hunt	26	3	29	42	0.90	0.47	0.93	0.45
PROFtmb	51	1	4	44	0.98	**0.93**	0.98	**0.90**
HHomp	51	2	4	43	0.96	**0.93**	0.96	0.88
CELLO v2.5	36	3	19	42	0.92	0.66	0.93	0.60
PSORTb 3.0	17	0	38	45	**1.00**	0.31	**1.00**	0.41

The number of true and false positives and negatives along with otherperformance measures are shown for several BBOMP predictors

We performed our comparison on six representative prediction methods listed in Table 1, each of which provides a web site which can accept multiple sequence input. Table 3 lists precision, recall, specificity, and MCC of the six prediction methods. The HMM profile-based predictions are clearly superior to the others, with the highest MCC (0.90) achieved by PROFtmb and the 2nd highest (0.88) by HHomp. One hundred percent precision and specificity were attained by BOMP and PSORTb, but their recall was lower (0.71 and 0.31, respectively). For the three experimentally verified BBOMPs, PROFtmb, TMB-hunt, and CELLO could predict all three, while HHomp missed one and PSORTb two. Taken together, these results suggest that HMM profile-based predictions are best for the detection of BBOMPs. However, the methods using N- and C-terminal targeting signals could be said to more closely model the underlying phenomenon of BBOMP sorting and assembly.

3.5. Improved Prediction of the Topology of Transmembrane β-Barrel Proteins

BBOMPs are formed by antiparallel β-strands that form a closed barrel-like central pore region. Residues in the β-strands tend to follow a dyad repeat pattern (68), but their less prominent hydrophobicity profile makes it difficult to detect β-strands based on hydrophobicity alone. Since only a limited number of BBOMP three-dimensional structures have been determined, computational methods to provide structural information are needed. Fortunately, the structures of the barrels are regular, so it should be possible to determine their overall structure, given accurate topology prediction (69).

The small number of known transmembrane β-barrel (TMB) structures also poses problems in the development of computational methods for BBOMP topology prediction and identification. Nevertheless, many BOMP topology prediction methods have been developed (61, 62, 70–74). These predictors employ a variety of computational methods, but a comparison carried out by (75) found that the HMM-based methods performed best.

Recently, we introduced BOCTOPUS (76), a computational method for BOMP topology prediction. BOCTOPUS uses a position-specific scoring matrix (PSSM) as its input and employs a combination of SVMs and an HMM to account for local and global residue preferences, respectively. BOCTOPUS is based on similar ideas to those used in two recent methods for topology prediction of α-helical membrane proteins: OCTOPUS (77) and MEMSAT-3 (78).

Based on a cross-validation estimation, the per-residue Q3 accuracy of BOCTOPUS is 91%. Furthermore, BOCTOPUS predicted the correct number of strands for 34 out of 36 proteins in the dataset.

3.5.1. BOCTOPUS: Dataset and Method

BOCTOPUS is trained on a nonredundant dataset of 36 BBOMP structures, homology reduced at 30% sequence identity. The 3-D structures were obtained from the OPM database (79). Tenfold cross validation was employed in the training of the SVMs used in BOCTOPUS, and to further avoid influence by distantly related homologs, the training was performed such that all proteins that belong to the same OPM family were put together in the same cross-validation partition. Based on the membrane boundaries included in the PDB files obtained from the OPM database (79) and the z-coordinate of the C-α atoms, all residues in the dataset were annotated as either "i" (inner loop), "o" (outer loop), or "M" (transmembrane β-strand).

The first step in a prediction is to obtain PSSMs with PSI-BLAST as the input for the SVMs. In the first stage, BOCTOPUS employs three separate SVMs (80) to recognize residues belonging to the "i," "o," and "M" classes. The second part of BOCTOPUS consists of an HMM-like module that uses the ioM-profile as its input, using the modhmm package (81). As shown in Fig. 3(B), the HMM describing the global topology contains a pre-barrel stage (P) that describes the protein region before the first transmembrane β-strand is detected. Further, a BBOMP is defined by 4 different states, representing the inner loop, outer loop, and the up and down transmembrane β-strands. As previously described by (77), the transition probabilities between all states are set to 0.0 or 1.0 between most states. Different combinations of weights for estimating the emission scores for the P and O loop states were tried. The up and down strand states can handle β-strands in the range of 6–15 residues. To be consistent with structural properties known from the 3-D structures available so far, all protein topologies start in the "P" or "i" state and end in the "M" (down strand) or "i" state. The Viterbi algorithm is used to predict the most likely topology based on the emission scores.

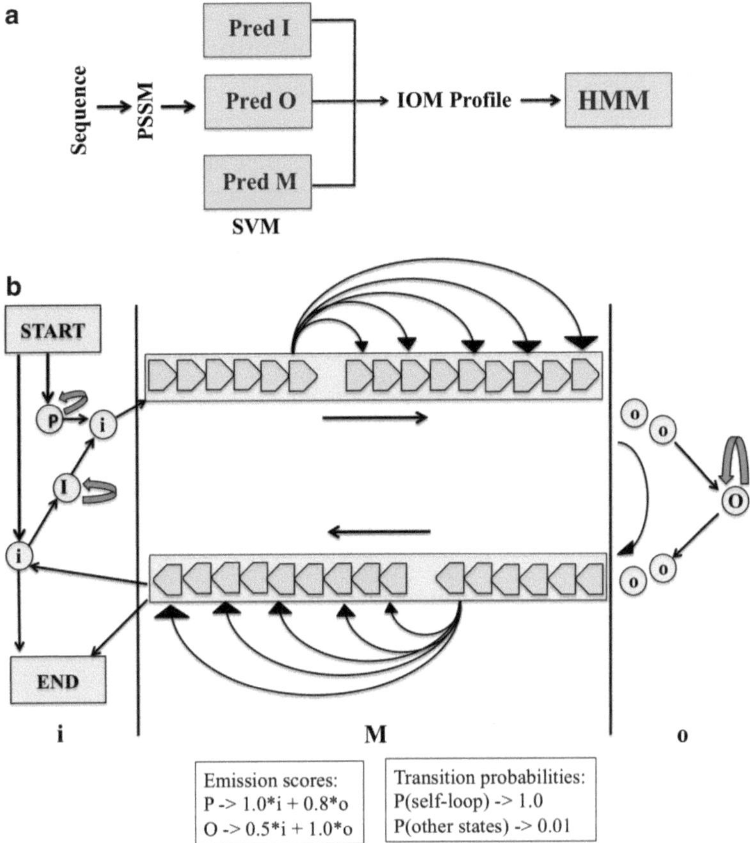

Fig. 3. BOCTOPUS pipeline. (1) PSI-BLAST is used to generate a PSSM for each given sequence. (2) Three separate SVMs are used to predict the residue-level preference for each amino acid to be in the i, M, and o regions, respectively. (3) An "ioM-profile" is generated from the probabilities obtained from the SVMs. (4) The "ioM-profile" is then used by an HMM to predict the global topology, calculated using the Viterbi algorithm.

The per-residue performance of BOCTOPUS was evaluated based on Q3 and segment overlap (SOV) measures (82). In addition, the number of strands predicted per protein and the number of strands with an overlap of at least two residues were used to evaluate the overall predicted topology. Correct predicted topology is defined as the correct number of predicted strands, where correct is defined as overlapping by at least two residues with an observed strand.

3.5.2. Topology Prediction Using BOCTOPUS

As shown in Table 4, BOCTOPUS predicts the correct number of strands in 34 out of 36 cases (94%). In the remaining two cases, BOCTOPUS under-predicts the number of strands (fimbrial usher porin from *E.coli* (2vqi) and anion-selective porin from *Comamonas acidovorans* (1e54)). Furthermore, BOCTOPUS also has the highest accuracy (83%) for predicting the correct topology. The BOCTOPUS web server takes a FASTA-formatted sequence as the input to predict the topology of putative BBOMPs. Figure 4 shows the

Table 4
Comparison of BBOMP topology predictor performance

Methods	Proteins with correct no. strands	Proteins with correct topology	UP	OP
BOCTOPUS	34(94%)	30(83%)	2	0
PRED-TMBB	22(61%)	15(42%)	5	9
PROFtmb	27(75%)	25(69%)	4	5
TMBETAPRED-RBF[a]	27(75%)	21(58%)	4	4
TMBpro	24(67%)	19(53%)	3	9

Correct no. strands—number of sequences where the number of predicted strandsis equal to the number of observed strands. Correct topology—number for whichthe number of strands is correct and the predicted strands overlap by at least 2residues. Underpredicted (UP)—sequences where the number of strands isunderpredicted, i.e., some strands are missed. Overpredicted (OP)—sequenceswhere the number of strands is overpredicted. BOCTOPUS-SVM shows theaccuracy measures without the HMM stage[a]TMBETAPRED-RBF (73) classified 2qomA as a non-BOMP protein

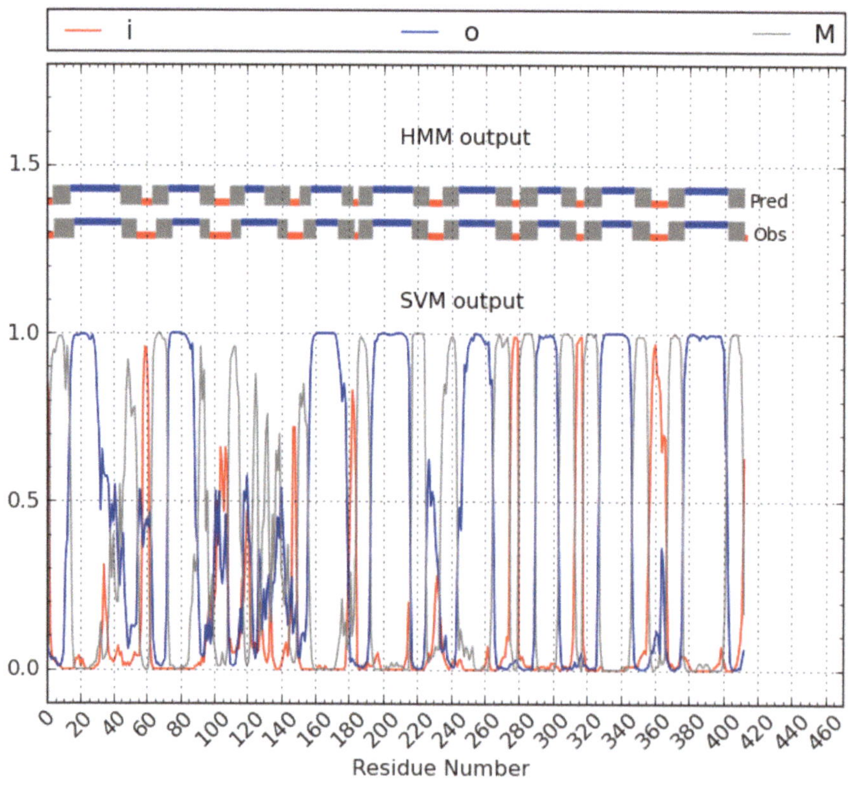

Fig. 4. BOCTOPUS output. Output for sucrose-specific porin ScrY from *Salmonella typhimurium* (PDB ID: 1a0s). The X-axis shows the residue number. Output probabilities from the SVMs are shown at the *bottom* (0 to 1). Final topology predictions are shown at the *top* with *horizontal bars*. Outer loops, inner loops, and the TM strands are shown in *blue*, *red*, and *gray* color, respectively.

Table 5
Comparison of predictor topology results on multichain BBOMPs

PDB-ID	Actual no. strands	Predicted no. strands BOCTOPUS	PRED-TMBB	PROFtmb
1tqq_A	4	4	2	6
1wpl_A	4	0	4	–
1yc9_A	4	0	2	4
1uun_A	2	4	4	2
3emo_A	4	4	2	4
7ahl_A	2	10	10	8

The prediction accuracy on multichain BBOMPs is shown. BOCTOPUS reports the correctnumber of β-strandsin 2 out of 6 cases

output from the BOCTOPUS web server for sucrose-specific porin ScrY from *Salmonella typhimurium* (PDB ID: 1a0s). The observed topology is overlaid to compare the observed and the predicted topologies.

BOCTOPUS has been trained on single-chain BBOMPs and does not take into account multichain BBOMPs, which are BBOMPs whose barrel comprises of β-strands from different chains (61, 63). Only a few multichain BBOMP structures are available: TolC protein from *E. coli* (1tqq), drug-discharge OMP, OprM from *Pseudomonas aeruginosa* (1wp1), and multidrug resistance (VceC) protein from *Vibrio cholerae* (1yc9) have long inner loops which are different from typical single-chain BBOMPs. To the best of our knowledge, most topology prediction methods in the literature do not consider multichain BBOMPs. Nevertheless, a comparison of two prediction methods on multichain BBOMPs is provided in Table 5. In the case of the α-hemolysin protein from *Staphylococcus aureus* (7ahl), BOCTOPUS overpredicts the number of strands, which could be due to the presence of multiple β-sheets in the outer-loop of 7ahl which are wrongly classified as transmembrane β-strands. The under-predicted number of strands in the case of 1wp1 and 1yc9 could be due to their large inner loops. It should be noted that 7ahl and porin MspA from *Mycobacterium smegmatis* (1uun) are found in Gram-positive bacteria. Given the role played by these atypical BBOMPs as toxins, it will be interesting to further investigate them in the future when more 3D structures are made available (83).

4. Structure-Based Sequence Analysis of Bacterial Proteins

This section focuses on general computational methods for protein structure prediction. Although amino acid sequence alignment methods are still commonly used techniques to infer protein structures based on known 3-D structures, profile–profile (and also HMM–HMM) alignment methods have become increasingly important techniques since their introduction in 2000 (84). Indeed, profile–profile alignment methods generally outperform conventional methods, such as PSI-BLAST, in terms of both sensitivity and alignment accuracy (85).

Several profile–profile methods have been developed for structural analysis (86). One well-performing method is FORTE (87), which performs profile–profile alignments for protein structure prediction. To predict the structure of a protein, FORTE uses PSSMs, which are calculated by PSI-BLAST iterations over the NCBI nonredundant database of amino acid sequences, with both the query and templates as profiles. FORTE utilizes massive sequence information, optimized gap penalties, and the Pearson's correlation coefficient of amino acid frequency as the scoring scheme to measure the similarity between two profile columns. Global–local algorithm is employed to build an optimal alignment between a query profile and a template one. The statistical significance of each alignment score is estimated by calculating Z-scores with a simple log-length correction. Candidate templates are sorted by Z-scores and presented to the user.

The FORTE server is available at http://www.cbrc.jp/forte/. The server holds a profile library of representative proteins mainly based on the SCOP database as templates. Users can submit a single protein sequence to explore the possibilities of similarity with known structures. FORTE has performed well in CASP (the Critical Assessment of protein Structure Prediction experiments) http://predictioncenter.org/ and has been employed in various studies (88–92). It is expected that the performance of profile–profile alignment methods will become even more effective as more protein sequence and structure information accumulate and computational methods are refined.

In recent CASPs, consensus prediction methods, also known as "meta-servers," which combine prediction results from many different methods, including profile–profile alignment methods have been the most successful. Generally speaking, these methods compare 3-D models constructed based on alignments of a query protein and the top hits from the involved methods, and then report the medoid structure as their prediction result. For example, Pcons (93), one of the earliest meta-servers, combines machine-learning approaches with consensus analysis to select the best possible model from a set of predicted structures.

According to a comprehensive study of structural annotation of genomes (94), we can expect that about 45% of genes in bacteria have at least one domain with known structure (see Figure 1 of (94)). By employing the sophisticated methods described above, one can expect even greater coverage. In the following section, we apply structure prediction methods to predicted *E. coli* OMPs.

4.1. Structural Predictions of E. coli BBOMPs

To estimate the number of BBOMPs in *E. coli* that could be reliably modeled using state-of-the-art methods, we first applied a localization prediction method (PSORTb 3.0) (20) on the complete proteome of *E. coli*. This resulted in 82 proteins predicted to be located in the outer membrane. Thereafter, we used the Pcons.net (93) structure prediction method on these proteins. However, Pcons. net can create models both of the barrel and the nonbarrel domains of the proteins. Therefore, hits to known BBOMPs were manually identified, and it was found that a good model could be made for 31 proteins, (see Table 6). For 27 proteins, a non-TMB domain could be modeled. Furthermore, we used BOCTOPUS to predict the topology of the 82 proteins, see Fig. 5. Here, it was found that 22 of the 82 proteins were predicted to have no β-strands, indicating that they most likely are not BBOMPs. This was thereafter verified by applying BOMP (59) and PROFtmb (61) on these proteins.

In summary, we found that a reliable structural model can be made for 31 of 60 (52%) of the identified BBOMPs in *E.coli*. Further, it can be seen that these proteins are predicted to contain between 2 and 30 β-strands, and the correct topology could be predicted by BOCTOPUS for 26 out of these 31 proteins. Some of the 29 proteins may contain some multichain BBOMPs which cannot be reliably identified using the BOCTOPUS topology prediction method.

5. Notes

Based on the methods described in this chapter, here, we propose a step-by-step procedure for how to identify and annotate BBOMPs within a Gram-negative bacterial proteome. We also discuss the possible pitfalls and hints that can aid in accessing the reliability of the generated models.

Step 1: The identification of OMPs by subcellular localization prediction is an important first step to obtain a set of likely BBOMPs. One of the better methods for this is to use BOMP (59), as it shows relatively good coverage (Table 3) and is much faster than PROFtmb (61) or HHomp (62). If obtaining maximal coverage is not of high concern, PSORTb (20) is convenient,

Table 6
Modeling of BBOMPs using the Pcons meta-server

	Uniprot ID	PCONS score	Template	%coverage	%identity	Observed no. of strands in template	Predicted no. of strands BOCTOPUS
1	P0A910	0.089	2K0L	57	82	8	10
2	P09169	0.273	1I78	94	98	10	10
3	P77211	0.172	1YC9	89	26	12(4)	0
4	P0A921	0.117	1QD5	89	100	12	12
5	P75733	0.212	3JTY	93	14	18	18
6	P75780	0.38	2W16	97	20	22	22
7	P0A915	0.14	2F1V	90	100	8	8
8	P06996	0.369	2J1N	94	100	16	16
9	P76045	0.132	2X9K	92	99	14	14
10	P0A927	0.207	1TLY	90	100	12	12
11	Q47706	0.201	2QTK	93	18	18	18
12	P31827	0.085	3FHH	96	17	22	22
13	P77747	0.39	2J1N	94	67	16	18
14	P02932	0.401	2J1N	94	63	16	16
15	P05825	0.271	1FEP	96	98	22	22
16	P17315	0.333	2HDI	95	99	22	22
17	P02930	0.136	1EK9	87	99	12(4)	4
18	P02931	0.423	1PHO	94	64	16	16
19	P76773	0.124	2WJQ	90	27	12	12
20	P37001	0.122	3GP6	87	99	8	6
21	P0A917	0.157	1ORM	87	98	8	8
22	P06971	0.301	1QFG	86	100	22	22
23	P02943	0.298	1AF6	94	100	18	18
24	P16869	0.351	2W16	98	33	22	22
25	P32714	0.145	1YC9	87	28	12(4)	2
26	P10384	0.23	3DWO	94	22	14	14
27	P06129	0.311	2HDI	96	25	22	22
28	P13036	0.385	1KMO	85	100	22	22
29	P69856	0.311	2WJR	90	99	12	12
30	P26218	0.185	1AF6	77	25	18	18
31	P76115	0.262	2W16	97	17	22	22

Using a Pcons score cutoff of 0.05, 31 out of 60 BBOMPs canbe modeled using Pcons. BOCTOPUS gets the number ofβ-strandscorrect in 26 out of these 31 cases. Contribution to the barrel by a single chain ina multichain BBOMP is shown in circular brackets

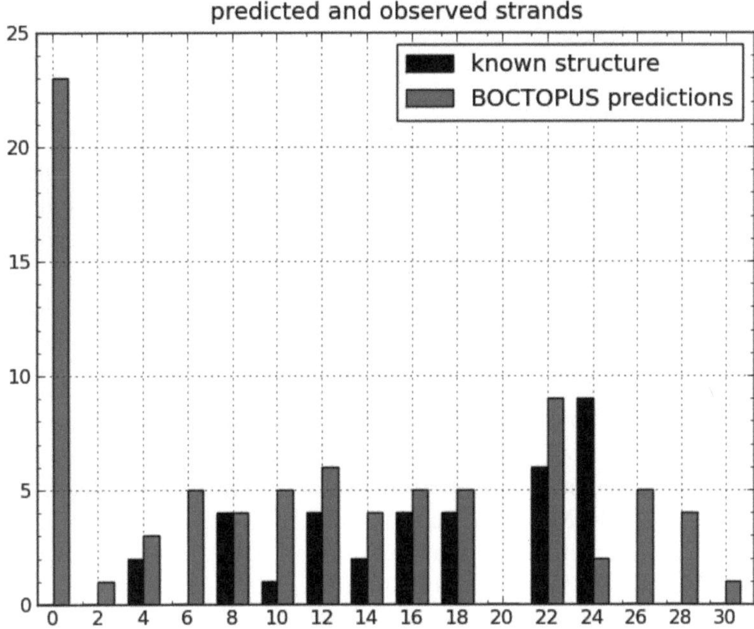

Fig. 5. BOCTOPUS predictions for *E. coli*. Predicted topologies of all 82 putative outer membrane proteins in *E. coli*. Predictions by BOCTOPUS are shown in *gray*. For 36 of these proteins, a significant hit to an OMP of known structure is found. The number of strands for these proteins is shown in *black*.

because their web site provides precomputed predicted subcellular localization for a number of genomes for download.

Step 2: The next step is to perform a topology prediction of the putative BBOMPs. One of the better methods for this is BOCTOPUS (http://boctopus.cbr.su.se/), but PRED-TMBB and other methods can also be used to determine the number of β-strands in all putative BBOMPs. Although the state-of-the-art methods perform quite well (> 80% accuracy), it should be noted that they are not optimized to correctly predict the topology of multi-strand BBOMPs.

Step 3: The predicted number of β-strands can theoretically be used to generate ideal BBOMP templates based on theoretically determined barrel radius and β-strand tilt (95). However, to obtain more accurate models, structure prediction servers such as Pcons.net (93), FORTE (87), or HHpred (96) can then be used to create homology models of putative BBOMPs. Of these methods, HHpred has the advantage to be significantly faster than the others. Here, a query sequence is aligned against a set of templates of known structure. However, a hit does not guarantee that the barrel part of a potential BBOMP can be modeled accurately. This is due to the fact that large proteins, such the AG43 protein in *E. coli* (Uniprot id: P39180), can contain several non-BBOMP domains whose structure might be accurately modeled. In the case of BBOMPs in *E. coli*, 27 non-TMB domains could be modeled, suggesting that either the query sequences had no reliable known structural template for their TMB domains or that those sequences do not contain a TMB domain and are in fact not BBOMPs. Thus, it is imperative to inspect what region of the sequence is represented by the template found. For this chapter, this was done both manually by visualizing the structures and by comparing the identified templates as BBOMPs or not.

Step 4: To verify that the putative BBOMPs without predicted strands and without a reliable hit to a BBOMP by Pcons.net are not BBOMPs, a more reliable identification can be done using various dedicated BBOMP identification methods (58, 59, 61, 62, 97, 98). These methods employ a variety of features such as statistical propensities and C-terminal pattern identification (BOMP) (59), amino acid composition (58, 97), and secondary structure element alignments (98). In this chapter, we applied BOMP (59) and PROFtmb (62) for this step.

6. Conclusion

In this chapter, we introduced and evaluated some tools for sequence analysis of bacterial proteins—with a special emphasis on OMPs. It can be seen that, while far from perfect, these tools can provide important hints regarding the localization and structure of bacterial proteins from amino acid sequence alone. We hope this chapter may serve as a starting point for readers planning to carry out this kind of analysis.

References

1. von Heijne G (1985) Signal sequences. The limits of variation. J Mol Biol 184:99–105
2. Bendtsen J, Nielsen H, von Heijne G, Brunak S (2004) Improved prediction of signal peptides: SignalP 3.0. J Mol Biol 340:783–795
3. De Buck E, Lammertyn E, Anné J (2008) The importance of the twin-arginine translocation pathway for bacterial virulence. Trends Microbiol 16:442–453
4. Bendtsen J, Nielsen H, Widdick D, Palmer T, Brunak S (2005) Prediction of twin-arginine signal peptides. BMC Bioinf 6:167
5. Voulhoux R, Ball G, Ize B, Vasil M, Lazdunski A, Wu L, Filloux A (2001) Involvement of the twin-arginine translocation system in protein secretion via the type II pathway. EMBO J 20:6735–6741
6. Reichow S, Korotkov K, Gonen M, Sun J, Delarosa J, Hol WGJ, Gonen T (2011) The binding of cholera toxin to the periplasmic vestibule of the type II secretion channel. Channels (Austin) 5:215–218
7. Jacob-Dubuisson F, Fernandez R, Coutte L (2004) Protein secretion through autotransporter and two-partner pathways. Biochim Biophys Acta 1694:235–257
8. Desvaux M, Parham N, Henderson I (2004) The autotransporter secretion system. Res Microbiol 155:53–60
9. Thanassi D, Stathopoulos C, Karkal A, Li H (2005) Protein secretion in the absence of ATP: the autotransporter, two-partner secretion and chaperone/usher pathways of gram-negative bacteria (review). Mol Membr Biol 22:63–72

10. Cescau S, Debarbieux L, Wandersman C (2007) Probing the in vivo dynamics of type I protein secretion complex association through sensitivity to detergents. J Bacteriol 189:1496–1504
11. Sory M, Boland A, Lambermont I, Cornelis G (1995) Identification of the YopE and YopH domains required for secretion and internalization into the cytosol of macrophages, using the cyaA gene fusion approach. Proc Natl Acad Sci USA 92:11998–2002
12. Anderson D, Schneewind O (1997) A mRNA signal for the type III secretion of Yop proteins by Yersinia enterocolitica. Science 278:1140–1143
13. Arnold R, Brandmaier S, Kleine F, Tischler P, Heinz E, Behrens S, Niinikoski A, Mewes H-W, Horn M, Rattei T (2009) Sequence-based prediction of type III secreted proteins. PLoS Pathog 5:e1000376
14. Samudrala R, Heffron F, McDermott J (2009) Accurate prediction of secreted substrates and identification of a conserved putative secretion signal for type III secretion systems. PLoS Pathog 5:e1000375
15. Vergunst A, van Lier M, den Dulk-Ras A, Stüve TAG, Ouwehand A, Hooykaas P (2005) Positive charge is an important feature of the C-terminal transport signal of the VirB/D4-translocated proteins of Agrobacterium. Proc Natl Acad Sci USA 102:832–837
16. Nagai H, Cambronne E, Kagan J, Amor J, Kahn R, Roy CR (2005) A C-terminal translocation signal required for Dot/Icm-dependent delivery of the Legionella RalF protein to host cells. Proc Natl Acad Sci USA 102:826–831
17. Hohlfeld S, Pattis I, Püls J, Plano G, Haas R, Fischer W (2006) A C-terminal translocation signal is necessary, but not sufficient for type IV secretion of the Helicobacter pylori CagA protein. Mol Microbiol 59:1624–1637
18. Records A (2011) The type VI secretion system: a multipurpose delivery system with a phage-like machinery. Mol Plant-Microbe Interact 24:751–757
19. Gardy J, Laird M, Chen F, Rey S, Walsh C, Ester M, Brinkman FSL (2005) PSORTb v.2.0: expanded prediction of bacterial protein subcellular localization and insights gained from comparative proteome analysis. Bioinformatics 21:617–623
20. Yu N et al (2010) PSORTb 3.0: improved protein subcellular localization prediction with refined localization subcategories and predictive capabilities for all prokaryotes. Bioinformatics 26:1608–1615
21. Yu C-S, Chen Y-C, Lu C-H, Hwang J-K (2006) Prediction of protein subcellular localization. Proteins 64:643–651
22. Imai K, Asakawa N, Tsuji T, Akazawa F, Ino A, Sonoyama M, Mitaku S (2008) SOSUI-GramN: high performance prediction for sub-cellular localization of proteins in gram-negative bacteria. Bioinformation 2:417–421
23. Hirokawa T, Boon-Chieng S, Mitaku S (1998) SOSUI: classification and secondary structure prediction system for membrane proteins. Bioinformatics 14:378–379
24. Shen H-B, Chou K-C (2009) Gpos-mPLoc: a top-down approach to improve the quality of predicting subcellular localization of Gram-positive bacterial proteins. Protein Pept Lett 16:1478–1484
25. Shen H-B, Chou K-C (2010) Gneg-mPLoc: a top-down strategy to enhance the quality of predicting subcellular localization of Gram-negative bacterial proteins. J Theor Biol 264:326–333
26. Altschul S, Madden T, Schaffer A, Zhang J, Zhang Z, Miller W, Lipman D (1997) Gapped BLAST and PSI-BLAST: a new generation of protein database search programs. Nucleic Acids Res 25:3389–3402
27. Li W, Godzik A (2006) Cd-hit: a fast program for clustering and comparing large sets of protein or nucleotide sequences. Bioinformatics 22:1658–1659
28. Tusnády G, Dosztányi Z, Simon I (2005) PDB_TM: selection and membrane localization of transmembrane proteins in the protein data bank. Nucleic Acids Res 33:D257–278
29. Matthews B (1975) Comparison of the predicted and observed secondary structure of T4 phage lysozyme. Biochim Biophys Acta 405:442–451
30. Dong C, Beis K, Nesper J, Brunkan-Lamontagne A, Clarke B, Whitfield C, Naismith J (2006) Wza the translocon for *E. coli* capsular polysaccharides defines a new class of membrane protein. Nature 444:226–229
31. Chandran V, Fronzes R, Duquerroy S, Cronin N, Navaza J, Waksman G (2009) Structure of the outer membrane complex of a type IV secretion system. Nature 462:1011–1015
32. Schulz G (2000) beta-barrel membrane proteins. Curr Opin Struct Biol 10:443–447
33. Ruiz N, Kahne D, Silhavy T (2006) Advances in understanding bacterial outer-membrane biogenesis. Nat Rev Microbiol 4:57–66
34. Bishop R (2008) Structural biology of membrane-intrinsic beta-barrel enzymes: sentinels of the bacterial outer membrane. Biochim Biophys Acta 1778:1881–1896

35. Nikaido H (2009) Multidrug resistance in bacteria. Annu Rev Biochem 78:119–146
36. Knowles T, Scott-Tucker A, Overduin M, Henderson I (2009) Membrane protein architects: the role of the BAM complex in outer membrane protein assembly. Nat Rev Microbiol 7:206–214
37. Hagan C, Silhavy T, Kahne D (2011) β-barrel membrane protein assembly by the Bam complex. Annu Rev Biochem 80:189–210
38. Wickner W, Schekman R (2005) Protein translocation across biological membranes. Science 310:1452–1456
39. Sklar J, Wu T, Kahne D, Silhavy T (2007) Defining the roles of the periplasmic chaperones SurA, Skp, and DegP in Escherichia coli. Genes Dev 21:2473–2484
40. Bitto E, McKay D (2003) The periplasmic molecular chaperone protein SurA binds a peptide motif that is characteristic of integral outer membrane proteins. J Biol Chem 278:49316–49322
41. Xu X, Wang S, Hu Y, McKay D (2007) The periplasmic bacterial molecular chaperone SurA adapts its structure to bind peptides in different conformations to assert a sequence preference for aromatic residues. J Mol Biol 373:367–381
42. Qu J, Mayer C, Behrens S, Holst O, Kleinschmidt J (2007) The trimeric periplasmic chaperone Skp of Escherichia coli forms 1:1 complexes with outer membrane proteins via hydrophobic and electrostatic interactions. J Mol Biol 374:91–105
43. Spiess C, Beil A, Ehrmann M (1999) A temperature-dependent switch from chaperone to protease in a widely conserved heat shock protein. Cell 97:339–347
44. Arnold T, Zeth K, Linke D (2010) Omp85 from the thermophilic cyanobacterium Thermosynechococcus elongatus differs from proteobacterial Omp85 in structure and domain composition. J Biol Chem 285:18003–18015
45. Kim S, Malinverni J, Sliz P, Silhavy T, Harrison S, Kahne D (2007) Structure and function of an essential component of the outer membrane protein assembly machine. Science 317:961–964
46. Struyve M, Moons M, Tommassen J (1991) Carboxy-terminal phenylalanine is essential for the correct assembly of a bacterial outer membrane protein. J Mol Biol 218:141–148
47. Robert V, Volokhina E, Senf F, Bos M, Van Gelder P, Tommassen J (2006) Assembly factor Omp85 recognizes its outer membrane protein substrates by a species-specific C-terminal motif. PLoS Biol 4:e377
48. Fairman J, Noinaj N, Buchanan S (2011) The structural biology of beta-barrel membrane proteins: a summary of recent reports. Curr Opin Struct Biol 29:1–9
49. Schulz G (2002) The structure of bacterial outer membrane proteins. Biochim Biophys Acta 1565:308–317
50. Koronakis V, Sharff A, Koronakis E, Luisi B, Hughes C (2000) Crystal structure of the bacterial membrane protein TolC central to multidrug efflux and protein export. Nature 405:914–919
51. Meng G, Surana NK, St Geme JW III, Waksman G (2006) Structure of the outer membrane translocator domain of the Haemophilus influenzae Hia trimeric autotransporter. EMBO J 25:2297–2304
52. Phan G et al (2011) Crystal structure of the FimD usher bound to its cognate FimC-FimH substrate. Nature 474:49–53
53. Hemmingsen J, Gernert K, Richardson J, Richardson D (1994) The tyrosine corner: a feature of most Greek key beta-barrel proteins. Protein Sci 3:1927–1937
54. Yau W, Wimley W, Gawrisch K, White S (1998) The preference of tryptophan for membrane interfaces. Biochemistry 37:14713–14718
55. Hiller S, Garces R, Malia T, Orekhov V, Colombini M, Wagner G (2008) Solution structure of the integral human membrane protein VDAC-1 in detergent micelles. Science 321:1206–1210
56. Bayrhuber M, Meins T, Habeck M, Becker S, Giller K, Villinger S, Vonrhein C, Griesinger C, Zweckstetter M, Zeth K (2008) Structure of the human voltage-dependent anion channel. Proc Natl Acad Sci USA 105:15370–15375
57. Ujwal R, Cascio D, Colletier J, Faham S, Zhang J, Toro L, Ping P, Abramson J (2008) The crystal structure of mouse VDAC1 at 2.3 Å resolution reveals mechanistic insights into metabolite gating. Proc Natl Acad Sci USA 105:17742–17747
58. Garrow A, Agnew A, Westhead D (2005) TMB-Hunt: an amino acid composition based method to screen proteomes for beta-barrel transmembrane proteins. BMC Bioinf 6:56
59. Berven F, Flikka K, Jensen H, Eidhammer I (2004) BOMP: a program to predict integral β-barrel outer membrane proteins encoded within genomes of Gram-negative bacteria. Nucleic Acids Res 32:W394–W399
60. Wimley W (2002) Toward genomic identification of beta-barrel membrane proteins: composition and architecture of known structures. Protein Sci 11:301–312

61. Bigelow H, Rost B (2006) PROFtmb: a web server for predicting bacterial transmembrane beta barrel proteins. Nucleic Acids Res 34: W186–W188
62. Remmert M, Linke D, Lupas A, Söding J (2009) HHomp prediction and classification of outer membrane proteins. Nucleic Acids Res 37:W446–W451
63. Remmert M, Biegert A, Linke D, Lupas A, Söding J (2010) Evolution of outer membrane β-barrels from an ancestral ββ hairpin. Mol Biol Evol 27:1348–1358
64. Yu C, Chen Y, Lu C, Hwang J (2006) Prediction of protein subcellular localization. Proteins 64:643–651
65. Tsirigos K, Bagos P, Hamodrakas S (2011) OMPdb: a database of beta-barrel outer membrane proteins from Gram-negative bacteria. Nucleic Acids Res 39:D324–D331
66. Hunter S et al (2009) InterPro: the integrative protein signature database. Nucleic Acids Res 37:D211–D215
67. Finn R et al (2010) The Pfam protein families database. Nucleic Acids Res 38:D211–D222
68. Seshadri K, Garemyr R, Wallin E, von Heijne G, Elofsson A (1998) Architecture of beta-barrel membrane proteins: analysis of trimeric porins. Protein Sci 7:2026–2032
69. Schulz G (2002) The structure of bacterial outer membrane proteins. BBA-Biomembranes 1565:308–317
70. Martelli P, Fariselli P, Krogh A, Casadio R (2002) A sequence-profile-based HMM for predicting and discriminating β barrel membrane proteins. Bioinformatics 18:S46
71. Deng Y, Liu Q, Li Y (2004) Scoring hidden Markov models to discriminate beta-barrel membrane proteins. Comput Biol Chem 28:189–194
72. Randall A, Cheng J, Sweredoski M, Baldi P (2008) TMBpro: secondary structure, β-contact and tertiary structure prediction of transmembrane β-barrel proteins. Bioinformatics 24:513–520
73. Ou Y, Chen S, Gromiha M (2010) Prediction of membrane spanning segments and topology in β-barrel membrane proteins at better accuracy. J Comput Chem 31:217–223
74. Singh N, Goodman A, Walter P, Helms V, Hayat S (2011) TMBHMM: a frequency profile based HMM for predicting the topology of transmembrane beta barrel proteins and the exposure status of transmembrane residues. Biochim Biophys Acta Protein Proteomics 1814:664–670
75. Bagos P, Liakopoulos T, Hamodrakas S (2005) Evaluation of methods for predicting the topology of β-barrel outer membrane proteins and a consensus prediction method. BMC Bioinf 6:7
76. Hayat S, Elofsson A (2011) BOCTOPUS: improved topology prediction of transmembrane β barrel proteins. submitted
77. Viklund H, Elofsson A (2008) OCTOPUS: improving topology prediction by two-track ANN-based preference scores and an extended topological grammar. Bioinformatics 24:1662–1668
78. Jones D (2007) Improving the accuracy of transmembrane protein topology prediction using evolutionary information. Bioinformatics 23:538–544
79. Lomize M, Lomize A, Pogozheva I, Mosberg H (2006) OPM: orientations of proteins in membranes database. Bioinformatics 22:623–625
80. Dimitriadou E, Hornik K, Leisch F, Meyer D, Weingessel A (2009) Misc functions of the department of statistics (e1071). TU Wien Technical report
81. Viklund H, Elofsson A (2004) Best alpha-helical transmembrane protein topology predictions are achieved using hidden Markov models and evolutionary information. Protein Sci 13:1908–1917
82. Rost B, Sander C, Schneider R (1994) Redefining the goals of protein secondary structure prediction. J Mol Biol 235:13–26
83. Iacovache I, Paumard P, Scheib H, Lesieur C, Sakai N, Matile S, Parker M, Van der Goot F (2006) A rivet model for channel formation by aerolysin-like pore-forming toxins. EMBO J 25:457–466
84. Rychlewski L, Jaroszewski L, Li W, Godzik A (2000) Comparison of sequence profiles. Strategies for structural predictions using sequence information. Protein Sci 9:232–241
85. Ohlson T, Wallner B, Elofsson A (2004) Profile-profile methods provide improved fold-recognition: a study of different profile-profile alignment methods. Proteins 57:188–197
86. Dunbrack R (2006) Sequence comparison and protein structure prediction. Curr Opin Struct Biol 16:374–384
87. Tomii K, Akiyama Y (2004) FORTE: a profile-profile comparison tool for protein fold recognition. Bioinformatics 20:594–595
88. Shiozawa K, Maita N, Tomii K, Seto A, Goda N, Akiyama Y, Shimizu T, Shirakawa M, Hiroaki H (2004) Structure of the N-terminal domain of PEX1 AAA-ATPase. Characterization of a putative adaptor-binding domain. J Biol Chem 279:50060–50068

89. Tomii K, Hirokawa T, Motono C (2005) Protein structure prediction using a variety of profile libraries and 3D verification. Proteins 61:114–121
90. Wang G, Jin Y, Dunbrack R (2005) Assessment of fold recognition predictions in CASP6. Proteins 61:46–66
91. Ogawa H et al (2010) Chondroitin sulfate synthase-2/chondroitin polymerizing factor has two variants with distinct function. J Biol Chem 285:34155–34167
92. Imai K, Fujita N, Gromiha M, Horton P (2011) Eukaryote-wide sequence analysis of mitochondrial β-barrel outer membrane proteins. BMC Genomics 12:79
93. Wallner B, Larsson P, Elofsson A (2007) Pcons.net: protein structure prediction meta server. Nucleic Acids Res 35:W369–W374
94. Yeats C, Lees J, Reid A, Kellam P, Martin N, Liu X, Orengo C (2008) Gene3D: comprehensive structural and functional annotation of genomes. Nucleic Acids Res 36:D414–D418
95. Murzin A, Lesk A, Chothia C (1994) Principles determining the structure of β-sheet barrels in proteins I. A theoretical analysis. J Mol Biol 236:1369–1381
96. Söding J, Biegert A, Lupas A (2005) The HHpred interactive server for protein homology detection and structure prediction. Nucleic Acids Res 33:W244
97. Gromiha M, Ahmad S, Suwa M (2005) Application of residue distribution along the sequence for discriminating outer membrane proteins. Comput Biol Chem 29:135–142
98. Yan R, Chen Z, Zhang Z (2011) Outer membrane proteins can be simply identified using secondary structure element alignment. BMC Bioinf 12:76

Chapter 11

Analysis Strategy of Protein–Protein Interaction Networks

Zhenjun Hu

Abstract

Protein interactions, as well as the networks they formed, play a key role in many cellular processes and the distortion of the protein interacting interfaces may lead to the development of many diseases. In this chapter, we will briefly introduce the background knowledge of the protein–protein interaction, followed by the detailed explanation of varied analysis—from basic to advanced, as well as related tools and databases. VisANT (http://visant.bu.edu)—a free Web-based software platform for the integrative visualization, mining, analysis, and modeling of the biological networks—will be used as a main tool for all examples used in this section.

Key words: Interaction, Multi-scale visualization, Large-scale network, Integration, Pathway, Systems biology, Expression, Enrichment analysis

1. Background

In the past few years protein–protein interactions (PPIs) have gained a strong interest in the fields of pharmacy, medicine, biology, and bioinformatics. Identifying and characterizing PPIs and their networks is essential to understand the mechanisms of biological processes on a molecular level. For example, signals from the exterior of a cell are mediated to the inside of that cell by PPIs of the signaling molecules. Protein interactions can be classified into different types depending on their strength (permanent and transient), specificity (specific or nonspecific), the location of interacting partners within one or on two polypeptide chains, and the similarity between interacting subunits. In many cases, the analyses of PPIs are usually coupled with other type of interactions, such as transcription binding (protein–DNA interaction) and synthetic lethal (gene–gene interaction). This is mainly because the complexity of the biological processes is

Hiroshi Mamitsuka et al. (eds.), *Data Mining for Systems Biology: Methods and Protocols*, Methods in Molecular Biology, vol. 939, DOI 10.1007/978-1-62703-107-3_11,

carried out by the combination of various types of interactions between different types of molecules. From this perspective, our discussion will also cover other biological interactions of living organisms, in addition to the PPIs.

1.1. Techniques for Identification of Protein Interactions

Although PPIs have been invaluable biological knowledge for quite a long time, its importance to the systems biology becomes more practical only after high-throughput methods (method that enables the screening of a large number of proteins), such as yeast two-hybrid system (Y2H), become available recently. In fact, there are a multitude of methods to detect interactions now (Table 1 for typical ones). Each of the approaches has its own strengths and weaknesses, especially with regard to the sensitivity and specificity of the method. A high sensitivity means that many of the interactions that occur in reality are detected by the method; a high specificity indicates that most of the interactions detected by the screen are also occurring in reality.

1.1.1. Experimental Method

A detailed classification, comparison, and description of the experimental methods to detect the interactions can be found in the review of Phizichy and Fields (1); while the review of Berggard et al. (2) emphasizes on the advantage/disadvantage of different methods. The technology development has mainly been fuelled by the advances in mass spectrometry (MS) (3, 4), which has made the identification of proteins, as well as macromolecular complexes such as ribosomes and exosomes, a relatively simple task. A number of large-scale studies have been presented, using e.g., Y2H screens and coaffinity purification followed by MS to detect PPIs on a genome-wide scale. However, only a small number of the interactions are supported by more than one method (5). Estimates of 40–80% false negatives and 30–60% false positives and have been assigned to high-throughput studies that have used two-hybrid techniques, affinity-based techniques or computational approaches (5–7). The poor overlap can be explained partly by the fact that many different methods have been used. However, even within subsets of PPIs identified using the same method, the overlap can be poor (e.g., compare the results from the yeast two-hybrid (Y2H) screens from Ito et al. (8) and Uetz et al. (9)). It has been estimated that due to a high false positive rate, current yeast and human interaction maps are roughly only 50 and 10% complete, respectively (10). It is clearly important to experimentally validate PPIs by several methods. However, very few databases or tools, except VisANT (11–16), provide convenient method-based query and filtering of the interactions.

In addition to the specificity and sensitivity, the method may also hint the type of the interaction as shown in Table 1. For example, the Y2H always identifies the physical interactions between proteins while the synthetic lethality tells the genetic

Table 1
Typical methods to identify the interactions

Method name	(H)igh/(L)ow throughput	VisANT method ID	Interaction type	Brief description
Experimental methods				
Mass spectrometry (11, 22)	H	M0028	Protein complex	Mass spectrometric approaches to the study of protein in complexes permits the identification of subunit stoichiometry and transient associations. By preserving complexes intact in the mass spectrometer, mass measurement can be used for monitoring changes in different experimental conditions, or to investigate how variations of collision energy affect their dissociation
Yeast two hybrid (33, 44)	H	M0034	Physical interaction	The classical yeast two-hybrid system is a method that uses transcriptional activity as a measure of protein–protein interaction. The DNA-binding domain serves to target the activator to the specific genes that will be expressed, and the activation domain contacts other proteins of the transcriptional machinery to enable transcription to occur
Copurification (55)	H	M0013	Protein complex	Approaches designed to separate cell components on the basis of their physicochemical properties. The copurified components are thought to form a molecular complex
Synthetic lethality (66)	H	M0047	Genetic interaction	Death phenotype observed on cells carrying combination of two independently silent mutations
Electron microscopy (77)	L	M0062	Protein complex	Electron microscopy methods provide insights into the structure of biological macromolecules and their supramolecular assemblies. Resolution is on average around 10 Angstroms but can reach the atomic level when the samples analyzed are 2D crystals
Computational methods				
Domain fusion (88, 99)	H	M0036	Functional association	The rosetta stone, or domain fusion, procedure is based on the assumption that proteins whose homologues in other organisms happen to be fused into a single protein chain are likely to interact or to be functionally related

(continued)

Table 1 (continued)

Method name	(H)igh/(L)ow throughput	VisANT method ID	Interaction type	Brief description
Phylogenetic profile (10, 11)	H	M0037	Functional association	The phylogenetic profile of a protein stores information about the presence and the absence of that protein in a set of genomes. By clustering identical or similar profiles, proteins with similar functions and potentially interacting are identified
Gene neighborhoods (12, 13)	H	M0038	Functional association	Gene pairs that show a conserved topological neighborhood in many prokaryotic genomes are considered by this approach to encode interacting or functionally related proteins. By measuring the physical distance of any given gene pair in different genomes, interacting partners are inferred

interactions between genes. As a matter of fact, these methods are classified hierarchically by the committee of Protein Standard Initiative-Molecular Interaction (PSI-MI) (17) and can easily be explored using MI Ontology Browser (http://www.ebi.ac.uk/ontology-lookup/browse.do?ontName=MI).

1.1.2. Computational Method

Although high-throughput experimental methods produce a large amount of interaction data, for many organisms they are far from complete. The low interaction coverage along with the experimental biases toward certain protein types and cellular localizations reported by most experimental techniques call for the development of computational methods to predict whether two proteins interact. These methods can be very useful for choosing potential targets for experimental screening. Some computational methods are evolution based, such as fusion and phylogenetic profiling listed in Table 1; many others use mathematical methods, such as one or another Bayesian variant or support vector machines, to carry out a principled and systematic integration of experimental and computational techniques to different extent and do not predict physical interactions directly but rather infer the functional associations between potentially interacting proteins (18, 19). Computational methods typically take as input weighted interactions of diverse types, and return confidence levels for inferred relations. They are essential for drawing network inferences, and knowing how likely an inference is to be valid. Furthermore, computational methods can greatly reduce the complexity of an integrated network by condensing diverse biological data into single all-encompassing relation.

1.2. Protein Interaction Databases

A large variety of databases exists to study binary protein interactions and the higher order interactions in protein complexes (20). Typical examples are BioGrid (21), IntAct (22), MINT (23), and MIPS (24). Some organism-specific database, such as Saccharomyces Genome Database (SGD) (25), FlyBase (26), and Human Protein Reference Database (HPRD) (27). Different databases contain interactions obtained by direct submission from experimentalists and by mining literature and other data sources; in some cases the data is verified using automated algorithms or manual curation. In addition, some databases, such as Predictome (28) and String (29), also provide the computationally predicted functional association between proteins.

2. Basic Analysis Strategy

2.1. Integrating Data from Diverse Sources

In spite of the interaction data diversity, there exist considerable overlaps in the datasets contained in the databases, making it difficult to recommend a single resource for a particular type of information. At the same time, this also indicates that you will need to visit all of these interaction databases and then integrate all the downloaded data to get a most complete interaction data set.

2.1.1. Name Unification

The first task in network integration is the unification of equivalent genes/proteins which have been labeled using different classifications, such that total knowledge available for each node can be determined. This very basic, yet vital, unification operation is nontrivial in practice. In VisANT, name unification is achieved using the *Name Normalization* function. Currently, this function can resolve the identities of genes or proteins, but not compounds.

Unification requires knowing whether the node refers to a gene or a protein. Although names of genes and proteins are often used interchangeably, gene IDs cannot reliably be used to represent proteins, primarily because splice variants exist. From this perspective, the integration of the interaction is often gene based, meaning that two proteins will be recognized as a single protein, if the same gene encodes them. This decision not to distinguish gene nodes from protein nodes was based primarily on the following observations:

- Independently created interaction databases often use different naming systems for proteins/genes. For example, MINT (23) uses a protein's UniProt (30) id in PPI for *Homo sapiens* which, however, is represented using gene's HGNC (HUGO Gene Nomenclature Committee) (31) id in BioGrid (21).
- Genetic associations very often need to be compared to PPIs (32–34).
- High-throughput gene expression results must often be mapped onto a PPI network or pathway.

Because genes often encode multiple proteins, using gene id to represent both gene and protein is a natural choice; otherwise, uncertainty may arise when integrating interaction data. For example, for a gene with two splice variants, the variant to use when integrating interactions between the MINT (protein based) and BioGrid databases (gene based) will be ambiguous. The shortcoming of this solution is that certain splice variants may only share a subset of interactions. This dilemma can be overcome if alternative splicing knowledge is represented using a metanode, as discussed in the Subheading 3.

The data used for name unification can be compiled from the Entrez Gene (35) and UniProt databases, as well as some organism-specific databases such as SGD (25), Flybase (26), and Wormbase (36). Careful attention, however, needs to be paid to assure that every ID/name will be mapped to only one gene for a given species.

2.1.2. Interaction Integration

Large-scale studies have resulted in networks composed of various biological interaction types, such as PPIs, genetic interactions and transcriptional coexpression. A systems-level understanding can be better achieved by integrating these diverse data types, such that not only the molecular assemblies involved can be deciphered, but also their functional connections. As discussed in the previous section, although the number of references reporting the interaction may be the straightforward way to roughly access the reliability of the integrated interactions from multiple resources, the number of methods detecting the interaction appears more promising due to the well-known systematic biasing of many experimental methods toward certain protein types and cellular localizations. All interaction data downloaded from public databases provides the corresponding experimental methods, some are compliant with PSI-MI and some not, it is a nontrivial task to integrate and categorize the interaction data according to the method hierarchy defined by PSI-MI.

2.1.3. Database of the Integration

The Predictome database stores relations based on some 90 different methods for 110+ species. The interaction statistics on the VisANT Web site (http://visant.bu.edu) lists the total number of interactions for each species, the methods contributing to these interaction totals, and the total number of interactions classified under each method. There total 918,312 interactions in the database with 300,297 being predicted computational and 618,015 identified experimentally. If an interaction is reported by multiple references using the same technology, it is counted as one interaction although all these references can be easily accessed through corresponding menus in VisANT. The numbers of the interactions grow all the time as the system is synchronized with major interaction databases, such as Biogrid, MINT, BIND, MIPS, IntAct, and HPRD monthly.

All interactions are classified based on the experimental/computational methods that report them. The name of the method is compatible with those defined in PSI-MI standard whenever possible. Different types of interactions can be distinguished using their method names. For example, interactions based on two-hybrid experiments are physical, while those with synthetic lethal are genetic, and do not necessarily imply a direct physical interaction. As shown in Fig. 1, edges for the interactions between proteins are colored based on the interaction method, and edges resulting from computational prediction are often associated with weights which can be used to quickly filter networks with different confidence scores (Fig. 1**III**, weight cutoff).

Among other issues illustrated in Fig. 1, are two major challenges which need to be addressed in the implementation of any tool for supporting the visualization of large-scale interaction networks, namely readability and performance. In general, a network becomes difficult to interpret with hundreds of nodes and edges; and for the majority of available tools, performance starts to deteriorate quickly when working with networks containing thousands of nodes and edges. Although VisANT has been highly optimized and can handle much larger networks (tested with 226 k+ edges and nodes in a PC with 1 GB memory and 2.33 GHz CPU), some functions, such as spring-force based layout, can become very slow when working with larger networks. In Subheading 3, we will discuss metagraph-based technology, which addresses this bottleneck in network integration. We will also introduce additional solutions in VisANT to support for computationally heavy tasks.

Figure 2 shows an example of the use of the *Name Normalization* function in VisANT. The original data (37) shows what is actually a single gene, but displayed as three different nodes each with its own label. This ambiguity is resolved by name normalization, as indicated in Fig. 2. Subsequent to node unification, the gene aliases and a brief functional annotation will be shown as node's tooltip (Fig. 2**II**), and HTTP links to the related nucleotide/protein sequences will be available in the corresponding menus. In addition, VisANT provides a special function to label a

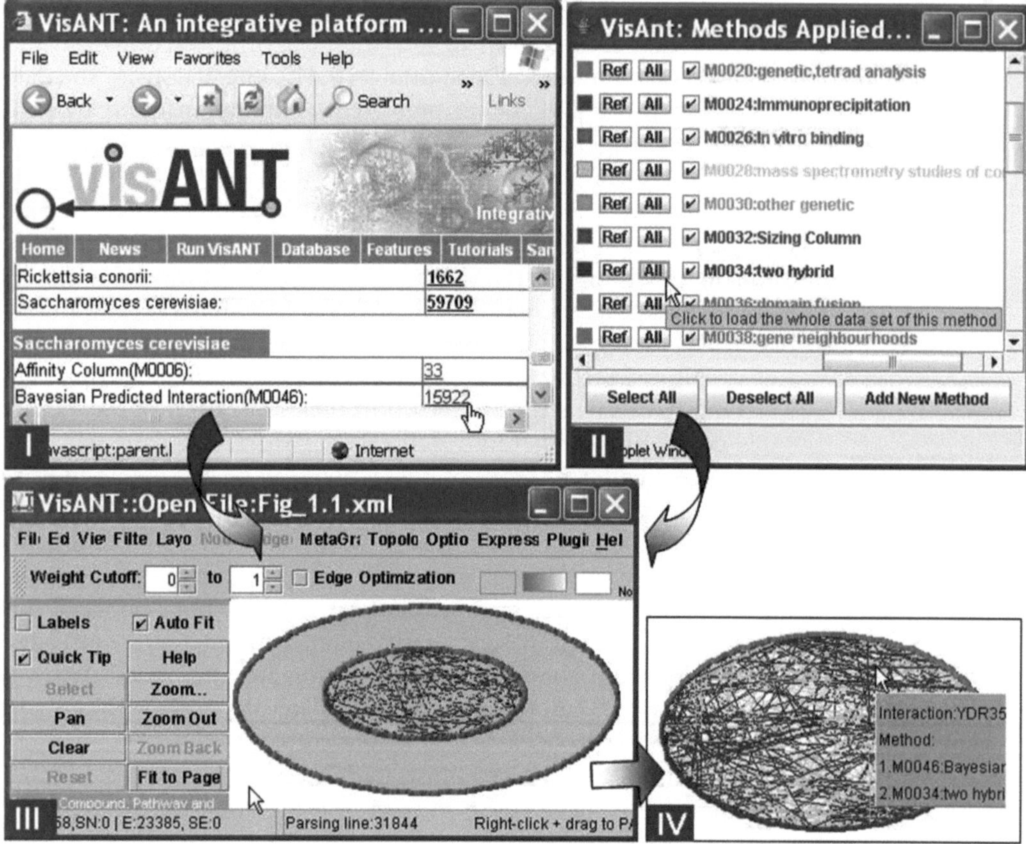

Fig. 1. Method-based loading and filtering of large-scale interaction data set. (**I**) Directly load interactions associated with method M0046 into VisANT through interaction statistics page. (**II**) Directly load interactions associated with method M0034 through methods table in VisANT. (**III**) The combined interaction network of M0034 and M0046 shown in VisANT. (**IV**) The interaction network with the overlap of the two data sets created using built-in filter.

gene with its official name (Fig. 2). This function will automatically be carried out when a gene's interactions are queried against the Predictome database. VisANT users can import their own mapping data using the ID-Mapping format, which is detailed in the VisANT user manual.

2.2. Interaction Visualization

Visualization of PPIs as a network is usually a start point to analyze the interaction data where nodes represent genes, proteins, or metabolites. and edges connect them in accordance with evidence for one or another type of interaction. Figure 3 shows a typical interaction network with hundreds of nodes and edges, as well as the various visual customization of both nodes and edges can be achieved in VisANT. Big nodes containing other nodes are meta-nodes that will be detailed later. While the edge color corresponds to the methods that identifies the interaction, the interaction detail (e.g., binding and inhibition) is represented the edge head/tail.

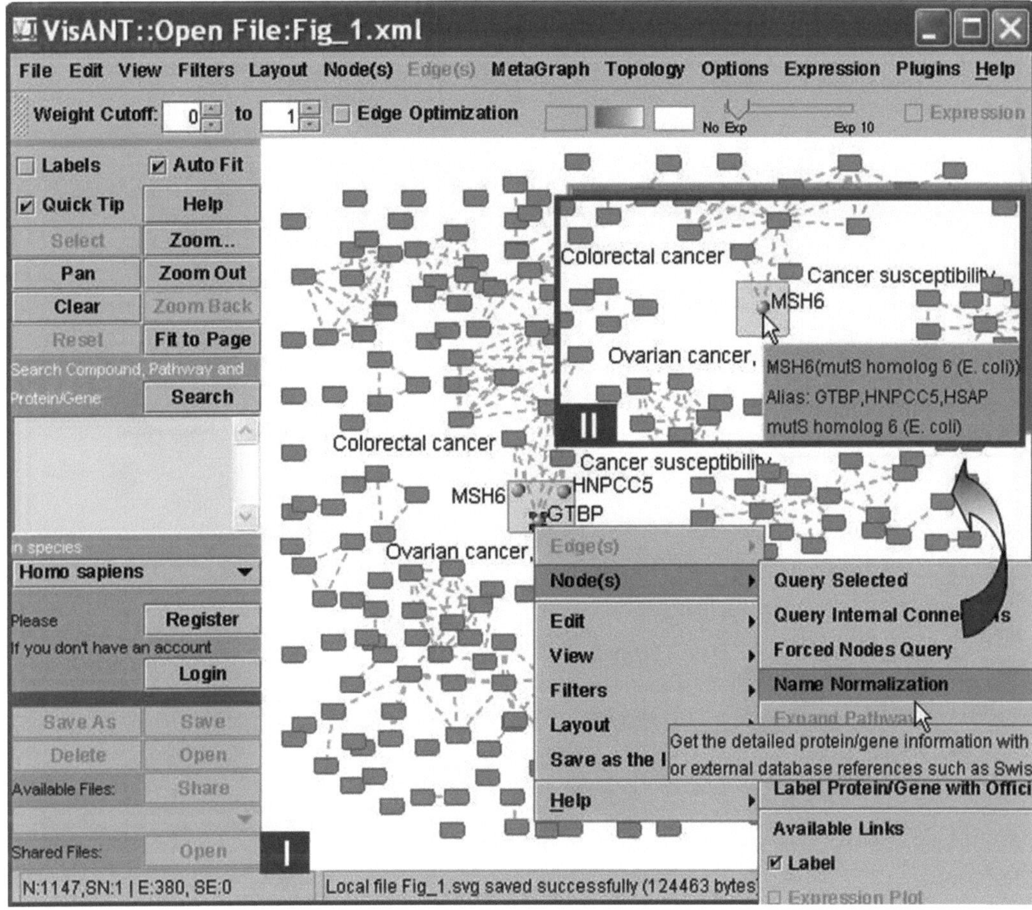

Fig. 2. Network of cancers rebuilt in VisANT using metagraph with subset data of cancer extracted from the work of Goh and coworkers (37). Each metanode (*gray box*) represents one type of cancer. The correlations between cancers are evaluated based on the number of shared genes. Mouse clicking a metanode will reveal its substructure: genes involved in the cancer and their correlations to one another if any. An example of an expanded node for ovarian and endometrial cancers is shown. The original data shows that ovarian cancer of endometrial type involves three different genes (MSH6, GTBP, HNPCC5) (**I**) which are actually all the same gene with official name MSH6 as discovered by the Name Normalization function (**II**).

Automatic layout usually helps to organism the interaction network, typically separate the network based on their connectivity. Figure 4 below shows the different visual presentation for a relative small interaction network with three different layout algorithms available in VisANT, while Fig. 5 presents a huge interaction network with 1.7+ million nodes and edges and its layout takes hours to finish using the batch mode of VisANT (http://visant.bu.edu/vmanual/cmd.htm).

Interactions, especially those predicted computationally, usually have an associated weight to indicate their reliability. The weight of the interaction is usually visualized using line thickness, or line color, or both, as shown in Fig. 6.

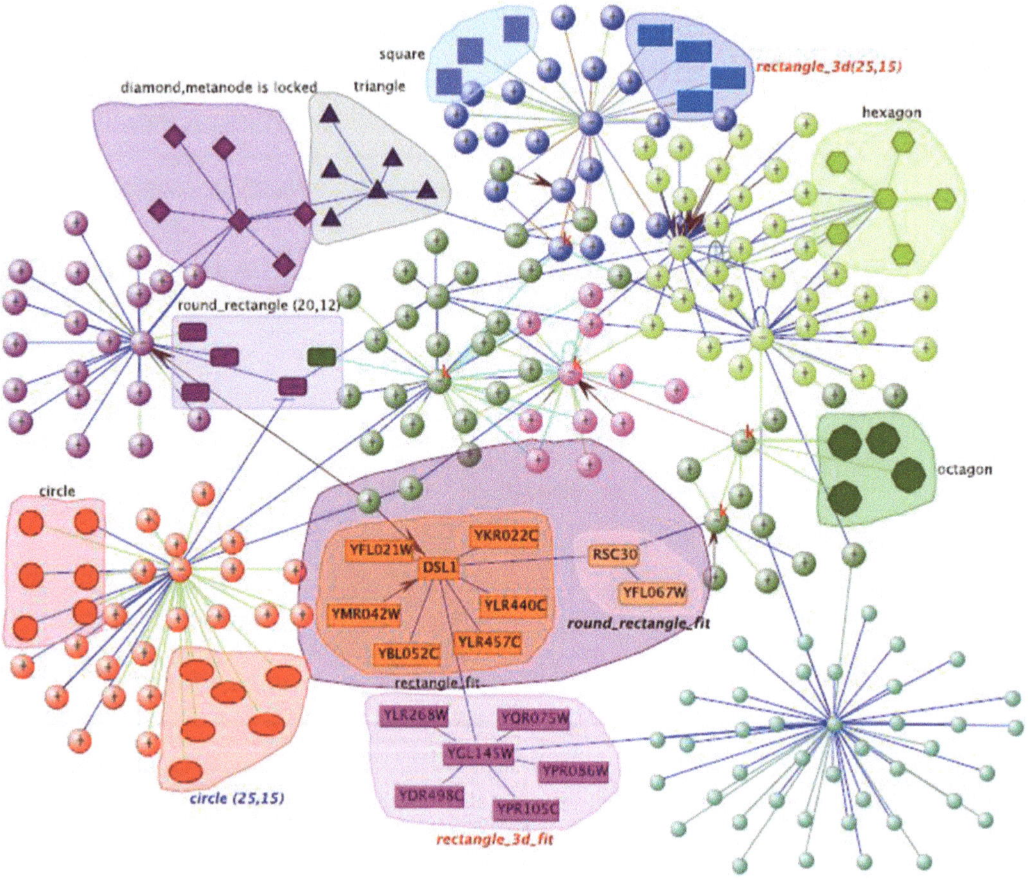

Fig. 3. Typical interaction visualization in VisANT.

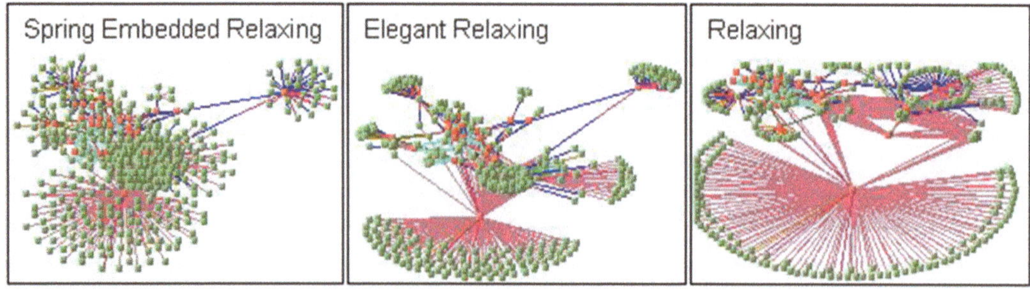

Fig. 4. The difference of three spring-forces-based layout.

2.3. Exploratory Navigation of the Interactions

Many researchers may not be interested in seeing the whole interactome (defined as the whole set of molecular interactions in cells), but instead are interested in performing an interaction walk starting from one protein of interest. From this perspective, a navigation technique named Exploratory Navigation is developed. Figure 7 shows how it works in VisANT.

Fig. 5. An interaction network with 1.7+ million nodes and edges laid out using Spring-Embedded-Relaxing in VisANT batch mode.

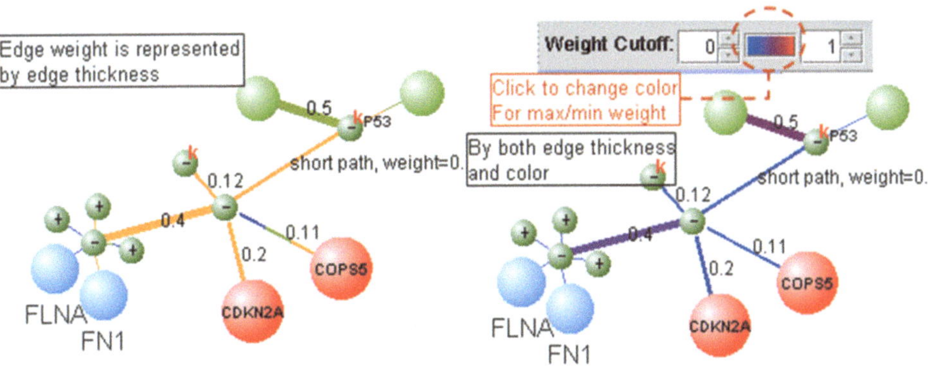

Fig. 6. Interaction (Edge) weight visualization in VisANT.

2.4. Interaction Filtering

Although the interaction data can be filtered based on many different rules, the most important factors that need to be considered are interaction type and reliability. As we discussed earlier, the method associated with the interaction not only informs the type of interaction it identifies, but may also its systematic accuracy. From this perspective, here we mainly focus on weight-based and method-based filtering.

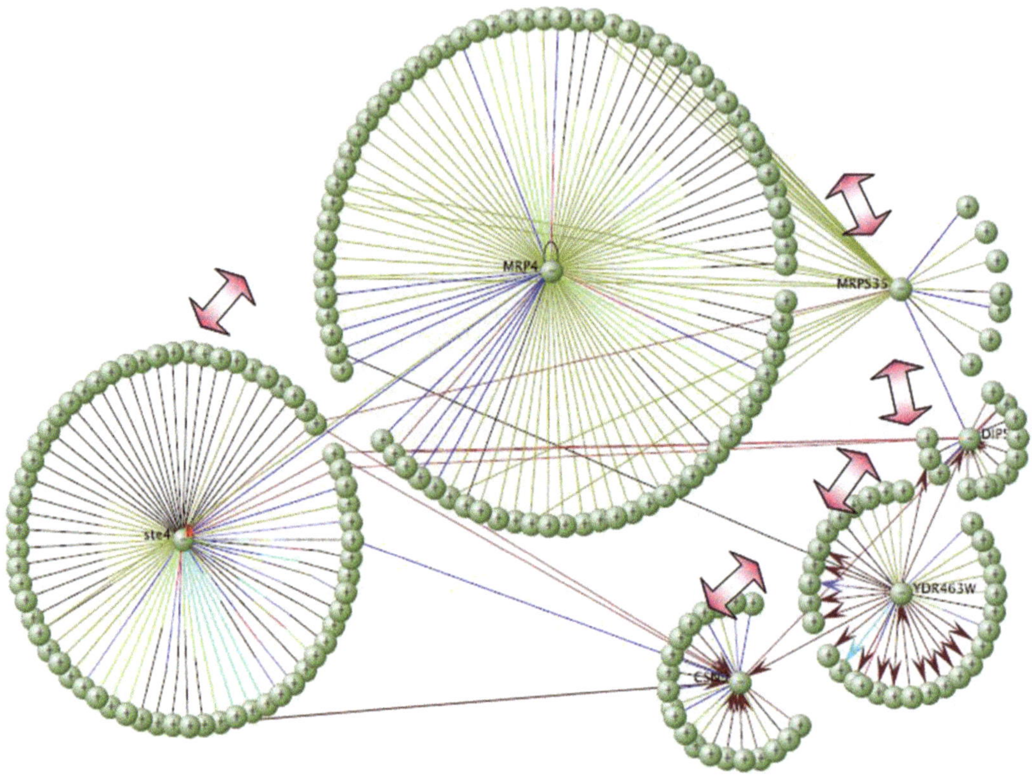

Fig. 7. Exploratory navigation of interactions. The navigation starts by querying the interaction of gene STE4 of *Saccharomyces cerevisiae* and followed by the order STE4 → MRP4 → MRPS35 → DIP5 → YDR463W → CSN9. The "−" sign of the node indicates its interactions have been expanded and the "+" sign indicates not. Be aware that you can always walk back from CSN9 → STE4.

Weight-based filtering. As shown in Fig. 8, increasing the weight cutoff not only results in an interaction network with higher confidence but may also reveal the network structure (Fig. 8**II**) that may be hidden in a denser (Fig. 8**I**) network.

Method-based filtering. As also shown in Fig. 1 (**I** and **II**), VisANT supports method-based fast loading of large numbers of interactions that can be obtained from the interaction statistics page of VisANT Web site (http://visant.bu.edu) or method table in VisANT. The latter also enables method-based filtering of the interactions: check/uncheck the checkbox of each method will make the corresponding interactions visible/invisible. In addition, it is somehow useful to filter out the interactions that are identified only by one method, as also shown in Fig. 1 (**III** and **IV**).

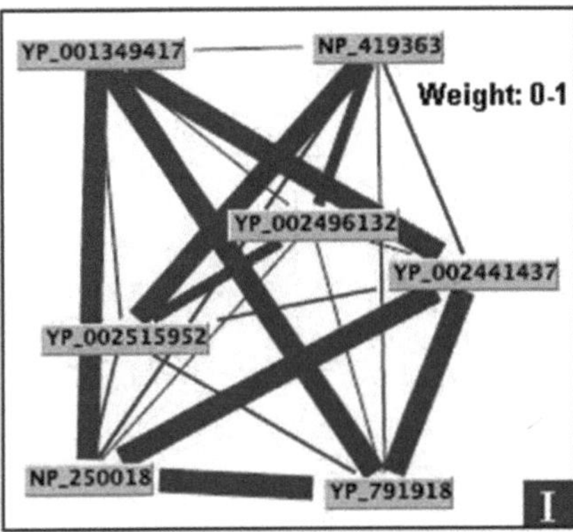

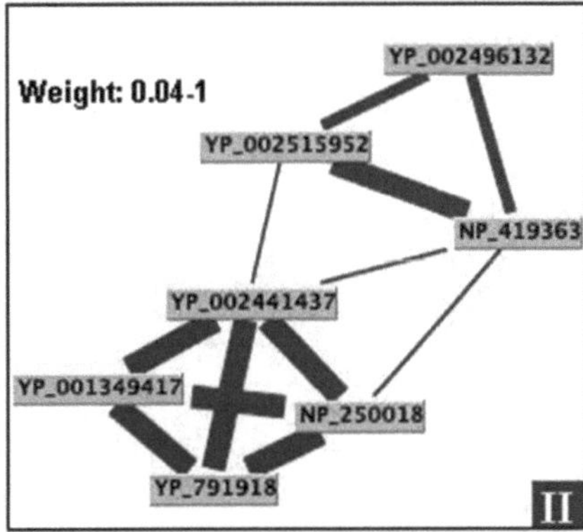

Fig. 8. Filter the weighted interactions. (**I**) A set of proteins in a protein cluster with weight of the interactions represents sequence similarities. (**II**) Same set of proteins with weight being filtered between 0.04 and 1, and two subclusters are clearly formed using the layout after the weight cutoff.

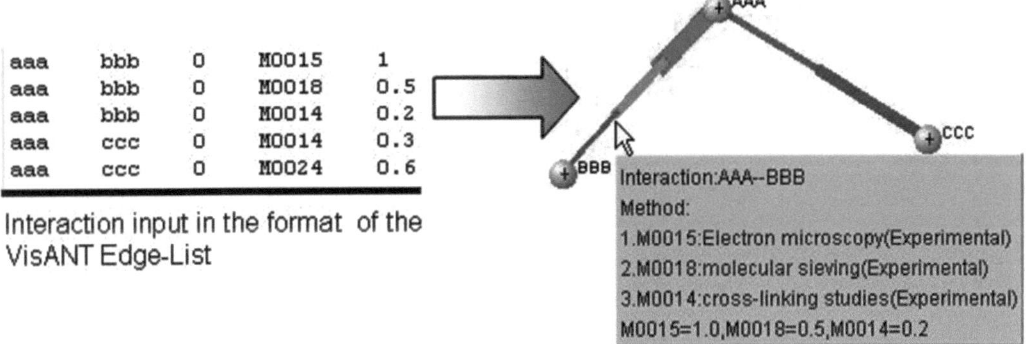

Fig. 9. Visual comparison of the weighted interactions.

2.5. Interaction Comparison

The network shown in Fig. 1 (**III** and **IV**) also indicates a convenience approach to compare the two interaction data sets, and is especially useful to find the overlap between two interaction data sets. In addition, the methods table in VisANT allows users to add, delete, or customize the methods as well, so that users can create methods for their own data for comparison purpose.

In the case that an interaction is identified by multiple methods and each method provides the weights, their weight can visually be compared as shown in Fig. 9.

2.6. Network Structure Analysis

Examines the topological and dynamical properties of the interaction networks may bring new insights to the corresponding biological processes. Integrated biological networks are often hybrid, meaning they contain both directed (e.g., transcription factor binding) and undirected (e.g., protein interaction) connections, and *compound* or *modular* (see Subheading 3 for detail),

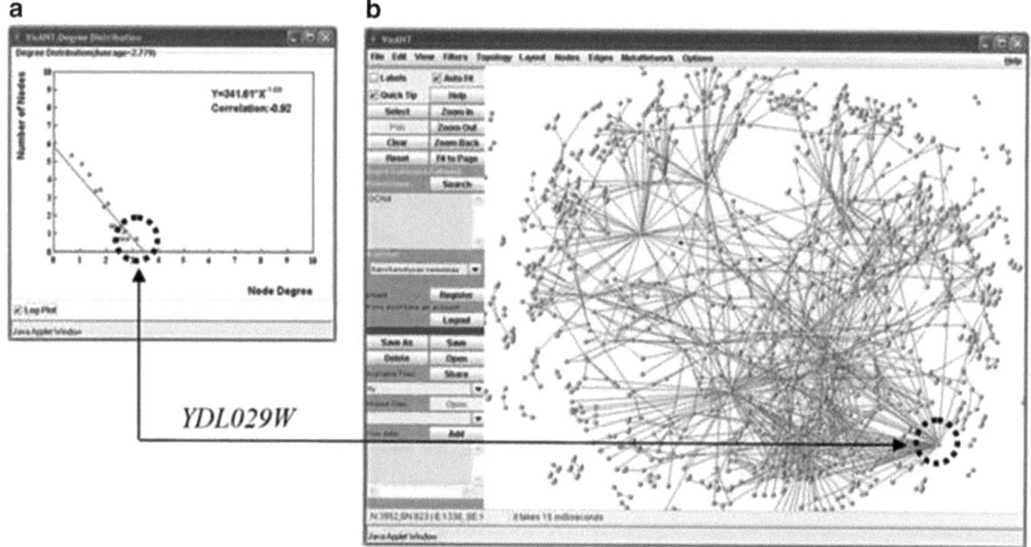

Fig. 10. Dynamic linking for finding topological features. The network of synthetic lethal genetic interactions (38) in yeast contains 823 genes and 3,952 interactions. A plot of the degree distribution of genes in this network allows users to quickly identify which genes have the highest (or lowest) connectivity. In this example, the gene shown is *YDL029W*, which has synthetic lethal interactions with 58 neighbors.

meaning that linked components can either be single or grouped. Unlike other software for biological network topological analysis, VisANT explicitly allows creation of mixed networks involving different types, with topological algorithms to support type. For example, if protein A inhibits B and B interacts C, then there will be no path from protein A to protein C (Fig. 10).

2.6.1. Node Degree and Distribution

The degree of a network component, k, is the number of connections it has with other components. The distribution of degrees among components is useful for characterizing the topology and scale of a network, and often has meaningful biological interpretation. In protein interaction and genetic interaction networks, for example, the degree of a hub is often its importance and essentiality for cell function. For directed networks, such as transcription factor binding networks, the degree is separated into the "in"-degree and "out"-degree, depending on the directions of interaction between two given components. Degree is also a feature which distinguishes hubs (highly connected nodes) from leaves or orphans (weakly or nonconnected nodes) in the network. In VisANT, users can see a scatter plot and log-linear regression fit of the degree distribution, $p(k)$, of the network. The degree exponent (γ) of the log-linear regression, where $p(k) = k^{-r}$, is a measure of the network's "scale-free" property (39, 40). The VisANT degree plot is dynamically linked to the network view (selection in one window maps to corresponding points in the other, see Fig. 3).

2.6.2. Clustering Coefficients and Distribution

The clustering coefficient ($C = 2n/[k(k-1)]$), where n is the number of links between k neighbors, measures the tendency of a network to have highly connected clusters. Fully connected sets of nodes have $C = 1$, because everything is connected to everything else. In large-scale mass spectrometric networks in yeast, this property can be used to identify groups of proteins involved in the assembly of the ribosome (41). The exponential degree of the log-linear fit $C(k) = k^{-r}$ can be used to characterize the hierarchical structure of a network (42). VisANT provides scatter plots and log-linear regression fit of the clustering coefficients in a network, allowing users to identify densely connected clusters of nodes (the direction in hybrid networks is ignored in calculating C). This plot, like the degree plot, is dynamically linked to the network view.

2.6.3. Shortest Path Lengths and Distribution

In studying the function of pathways, the property of interest is often how a given gene or protein is related to (or responds to) an up- or downstream signal. Given a large data set of interactions, it may be useful in some contexts to find the most direct path between two genes, proteins, complexes, or pathways; e.g., the overall lengths of such pathways may be related to the immediacy or breadth of signal response (43). The average shortest path also indicates the well-known "small-world" property of many real-life networks (44). Networks in VisANT are analyzed by depth-first-searching (DFS) for both the shortest path between two given components as well as the distribution of shortest paths between all components. In the consideration of the direction of the interaction, as well as some detailed interaction type (e.g., inhibition/repression) being treated as the stop of the path, a shortest path from A → B may not be the same as the one from B → A and VisANT will list both. VisANT also lists all the possible shortest paths that may be available.

2.6.4. Detection of Network Motifs

Certain patterns and motifs have been shown to occur with more frequency in biological networks than would be expected by chance alone (45, 46). This leads to the hypothesis that such motifs, for example feed-forward loops, have functional characteristics that correspond to their structure (47). Identifying topological features in networks is an important part of understanding the relationship between structure and function of these motifs. VisANT supports searching of basic motif types, such as feedback and feed-forward loops, and development is in progress for detection of other arbitrary motifs, and for assessing the statistical significance of these patterns in large networks. Motif detection is performed with exhaustive breadth-first-search (BFS) over nodes in the network.

2.6.5. Network Randomization

It is important to test whether some measured structure features are statistically significant. In such case, we will test these features in the corresponding random networks. The randomization criteria are different for directed and undirected networks. Randomization of undirected networks preserves the same number of nodes and edges, but edges are randomly distributed, and the resulting network therefore follows a Poisson distribution. Directed network randomization preserves the same number nodes and same in-degree and out-degree for each node, but the directed edges are randomly distributed among the nodes.

2.7. Integration of Interaction Network and Expression Data

Both expression and interaction data may be noisy and integration of two biologically relevant signals supported by both data types are more likely to be correct than those supported from either data source alone. VisANT provides two methods to visualize expression data over the pathways: either the node color is used to represent the expression value of the current experiment, or the plot of expression profile is embedded in the node, as shown in Fig. 11.

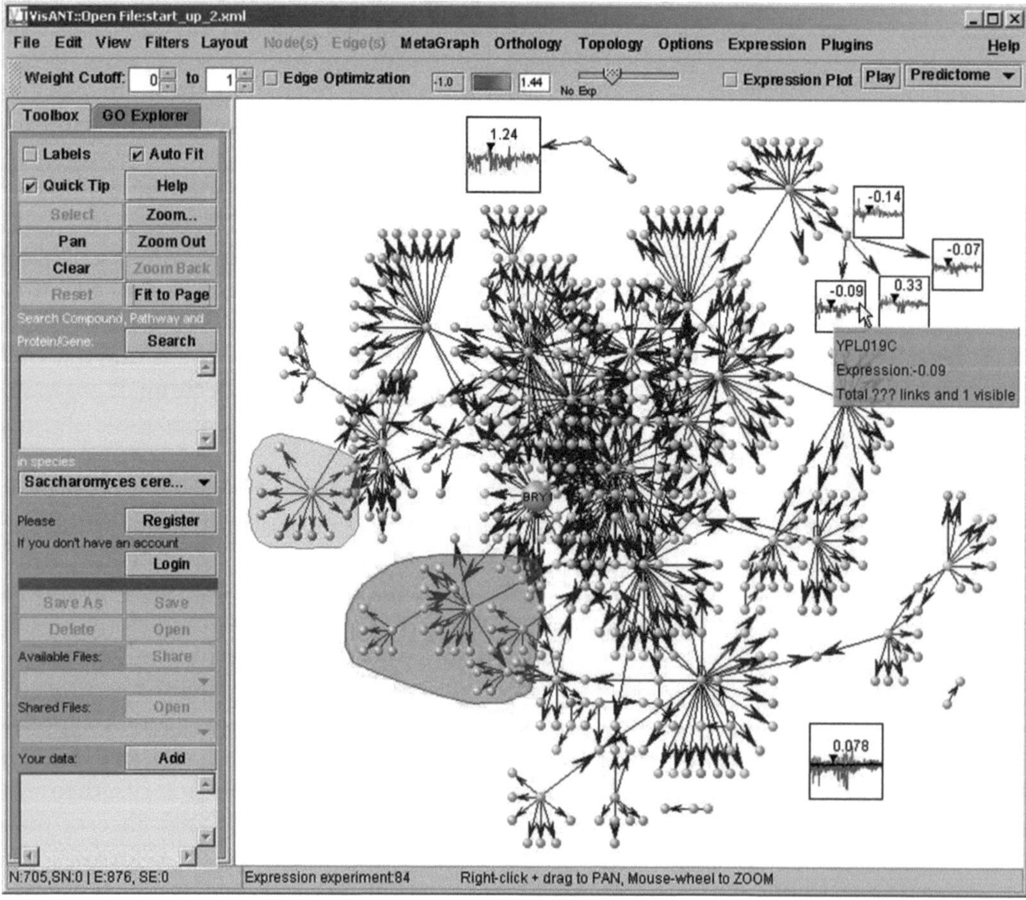

Fig. 11. Visual integration of expression and transcriptional interaction data. Node color here represents the value of expression and line arrow represents the direction of the transcription factor binding. The expression can also be shown as a plot if there are expression data for multiple experiments. At the right-bottom corner is the expression plot for a collapsed interaction module where average expression of all nodes inside the module is shown using black color.

The two methods can be toggled either for individual nodes, or for the whole network. Different experiments can be navigated using a sliding bar and the navigation process can be animated. When the expression profile is shown for the node, there will be a cursor to indicate the position of current experiment, as well as corresponding expression value.

It is also very convenient to test whether the genes in the same interaction module are coexpressed as all the expression profiles of the nodes contained in the module, as well as average profile, will be drawn together as one plot with average profiles in black.

2.8. Annotation of the Interaction Network Using Gene Ontology

One of the most widely used bioinformatics resources is the Gene Ontology (GO) (48), which provides hierarchically organized information about gene products, their activity, biological functions, and cellular locations. Tools for network visualization and analysis often provide functions to annotate gene functions using GO. However, the different relational meanings of GO and network edges, and the ontological structure of GO, make the integration between GO and interaction network more challenging. GO terms are structured as a Directed Acyclic Graph (DAG), where nodes represent terms, and edges represent inclusive relationships between terms. A key characteristic of such representation is that a term in a DAG can have multiple parents. As a result, genes are associated with multiple biological terms and individual biological terms can also be associated with multiple genes. These “many-genes-to-many-terms” (49) associations reflect the complex nature of biological processes and make visualization and modeling of the integrated network difficult (50).

VisANT provides four basic options to annotate genes using GO annotations with corresponding menus shown in Fig. 12. Options 1–3 listed below can also be applied to the selected branches. These options provide users great flexibility to test various

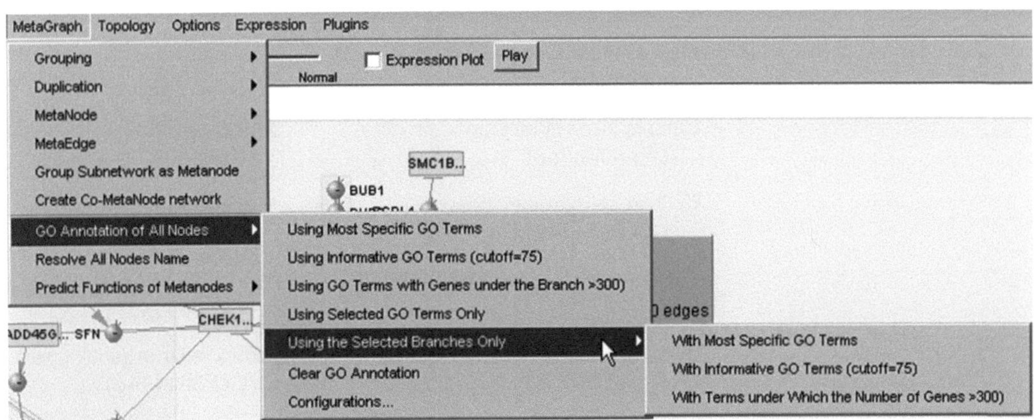

Fig. 12. Menus for GO annotations under the MetaGraph menu, which will annotate all the genes, including those hidden in the collapsed metanodes. The same list of menus is also available under *Nodes* menu which should be used to annotate the selected nodes.

hypotheses. To save the space, we use the human gene ACN9 that is involved in the predisposition to alcohol dependence to illustrate these options (indicated below):

1. *Using most specific GO terms.* Genes are annotated with the most specific functional descriptions available at Entrez Gene database. The table below lists the GO annotation of ACN9 with this option.

Biological process	Cellular component
Gluconeogenesis (GO:0006094) [ISS]	Mitochondrion (GO:0005739) [IEA] Mitochondrial intermembrane space (GO:0005758) [ISS]

2. *Using informative GO terms.* Genes are annotated using GO terms (i) having more than a user-specified number of genes and (ii) each of whose descendent terms have less than the specified number of genes. Let us use 145 as the cutoff (click the button near Search button of GO explorer, and enter 145 in the corresponding field and press Enter key), the informative GO annotations for ACN9 is shown below:

Biological process	Cellular component
Hexose metabolic process(GO:0019318)	

3. *Using GO Terms with Genes under the Branch > cutoff.* A term must have more than a user-specified number of genes. Let us again use 145 as the cutoff, and here are the results:

Biological process	Cellular component
Hexose metabolic process (GO:0019318)	
Cellular alcohol metabolic process (GO:0006066)	Mitochondrion (GO:0005739)
Cellular biosynthetic process (GO:0044249)	Mitochondrial envelope (GO:0005740)
Biosynthetic process (GO:0009058)	Mitochondrial part (GO:0044429)
Carbohydrate metabolic process (GO:0005975)	Intracellular organelle part (GO:0044446)
Monosaccharide metabolic process (GO:0005996)	Membrane-enclosed lumen (GO:0031974)
Monocarboxylic acid metabolic process (GO:0032787)	Organelle envelope (GO:0031967)

4. *Using selected GO terms only.* Genes are annotated using only selected GO terms. Following figure shows the selected terms and resulting annotation for ACN9:

Options 2 and 3 are frequently used when predicting gene functions using functional linkages. Annotations resulting from different options can coexist as node descriptions in VisANT for comparison purposes.

3. Advanced Analysis Strategy

The increasing importance of network models in biology stems from the emergence of systems biology, and the promise of representing the cell as a computable network of genes and proteins (51, 52). This in turn requires software for computable representations, such as the proposed "biological information system" of Endy and Brent (53), which can extend current databases to embody new types of mechanistic knowledge. Still, while networks will likely play a critical role as the basic data structures that represent the fundamental capabilities of such systems, common network models, as explained below, have limited utility to incorporate graphical representations of the interactions with the wealth of annotation, in part because they cannot readily represent overlapping functions between agents (molecules or sets of molecules) or multiple functions of a single agent (54).

3.1. Hierarchy and Complexity in Biological Networks

3.1.1. Abstraction

Biological networks differ from general networks because such networks abstract common features from different situations, at the expense of concreteness. Protein *p53*, for example, in PPI networks is an abstraction of many *p53* proteins in different cells under different conditions; their associated interactions are combined regardless of temporal, spatial, or conditional dependency. This is also true of conditional dependencies between proteins and DNA-binding sites or, in genetic interactions, between genes themselves. Clearly such network representations of protein–protein and protein–DNA interactions invariably omit environmental and temporal conditions; they display a static repertoire of potential interactions, but do not consider environmentally mediated combinatorial selection of subsets of the molecules. To have a computable model suitable for simulation, adaptability needs to be taken into account, and that will often require the same node to have multiple instances in one network. This is very different from a physical network such as a logic circuit in which each node represents a single physical component with invariant properties; consequently biological networks require considerably more sophisticated visualization systems than those currently available.

3.1.2. Multiple Scales

The scale of a network is related to the number of elements it contains (e.g., number of genes and proteins) and the number of states available to each element (e.g., posttranslational modifications of the proteins, promoter site occupancy, and conformation-dependent activity). Since each element generally has multiple states (think of the combinatorial possibilities for just a single type of modification—phosphorylation—of just a single protein) the combinatorial complexity of the network is enormous; considerably greater than that of a very large scale integrated (VLSI) circuit, which grows combinatorially, but only as a power of 2, and much greater still than geological maps, which are static for common purposes (53). This complexity would be practically impossible to represent or to relate—in even a very incomplete manner—to phenotype, were it not for the fact that the cell is organized into functional modules. In that respect it is reminiscent of a VLSI chip. Meanwhile, the size of the network is a key issue in its visualization. Not only because the increasing number of network elements can easily comprise the performance of the viewing platform but also because the large number of network elements will impact its viewability and usability: it becomes impossible to discern between nodes and edges.

The challenge is to represent information at different levels of detail and in different contexts while maintaining continuity between levels of what will likely be a hierarchy of modules. Thus, a protein complex in a pathway may be represented by a simple nod; however, the same complex may need to be represented as a network if the detailed associations among subunits needs to be explored, such as in a spliceosome in alternative splicing (54). In addition, a protein is not well represented by a simple node if its state, determined by condition-dependent posttranslational modifications, changes its function or localization.

When analyzing a large network, a user often needs to magnify it, or to focus on a subset of the network at a finer resolution, to see greater detail among a particular set of molecules. It is helpful to make a distinction between two kinds of magnification: *geometric zooming*, in which a region of the network is enlarged; and *semantic zooming*, in which additional properties are to be introduced with enlargement. Geometric zooming changes only the size of objects (nodes and links) (55) and therefore the field of view, but objects themselves are unaltered. Semantic zooming, on the other hand, changes resolution: objects can appear and disappear depending on context, allowing the kind and number of objects included in the network to change, in addition to the size of the objects (48, 55). For example, Google Maps implements a type of semantic magnification in which the annotation changes automatically as the zoom level changes (e.g., at level 12, states are seen as black boxes; at level 6 major roads within a state become visible; at level 1 local streets can be seen). Semantic zooming makes use of the viewing context to decide, in real time, what kind of information is appropriate to include.

3.1.3. Modularization and Hierarchy

The expression "biological module" is not uniquely defined, but is used in different ways by different researchers. Most modules are either computationally inferred or manually curated. A protein complex is the cleanest example of a module since it is detected directly by experiment. Manually curated pathways are also considered modules, though pathway boundaries are somewhat arbitrary (56). Additionally, genes subsumed by GO (48) biological process terms can also be thought as a functional module, although the interactions between genes can be further defined with other methods. Computational inference of modules usually adopts a bottom-up approach (57); typical methods include classifying nodes into universal roles according to their pattern of intra- and inter-module connection in an interaction network (58), iterative searching for coexpressed clusters from genome-wide expression data (59), and integrative clustering with additional evidence such as CHIP-chip data (60). The correlation threshold with which a link is defined plays an important role in the module definition, usually in a statistically defensible way. For example, when attempting to discover functional modules based on co-regulated elements, it would be very important to quantify the strength of postulated relationship based on sound statistical principles. Modules detected using different methods are usually cross-validated using existing definitions of curated pathways, or other "known" modules (61). Despite the different definitions of modules, their use in large-scale networks reduces visual complexity and increases the performance of graph drawing tools and algorithms.

Software tools with module visualization capability should not be tied to a particular definition of modules; the user defines the module for a particular biological context, and the tool mines and displays it. For example, the cellular network of protein complexes in yeast can be modeled by interactions between complexes based on shared components (Fig. 13**I**), or by the direct interactions between protein members of different complexes (Fig. 13**II**). It is also evident from Figs. 2 and 4 that modularization, if it is to be useful, must be connected with semantic zooming. This relation becomes even clearer in a more general context, in which the cell's network repertoire (i.e., the entire subset of connections, as opposed to the particular subset selected by a specific environment) is coarse grained as a hierarchical pyramid (62) with protein domains at the base and modules at the apex (modules ↔ pathways ↔ proteins ↔ domains) where resolution and annotation changes by zooming from one level to the next.

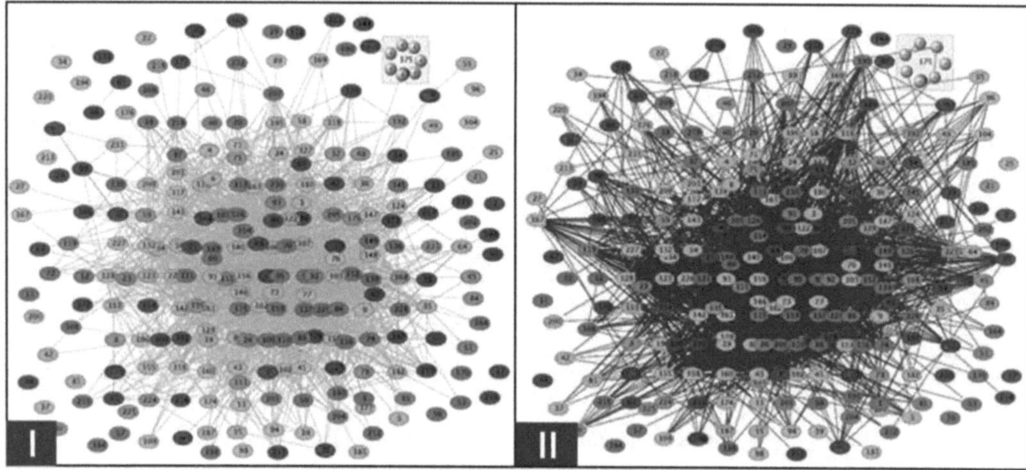

Fig. 13. The network of protein complexes inferred by bottom-up method. (**I**) A metagraph of the network of protein complexes discovered via TAP-MS-tagging by Gavin et al. (41) The complexes were determined by tandem affinity mass spectrometry and were colored to indicate subsequent functional assignments (41). The *gray* edges connect complexes that share protein components. The components of each complex are included in the model and are visible when the metanode representing the complex is expanded. As an example, the complex labeled as 175 is expanded so that its components, as well as the one being shared (shadowed) with complex 36, are visible. (**II**) A network using the same set of complexes as (**I**), only with edges representing interactions derived by large-scale two-hybrid assays (13). The integration of interaction data enables the internal connectivity of the complexes to be revealed when a node is expanded. An expansion of node 175 would reveal an interaction between Exo84 and Sec5, and between Sec8 and Sec10.

3.2. Metagraphs: Representing Networks of Networks

A metagraph (12, 13, 50) is an advanced graph type developed in our Lab to integrative inclusive or partially inclusive relationships and the adjacent relationships into one single network, as illustrated in Fig. 14. The inclusive relationship in a metagraph is represented by a metanode which is a special type of node that contains associated sub nodes, much as a Gene Ontology (GO) term contains its subterms or associated genes. A metanode has two states, expanded or collapsed; the expanded state manifests the internal subgraph (that is, places all descendent nodes with their connections into the graph) while the collapsed state replaces this subgraph with the single node. Networks represented by a metagraph are usually termed metanetworks, and such visualization technology is often referred to multi-scale visualization because information at different abstraction scales is presented in one network. Detailed mathematical definition of the metagraph can be found in Appendix.

In this session we focus on the advanced analysis using VisANT's functions associated with the work flows shown in Fig. 15. Users are advised to visit http://visant.bu.edu for the rest functions of VisANT. Both work flows shown in Fig. 15 are usually aimed to find network modules that may account for the differential RNA expression patterns (e.g., tumor vs. normal) determined by genome-wide association studies. The first work flow starts with the modules whose functions are unknown, therefore the task is to determine their functions; while the other starts with the modules

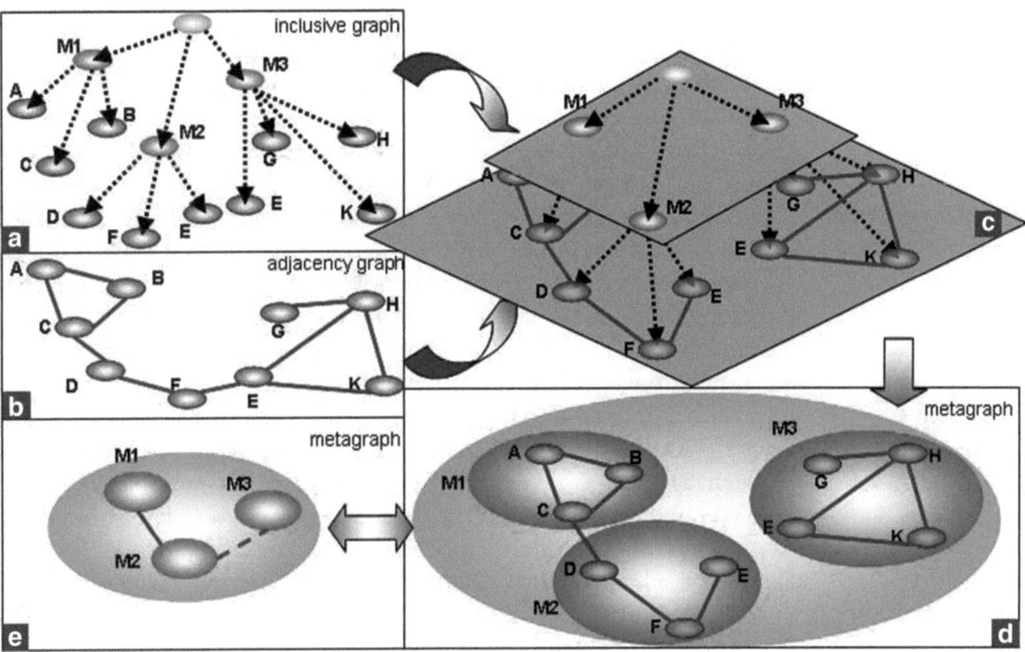

Fig. 14. Illustration of the multi-scale visualization using metagraph. Note that node E has two instances in the inclusive tree. (**a**) A network where an edge represents the inclusive relationship such as F belongs to M2, E is part of M2 and M3. (**b**) A network with adjacency relations. (**c**) Integration of inclusive relations (*dashed lines*) and adjacency relations (*solid lines*). (**d**) The intgrated network using metagraph (also referred as meta-network) where node E belongs to both metanode M2 and M3. (**e**) The same meta-network with three metanode (M1–3) collapsed, the *dashed line* between M2 and M3 indicates there is a shared node between two metanodes.

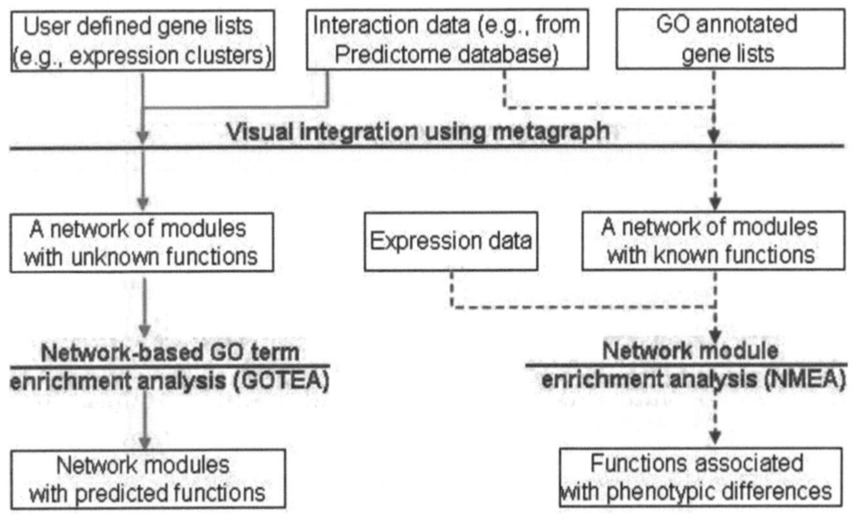

Fig. 15. Different work flows focused in this chapter. The *solid lines* represent the work flow of functional profiling where GO annotations are used to interpret the roles of a given gene set. The *dashed lines* represent the work flow of the gene set/network module enrichment analysis, where GO terms and associated genes may be used to construct the functional modules.

whose functions are already know and the task is to determine whether they are enriched in the expression pattern. Finally, we illustrate the automatic creation of the cancer gene network based on the cancer network shown in Fig. 2 using the built-in VisANT function “Create the co-metanode network.”

3.3. Network-Based Functional Profiling

Functional profiling (63), or GO term enrichment analysis (GOTEA), aims to determine whether particular GO terms inform the difference of molecular phenotypes in any set of user-specified genes, typically the coexpression modules (Fig. 15, solid lines). In a network context, the goal is to identify biological functions for a given subnetwork, or for a network module. Although many algorithms and tools (49, 64–75) have been developed for GOTEA, they generally omit correlations based on disparate and varied datasets, such as yeast two-hybrid, genetic interaction, mass spectrometry (MS) and so on. Such relations may help to overcome some drawbacks in the current enrichment analysis. For example, one drawback is that all terms are weighted equally (76), while in a network module, terms annotated for highly connected genes will have more weight than those annotated for the loosely connected genes. Accuracy may also be improved if network type is considered; e.g., for a regulatory network, we probably can exclude those annotations of metabolic processes. From this perspective, flexible annotation schema will be needed to enable users to select subsets of GO annotations as discussed in the section. Such flexibility could help determine the functions of genes in a specified network.

3.3.1. Construct a Network of Modules

Assume we have three coexpression clusters named CLUSTER_A, CLUSTER_B, and CLUSTER_C, each contains a number of genes as listed below, copy and paste (either through the pop-out menu, or the key combination CTRL-C and CTRL-V) the following text into the Add text box of VisANT’s toolbox and click the Add button (left-bottom corner indicated by the mouse cursor in Fig. 5), three metanodes will be created (Fig. 5). The Add textbox can be used to add any type of the data whose format is supported by VisANT (Table 1).

#group Cluster_A
KRT1 SIGIRR MYD88 MASP2 C1QA
MASP1 IL1R1 TLR4 TLR2 TLR1
TIRAP TBK1 IL1RAP TBKBP1 MBL2
SERPING1 CR2 C1S C1R

#group Cluster_B
NA SNAPAP BLOC1S3 BLOC1S2 DTNBP1 BLOC1S1
MUTED SNAP25 PLDN TRPV1 EBAG9 STX12

#group Cluster_C
HYAL2 CLP1 TEP1 RPP40 TSEN15
ERVWE1 RPP38 POP1 LOC100128314 TSEN34
RPP30 TSEN2 TSEN54 TERT

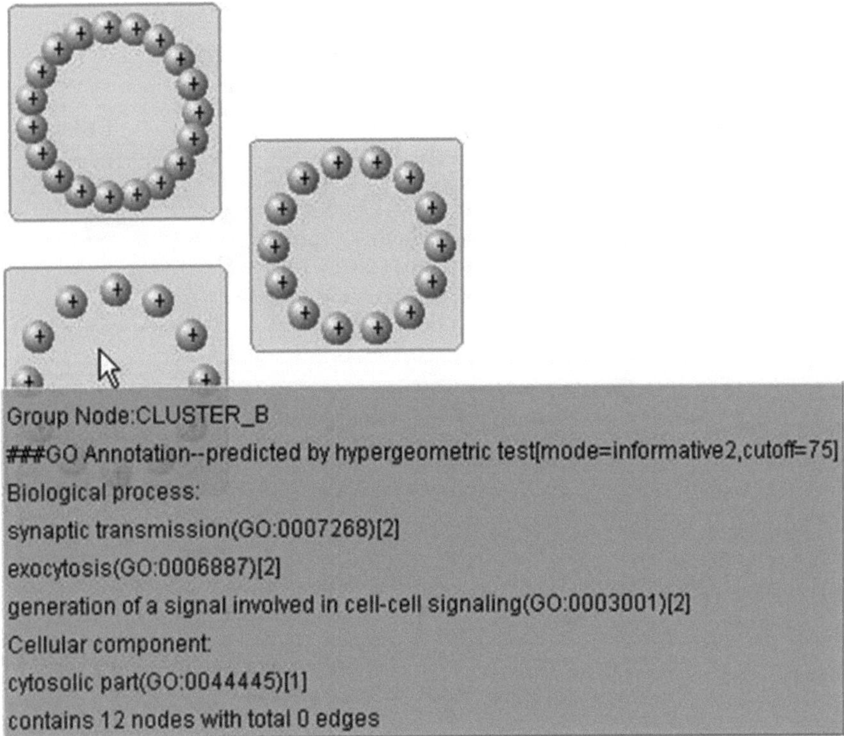

Fig. 16. Network of three clusters created using Edge-List format and laid out using Circle Layout. Functions of the cluster is predicted using hypergeometric testing.

Above text uses VisANT's extended Edge-List format[1] to create the network, which is the simplest format supported in VisANT. It can also be used to easily add nodes (each line with the name of one single node) or edges (each line with the name of the two nodes separated by space or tab). Alternatively, users can load this edge-list from URL through File → Open URL menu and enter the URL http://visant.bu.edu/other_formats/edge_list_3_clusters.txt (depending on the type of browsers, you may be able to paste the above URL using the key combination CTRL-V), and follow the instruction to achieve the same result. Once laid out using Circle Layout, the network shall look similar as the one shown in Fig. 16

3.3.2. Predict the Functions of Modules Using Hypergeometric Test

This method predicts the overall functions for a given set of genes by checking the overrepresented GO terms associated with the genes. Therefore the first step is to annotate the functions of the genes of each cluster through the menu "MetaGraph → GO Annotation of All Nodes → Using Most Specific GO Terms." VisANT will

[1]When you are uncertain about the format of edge-list, you can always export the network in the format of edge-list with the menu File→Export as Tab-Delimited File→All and follow the exported examples.

Table 2
Cluster functions predicted with hypergeometric-based test

	Molecular function	Biological process	Cellular component
Cluster_A	Cytokine binding (GO:0019955) Growth factor binding (GO:0019838) Serine-type endopeptidase activity (GO:0004252)	Positive regulation of immune response (GO:0050778) Innate immune response (GO:0045087) Acute inflammatory response (GO:0002526) ...	
Cluster_B		Synaptic transmission (GO:0007268) Exocytosis (GO:0006887) Generation of a signal involved in cell–cell signaling (GO:0003001)	
Cluster_C	Endonuclease activity (GO:0004519) Nucleotidyltransferase activity (GO:0016779)	tRNA metabolic process (GO:0006399)	Nucleolus (GO:0005730)

automatically resolve the node names when annotating the nodes. The same annotation menus are also available under the menu "Nodes" which are only used to annotate the selected nodes. In the case you have collapsed metanodes, such as for KEGG pathways, always use the annotation options under "MetaGraph" menu. Once the genes have been annotated using GO terms, we can easily predict the functions using hypergeometric test through the menu "MetaGraph → Predict Functions of Metanodes Using GO → Detect Overrepresented GO Terms Using Hypergeometric Test → Start Hypergeometric Test over GO Database." VisANT will perform the prediction for all non-embedded metanodes. For more information, please reference the manual at http://visant.bu.edu/vmanual/ver3.50.htm#hyper. The prediction results will be added to the metanode as part of its description that are available as tooltips when mouse-over the node (Fig. 16). Table 2 lists all predictions of three clusters based on the reported created by VisANT: http://visant.bu.edu/misi/hyper_3_cluster.htm

Predictome database maintains a local copy of GO database and the gene–GO associations are extracted from Entrez Gene database. Both data sets are being updated constantly therefore the actual prediction results may be a little different from the results shown in the link above. This also applies to the GOTEA algorithm that will be illustrated later because the interactions are also being updated from a list of interaction databases.

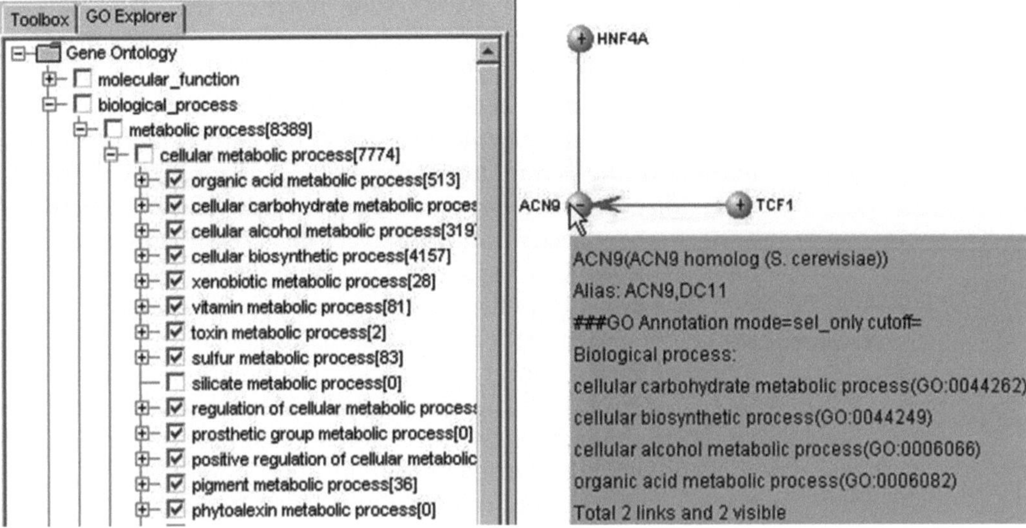

Fig. 17. Annotate the gene using the selected GO terms only. Four among the total thirteen terms are annotated for ACN9 because GO term hexose metabolic process (GO:0019318) are their child term, which will be very clear when the hierarchy of GO:0019318 is shown in the GO explorer. Please reference http:/visant.bu.edu/vmanual/ver3.50.htm for the information of GO hierarchy visualization.

3.3.3. GOEA

Although it is common and fast to use hypergeometric test to predict module's function, the algorithm, however, does not take into the account the interaction information for a given network module. From this perspective, a new algorithm has been developed and implemented as a VisANT plugin to find overrepresented GO terms in user-specified network modules (represented as metanodes in VisANT). The function is available under the "MetaGraph" menu. By default, the analysis will be performed for all non-embedded metanodes; i.e., it is not performed for descendent metanodes unless they are specifically selected. Similarly, overrepresented GO terms will be shown as a quick tip when the mouse is passed over a node, and clicking on a node will display the hierarchy of GO annotations in GO Explorer (Fig. 17). GOTEA also requires genes in the modules to be annotated prior to the analysis.

For a given target GO term, the algorithm first computes the density score of each node based on the path distance (number of links) to other nodes in the same module, and the similarity between its associated GO terms and the target term. The use of a similarity score rather than an exact match enables the algorithm to give the target term a high score so long as it is functionally similar to the annotations of the genes in a module. The similarity score between two terms is calculated by aggregating the semantic contributions of their ancestor terms in the GO graph (77). The enrichment of target term is determined using statistical measurement through permutation test over the subset of same number of genes extracted from all known genes annotated by Entrez Gene database (78) with appropriate false discovery rate (FDR) (79) cutoff. Details of the algorithm can be found in the Appendix. Related parameters, such as the cutoff and the iteration number of the permutation test can be configured. By default, all terms that have the associated genes for the current species will need to be tested; users however, may select subset of term branches in the GO Explorer to speed up the analysis.

The advantage of the algorithm over similar algorithms (such as hypergeometric test) is reflected in the computation of the density score, where the impact of one gene on another is a function of the GO term similarity, and the number of links between the genes. GO term similarity is calculated using a fuzzy search rather than a conventional exact match (77). With such a density score, a gene having many neighbors with similar GO terms will have more significant contributions to the enrichment outcome; the algorithm therefore leverages network topology, as well as the GO hierarchy. In addition, metagraphs provide a flexible visual context to perform analysis for hierarchically organized network modules. The function is designed for work flow shown as the solid red line in Fig. 15; network modules need not be limited to expression profiling.

Permutation-based algorithms tend to be computationally intensive and therefore time-consuming. In addition to the hypergeometric test-based algorithm, VisANT provides two options to address this shortcoming. First, VisANT provides an option "Fast GOTEA," which only scans related GO terms for a given network module (GO terms annotated for the genes in the module and corresponding ancestor terms); and second, macro commands have been created to allow the time-consuming GOTEA tasks be carried out in the background with the command-line mode of VisANT.

Continue with the same example as in the previous session, and load all interactions detected by the affinity technology (M0045) in Predictome database (when VisANT is run as Applet, this can be achieved through Interaction Statistics page as shown in Fig. 18). Otherwise, they can also be loaded through metapod table in VisANT (Fig. 1).

Once all interaction has been loaded, filter out all nodes that are not in the three clusters will results in a network similar to the one shown in Fig. 19. VisANT automatically adjust the global zoom level when loading large interaction set. To resume the zoom level, simply click first the Zoom Out button and then "Reset" button in VisANT's toolbox.

GOTEA can be performed through the menu "MetaGraph → Predict Functions of Metanodes Using GO → Network-based GOTEA → Fast GOTEA menu" and the iteration number is set to 20,000 using the menu "MetaGraph → Network-based GOTEA → Configure GOTEA." The prediction results will be visANT

Start VisANT

Integrative Visual Analysis Tool for Biological Networks and Pathways

Home | News | Run VisANT | Database | Features | Tutorials | Samples | Documents | VisANT Community | Interaction Statistics

Homo sapiens

affinity technology(M0045):	10672	anti tag coimmunoprecipitation(M0065):	3090
Biochemical Activity(M0083):	701	chromatography technology(M0085):	4
Co-fractionation(M0079):	167	coimmunoprecipitation(M0010):	4488
colocalization by immunostaining(M0025):	276	colocalization/visualisation technologies(M0063):	316
Competition binding(M0012):	121	copurification(M0013):	688
cosedimentation(M0033):	91	cross-linking studies(M0014):	3901
electron microscopy(M0062):	1	elisa:enzyme-linked immunosorbent assay(M0051):	79
enzymatic study(M0066):	118	far western blotting(M0060):	303
filter binding(M0053):	165	fluorescence technology(M0052):	114

Fig. 18. Total interactions available in Predictome database for *Homo sapiens*. Click on the number will load the corresponding interactions in VisANT.

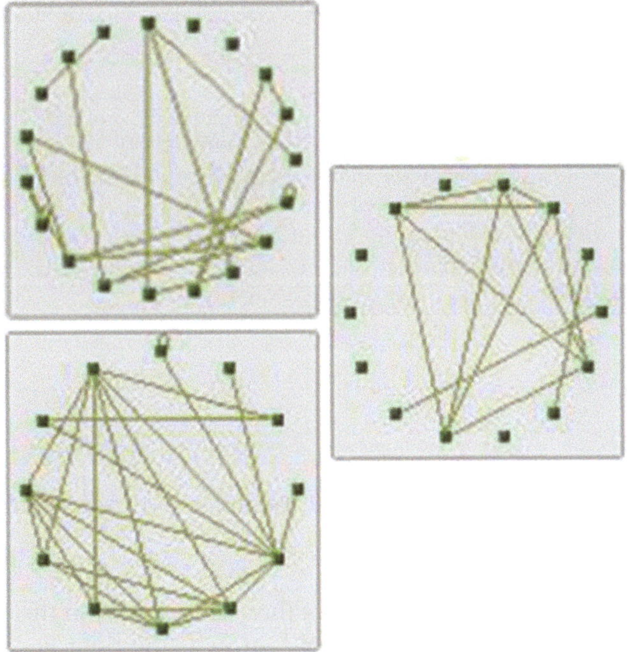

Fig. 19. Network modules for the three clusters with integrated interactions of M0045.

added to the metanode as part of its description that are available as tooltips. Table 3 lists top three GO terms resulted from GOTEA analysis for three clusters. The complete report can be found at: http://visant.bu.edu/misi/gotea_M0045_3_cluster.htm.

It is obvious that GOTEA finds more enriched GO terms for each cluster than hypergeometric test, which is mainly because GOTEA uses a fuzzy searching algorithm to find those GO terms that are semantically similar. As a result, GOTEA is much slower than hypergeometric test, and takes about half hour to finish the

Table 3
Cluster functions predicted by GO with integrated interaction of M0045

	Molecular function	Biological process	Cellular component
Cluster_A	Cytokine binding (GO:0019955) Growth factor binding (GO:0019838) Sugar binding (GO:0005529) ...	Cytokine biosynthetic process (GO:0042089) Positive regulation of immune response (GO:0050778) Innate immune response (GO:0045087) ...	Extracellular space (GO:0005615) Receptor complex (GO:0043235) Secretory granule (GO:0030141) ...
Cluster_B	Calmodulin binding (GO:0005516) ATP binding (GO:0005524) Calcium channel activity (GO:0005262)	Synaptic transmission (GO:0007268) Neurotransmitter transport (GO:0006836) Generation of a signal involved in cell–cell signaling (GO:0003001) ...	Clathrin-coated vesicle (GO:0030136) Neuron projection (GO:0043005) Cytoplasmic vesicle membrane (GO:0030659) ...
Cluster_C	Endonuclease activity (GO:0004519) Nucleotidyltransferase activity (GO:0016779) ATP binding (GO:0005524) ...	tRNA metabolic process (GO:0006399) DNA recombination (GO:0006310) Cellular carbohydrate catabolic process (GO:0044275) ...	Nucleolus (GO:0005730) Anchored to membrane (GO:0031225) Soluble fraction (GO:0005625)

N:48,SN:48 | E:56, SE:0 GOTEA for CLUSTER_C:scan GO term-->GO:0000171[8 of 206]-->permutation test[1396 of 20000]

Fig. 20. Use the *red* cancel button on VisANT status bar to cancel the computational heavy analysis.

analysis of three clusters. From this perspective, VisANT provides a red cancel button at the right end of the status bar to cancel the analysis, as shown in Fig. 20:

More information about GOTEA in VisANT can be found at http://visant.bu.edu/vmanual/ver3.50.htm#gotea.

3.4. Network-Based Expression Enrichment Analysis

Another typical application of the enrichment analysis is the study of differential RNA expression patterns (e.g., tumor vs. normal) determined by genome-wide association studies, to determine if one or more specified gene sets (e.g., KEGG pathways) might account for some of the differences (Fig. 2, dashed lines) (80–83). Gene Set Encrichment Analysis (GSEA) (82) is probably the most used algorithm in such analysis which does not take account of prior network knowledge. Here we introduce the

Network Module Enrichment Analysis (NMEA) to test whether the modules are enriched with transcriptional changes between the control and the sample. NMEA is basically an extension of GSEA but takes advantage of the extra information provided by network connectivity. In VisANT, a network can be constructed using the data from any combination of 70-odd methods (e.g., Y2H, ChIP-Chip, MS, and knockouts) for the interested gene lists. And modules can be easily constructed as metanodes through corresponding menus, simple drage&dop operation from GO explorer, and extended edge-list (http://visant.bu.edu/import#Edge) of user's own data.

Here we use the GO term to create the network modules and then perform NMEA over them. This example can be carried out in the following steps

1. Start VisANT as a local application (see Appendix for more detail) and have an empty network for *Homo sapiens.*
2. Resume the zoom level by clicking first the "Zoom Out" button and then "Reset" button in VisANT's toolbox.
3. Click on the GO Explorer tab in VisANT's control panel, enter GO:0000077 in the search box at the bottom of GO explorer, and click the "Search" button. Drag and drop the highlighted term DNA damage checkpoint to the network to create the metanode for GO:0000077 (Fig. 12).
4. Repeat step 3 for GO:0051320, GO:0007127 and GO:0051318. All three metanodes have overlaps with the first metanode of GO:0000077, move the overlapped genes to the center of each metanode, and a metanetwork shall appear similar to the one shown in Fig. 21 except there is no edge.
5. Mapping the expression profiles by opening the expression data from the following address: http://visant.bu.edu/sample/exp/p53_visant.dat using File → Open URL menu.

   The expression data shown in above link contains 22 microarray samples with mutations in P53 and 17 wild-type samples. The data is downloaded from GSEA Web site (http://www.broad.mit.edu/gsea/). Please reference http://visant.bu.edu/vmanual/ver3.50.htm#Expression for the format of expression data supported by VisANT.

   An alternative way to load the expression data is copy/paste expression data in the Add textbox of the toolbox.
6. Change the color mapped for the minimal and maximum expression values to the light green and darker green, respectively, by clicking left/right side the color map shown in the toolbar (Fig. 21). The color map will also be used to indicate the relative contribution to the enrichment score within each metanode.

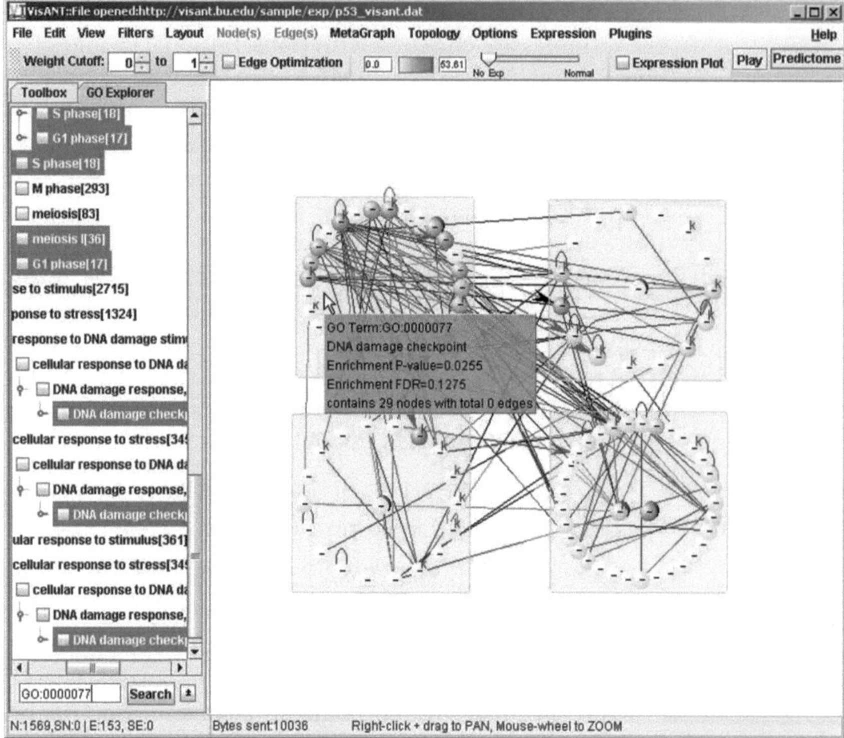

Fig. 21. NMEA analysis for four GO modules in VisANT.

7. Select all nodes using Edit → Select All Nodes menu
8. Query the interactions between selected nodes from Predictome database using Node(s) → Query Internal Interactions menu. The edges between the nodes appear as in Fig. 21.

   In comparison to the network modules shown in Fig. 19 where only a portion of the interactions are used to construct the network modules, here we query all possible interactions in the Predictome database.
9. Clear all selection with left-mouse clicking on empty space of the network panel.
10. Start NMEA using Expression → NMEA → Start NMEA Analyze menu. Once finished, *p*-value and FDR score will be added to each metanode's description (Fig. 21) and an html report will generated similar to the one at: http://visant.bu.edu/misi/nmea_go_modules.htm.

From the report it is clear that only the process DNA damage checkpoint (GO:0000077) exhibits the phenotypic difference in the expression of genes between mutated and wild-type samples, probably due to the fact that P53 plays a role in the process. As mentioned in step 6, nodes with the darker color have more contribution to the enrichment score.

More information about NMEA in VisANT can be found at http://visant.bu.edu/vmanual/ver3.50.htm#nmea.

3.5. Using Top-Down Method to Model the Cancer Gene Interaction Network

In this session we will illustrate how to use metagraph to build a network of cancers based on the simple cancer-gene association, and how this cancer network can be used to create the cancer gene network. Follow instructions shows the detailed step how this analysis can be carried out.

1. Construct the cancer network
   1.1 Clear the network by clicking Clear button
   1.2 Load the edge-list for the cancer network from http://visant.bu.edu/other_formats/edge_list_cancers.txt using the File → Open URL menu. Once finished, click the Fit to Page button on the toolbox.

       The data shown in the above URL is extracted from the work of Goh and coworkers (37). The disease is represented by disease ID and not very informative. From this perspective, we use the ID-Mapping format (Table 1) to add informative description for each cancer. The first few lines of the file are shown below:

       #!ID Mapping AddNewNode = false
       #VisANT_ID description
       DOR2212 Rhabdomyosarcoma, alveolar, 268220 (3) [DOR2212]
       DOR2211 Rhabdomyosarcoma, 268210 (3) [DOR2211]
       DOR2210 Rhabdoid tumors (3) [DOR2210]
       DOR1804 Nasopharyngeal carcinoma, 161550 (3) [DOR1804]

   1.3 Similar to above step, load the ID-Mapping file from URL: http://visant.bu.edu/other_formats/IDMapping_cencers.txt
   1.4 Collapse all metanodes using MetaGraph → MetaNode → Collapse All menu.
   1.5 A dashed edge between two cancers will be created automatically if they share at least one gene.
   1.6 Click the Zoom Out button on the toolbox 6 times and then click the Fit to Page button to reduce the node size and make it easier to examine the connections between diseases
   1.7 Layout the cancer network using the Layout → Spring-Embedded Relaxing menu. Click the Stop Animation button whenever appropriate (Fig. 14). The cancer network shall look similar to the one shown below:

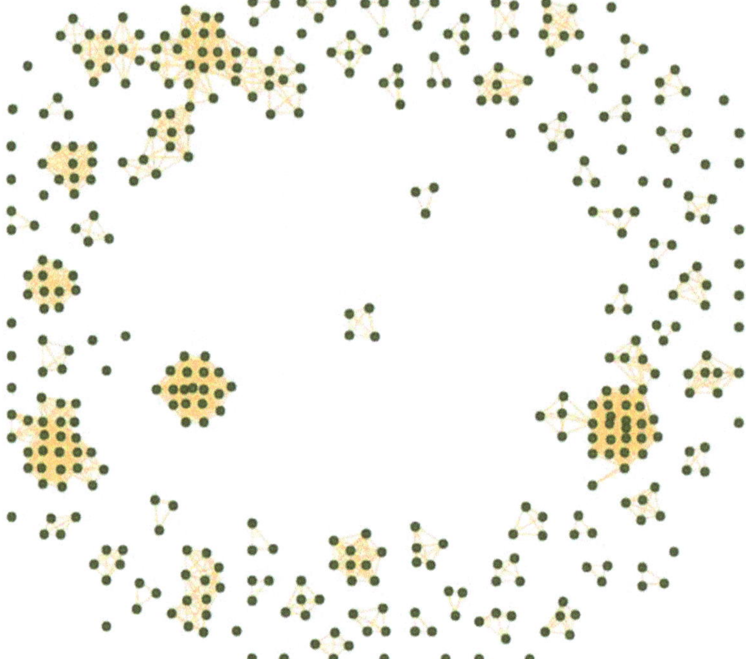

Fig. 22. Cancer gene network automatically created by VisANT based on the cancer network. The *dark dot* represents the gene and the edge represents that the two genes are associated with the same cancer.

2. Create the cancer gene network
   Apply the similar concept as "disease gene network" (37), i.e., two genes are connected if they are associated with the same disorder; we can easily create cancer gene network in VisANT.
   2.1. Use the MetaGraph → Create Co-Metanode Network menu to create the cancer gene network.
   2.2. Repeat step 4.1.6 to apply spring-embedded relaxing layout.
   2.3. Change node shape, color, and size of the cancer gene network by copy/paste following macros into the Add textbox (clear the textbox if necessary using the key CTRL-A and then Backspace):

   #!batch commands
   select_all_node
   set_node_property=node_size:7
   set_node_property=node_shape:circle
   clear_selection

   Please reference http://visant.bu.edu/vmanual/cmd.htm for more information about macros.
   2.4. The cancer gene network shall look similar to the one shown in Fig. 22:

Appendix

Mathematical Definition of Metagraph

A metagraph $G_m = \{V, E\}$ consists of a finite set V of the nodes and a finite set E of the edges. Nodes in a metagraph can be denoted as $V = \{V_s, V_m\}$ where V_s represents simple nodes as generally defined in simple graph and V_m represents the metanodes. The subscription s represents the simple node/edge and the subscription m represents metanode/metaedge. Each metanode $v_m \in V_m$ contains a subgraph consisting of child nodes and connected edges. In addition, each node $v \in V$ represents a set of its instance nodes, i.e., $v = \{v_i | i>0\}$ where v_i is the instance node of v. Instance nodes remains exact same identity between them but can have individual-specific properties. The statement that two metanodes share a node implies that each metanode contains an instance of the same node.

A metanode v_m has two states, *expanded* or *contracted*; the expanded state manifests the internal subgraph (that is, places all children nodes with their connections into the graph) while the contracted state replaces this subgraph with the single node. The combination of different states of the metanodes for a given metagraph results in multiple *views* that are abstract representations of the same underlying data. The change of the views for a given metagraph is defined as the dynamics of the metagraph, as shown in Fig. 1D, E.

Edges in a metagraph can be denoted as $E = \{E_s, E_m\}$where $E_s$represents simple edges that are generally defined in the simple graph and E_m represents metaedges. Each metanode edge $e_m \in E_m = e_{v_m, v}$is associated with at least one contracted metanode $v_m$and is transient: it appears when the metanode is contracted and disappears when one or two connected metanode nodes expanded, i.e., the metaedge is derived from the properties of two connected nodes. The most common derivation of the metaedge is the connection transfer. For example, when metanodes *M1* and *M2* are contracted in Fig. 1E, the connection between *C* and *E* is transferred to *M1* and *M2*. However, metaedge can also be derived from other properties of the metanode. The metaedge shown in Fig. 1E is derived because two metanode *M2* and *M3* share the same node *E*. The derivation of the metaedge can be generalized as $e_{v_{m,v}} = g(v_m, v)$, where g is the aggregation function and $v \in V$ can either be a metanode node or a simple node.

Download and Run VisANT as a Local Application

VisANT has four running modes in total, and two of them require a local copy of VisANT. Please visit http://visant.bu.edu and click the link "Run VisANT" for detailed instruction of other modes. It is recommended to run VisANT as a local application when handling large-scale network, such as the network with more than 100,000 nodes and edges because you will have the option to specify the memory size that VisANT can use. In addition, a local application allows VisANT to access local resources, such as load/save network files, directly; it also allows the user to develop VisANT plugins, as well as run a list of batch commands in the background without any user interface (batch mode).

The only drawback to run VisANT as a local application is that it easily becomes out of date because VisANT is under active development. Fortunately, VisANT provides a function to checks the update automatically and an icon will be shown near the Help menu if the update is available. Users can either click the icon, or corresponding menu to upgrade the VisANT to the latest version, as shown below:

1. If not already installed, download and install the Java 2 Platform, Standard Edition, version 1.4 or higher (http://java.sun.com/javase/downloads/index.jsp).
2. Go to http://visant.bu.edu and click on the link "Download," then click the link "Latest Version of VisANT."
3. Select a directory to save the file "VisAnt.jar"

   The VisAnt.jar is only about 400 K in size and the download shall take less than 1 min to finish. No installation is needed to run the VisANT.
4. To launch VisANT, double-clicking VisAnt.jar
5. To launch VisANT by an alternative mean: Open a Dos window in Win OS, or a shell window in other operation systems, and go to the directory where VisAnt.jar locates, and run the command:

   java -Xmx512M -classpath VisAnt.jar cagt.bu.visant.VisAntApplet

   where 512 M indicates the maximum size of the memory that VisANT can use. Increase this number if you have a large network or you get the "run out of memory" error.
6. The VisANT main window will appear (Fig. 4).
7. To exit VisANT, close the VisANT main window, or use the File → Exit menu option, or press the key combination ALT + X.

GO Term Enrichment Analysis

The four steps here describe how GOTEA works in VisANT. For illustration purposes, the following steps take only one metanode, *G*, into account and calculate only the enrichment score of one target GO term, *T*.

Step 1: Fully annotate all of the nodes in *G* with gene names and GO terms.

Step 2: Calculate density scores for each node based upon the topology and the GO term similarity to *T*. A vector D^G of density scores of each gene in *G* is computed, with the element of D^G for the *i*th gene denoted D_i. The density score is used to evaluate the impact of other genes in *G* on the *i*th gene, according to both the GO term similarity and the topological distance to the *i*th gene. D_i is defined as:

$$D_i = \sum_{j \in G} \log_2 \left[\left(\frac{M_j}{\alpha} \right) \Theta(M_j - \alpha) + \Theta(\alpha - M_j) \right] e^{-\beta d_{ij}},$$

where the step function,

$$\Theta(x - y) = \begin{cases} 1 & x \geq y \\ 0 & x < y, \end{cases}$$

ensures that $D_i \geq 0$. M_j is a measure of the GO term similarity calculated based upon the graph structure of the GO term hierarchy [85]. A significance threshold, α, is used to control the contribution that gene *j* makes to D_i. For larger α, a greater number of less statistically significant (with $M_j < \alpha$) genes are filtered and they do not contribute to D_i. The shortest distance between genes *i* and *j* given the topology of *G* is denoted d_{ij} and was calculated with the Floyd–Warshall algorithm. We assume that shorter distances make an exponentially greater contribution to the density than do longer distances, with the steepness of the exponential determined by the parameter $\tilde{\beta}$When a bigger β is chosen, more distant genes can contribute to the density. Taken together, the parameters α and β are used to control the sensitivity and selectivity of the density.

Step 3: Another vector of density scores, D^{NG}, is computed based upon a randomly chosen subset of genes representative of the background distribution. The background consists of all genes annotated by NCBI.

Step 4: Statistical significance for rejecting the null hypothesis is determined by a permutation test. For statistical robustness, step 3 is repeated *n* times. The number of times the average density score of randomly chosen genes is found to be larger than the average density score of genes in *G* is counted after n iterations and used to compute the final *p*-value (Fig. 23).

These four steps can be carried out for multiple testing by using multiple metanodes and multiple targeting GO terms. In this case, the *p*-values are corrected using FDR methods (79).

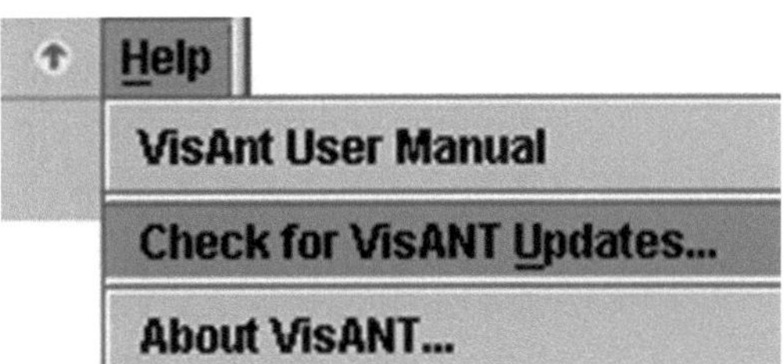

Fig. 23. VisANT upgrade.

Specifically, FDR $= p \times m/k$, where m is the total number of GO terms tested and k is the rank of the GO terms under consideration. There is also an option for GOTEA to identify representative GO terms from all its discoveries based upon approaches that identify the most informative GO term (84).

Network Module Enrichment Analysis

NMEA is implemented in a manner similar to GOTEA. Where GOTEA used GO term similarities, NMEA uses p-values from T-tests on the expression values of two phenotypes.

Step 1: Fetch the expression profile of each gene in a given module (i.e., metanode, denoted M in the following context) from formatted user input. The input should include an adequate number of samples with comparable phenotypes (e.g., normal and disease).

Step 2: A vector D^M of density scores of each gene is computed, with the element of D^M for the ith gene denoted as D_i. D_i is defined as:

$$D_i = \sum_{j \in G} \log_2 \left[\left(\frac{\alpha}{M_j} \right) \Theta(\alpha - M_j) + \Theta(M_j - \alpha) \right] e^{-\beta d_{ij}},$$

where the step function,

$$\Theta(x - y) = \begin{cases} 1 & x \geq y \\ 0 & x < y, \end{cases}$$

ensures that $D_i \geq 0$. M_j is the p-value from a two-tailed t-test of differential expression between two phenotypes (for example, normal and disease). The parameters α and β are used to control the sensitivity and selectivity of the density as described in the previous section.

The density score is used to evaluate the impact of other genes in M on the ith gene, according to both the p-value calculated by T-test (an indicator of differential expression) and their topological distances to the ith gene.

Step 3: Another vector of density scores, D^{NM}, is computed by randomly shuffling the phenotypes to obtain a representative sampling of the background distribution.

Step 4: Statistical significance for rejecting the null hypothesis is determined by a permutation test. For statistical robustness, step 3 is repeated n times. The number of times the average density score of randomly chosen genes is found to be larger than the average density score of genes in M is counted after n iterations and used to compute the final p-value.

When applying NMEA to multiple metanodes, the p-value must be corrected by FDR in a manner similar to what was described above for GOTEA. In this case, FDR $= p \times m/k$ as before, but m is the total number of metanodes and k is the rank of the metanodes under consideration.

References

1. Phizicky EM, Fields S (1995) Protein-protein interactions: methods for detection and analysis. Microbiol Rev 59(1):94–123
2. Berggard T, Linse S, James P (2007) Methods for the detection and analysis of protein-protein interactions. Proteomics 7(16): 2833–2842
3. Sobott F, Robinson CV (2002) Protein complexes gain momentum. Curr Opin Struct Biol 12(6):729–734
4. McCammon MG et al (2002) Screening transthyretin amyloid fibril inhibitors: characterization of novel multiprotein, multiligand complexes by mass spectrometry. Structure 10 (6):851–863
5. von Mering C et al (2002) Comparative assessment of large-scale data sets of protein-protein interactions. Nature 417(6887):399–403
6. Lu L, Arakaki AK, Lu H, Skolnick J (2003) Multimeric threading-based prediction of protein-protein interactions on a genomic scale: application to the *Saccharomyces cerevisiae* proteome. Genome Res 13(6A): 1146–1154
7. Aloy P, Russell RB (2004) Ten thousand interactions for the molecular biologist. Nat Biotechnol 22(10):1317–1321
8. Ito T et al (2001) A comprehensive two-hybrid analysis to explore the yeast protein interactome. Proc Natl Acad Sci USA 98(8): 4569–4574
9. Uetz P et al (2000) A comprehensive analysis of protein-protein interactions in *Saccharomyces cerevisiae*. Nature 403(6770):623–627
10. Hart GT, Ramani AK, Marcotte EM (2006) How complete are current yeast and human protein-interaction networks? Genome Biol 7 (11):120
11. Hu Z, Snitkin ES, DeLisi C (2008) VisANT: an integrative framework for networks in systems biology. Brief Bioinform 9(4): 317–325
12. Hu Z et al (2007) VisANT 3.0: new modules for pathway visualization, editing, prediction and construction. Nucleic Acids Res 35(Web Server issue):W625–W632
13. Hu Z et al (2005) VisANT: data-integrating visual framework for biological networks and modules. Nucleic Acids Res 33(Web Server issue):W352–W357
14. Hu Z, Mellor J, Wu J, DeLisi C (2004) VisANT: an online visualization and analysis tool for biological interaction data. BMC Bioinformatics 5:17
15. Hu Z, Mellor J, DeLisi C (2004) Analyzing networks with VisANT. In: Baxevanis A, Davison D, Page R, Petsko G, Stein L, Stormo G (eds) Current protocols in bioinformatics. Wiley, Hoboken
16. Hu Z et al (2009) VisANT 3.5: multi-scale network visualization, analysis and inference based on the gene ontology (translated from eng). Nucleic Acids Res 37(Web Server issue): W115–W121 (in eng)
17. Hermjakob H et al (2004) The HUPO PSI's molecular interaction format—a community standard for the representation of protein interaction data. Nat Biotechnol 22:177–183
18. Linghu B, Snitkin ES, Hu Z, Xia Y, Delisi C (2009) Genome-wide prioritization of disease genes and identification of disease-disease associations from an integrated human functional linkage network. Genome Biol 10(9):R91
19. Linghu B et al (2008) High-precision high-coverage functional inference from integrated data sources. BMC Bioinformatics 9:119
20. Bader GD, Cary MP, Sander C (2006) Pathguide: a pathway resource list. Nucleic Acids Res 34(Database issue):D504–D506
21. Breitkreutz BJ et al (2008) The BioGRID Interaction Database: 2008 update. Nucleic Acids Res 36(Database issue):D637–D640
22. Aranda B et al (2010) The IntAct molecular interaction database in 2010. Nucleic Acids Res 38(Database issue):D525–D531
23. Zanzoni A et al (2002) MINT: a Molecular INTeraction database. FEBS Lett 513(1): 135–140
24. Mewes HW et al (2008) MIPS: analysis and annotation of genome information in 2007. Nucleic Acids Res 36(Database issue): D196–D201
25. Cherry JM et al (1998) SGD: Saccharomyces Genome Database. Nucleic Acids Res 26(1): 73–79
26. Wilson RJ, Goodman JL, Strelets VB (2008) FlyBase: integration and improvements to query tools. Nucleic Acids Res 36(Database issue):D588–D593
27. Keshava Prasad TS et al (2009) Human Protein Reference Database—2009 update. Nucleic Acids Res 37(Database issue):D767–D772
28. Mellor JC, Yanai I, Clodfelter KH, Mintseris J, DeLisi C (2002) Predictome: a database of putative functional links between proteins. Nucleic Acids Res 30(1):306–309
29. von Mering C et al (2007) STRING 7—recent developments in the integration and prediction

of protein interactions. Nucleic Acids Res 35 (Database issue):D358–D362
30. UniProt Consortium (2008) The universal protein resource (UniProt). Nucleic Acids Res 36(Database issue):D190–D195
31. Bruford EA et al (2008) The HGNC Database in 2008: a resource for the human genome. Nucleic Acids Res 36(Database issue): D445–D448
32. Schuster-Bockler B, Bateman A (2008) Protein interactions in human genetic diseases. Genome Biol 9(1):R9
33. Yeger-Lotem E et al (2004) Network motifs in integrated cellular networks of transcription-regulation and protein-protein interaction. Proc Natl Acad Sci USA 101(16):5934–5939
34. Zhang LV et al (2005) Motifs, themes and thematic maps of an integrated *Saccharomyces cerevisiae* interaction network. J Biol 4(2):6
35. Benson DA, Karsch-Mizrachi I, Lipman DJ, Ostell J, Wheeler DL (2008) GenBank. Nucleic Acids Res 36(Database issue): D25–D30
36. Rogers A et al (2008) WormBase 2007. Nucleic Acids Res 36(Database issue): D612–D617
37. Goh KI et al (2007) The human disease network. Proc Natl Acad Sci USA 104(21): 8685–8690
38. Tong AH et al (2004) Global mapping of the yeast genetic interaction network. Science 303 (5659):808–813
39. Jeong H, Tombor B, Albert R, Oltvai ZN, Barabasi AL (2000) The large-scale organization of metabolic networks. Nature 407 (6804):651–654
40. Albert R, Jeong H, Barabasi AL (2000) Error and attack tolerance of complex networks. Nature 406(6794):378–382
41. Gavin AC et al (2002) Functional organization of the yeast proteome by systematic analysis of protein complexes. Nature 415(6868): 141–147
42. Ravasz E, Somera AL, Mongru DA, Oltvai ZN, Barabasi AL (2002) Hierarchical organization of modularity in metabolic networks. Science 297(5586):1551–1555
43. Babu MM, Luscombe NM, Aravind L, Gerstein M, Teichmann SA (2004) Structure and evolution of transcriptional regulatory networks. Curr Opin Struct Biol 14(3):283–291
44. del Sol A, O'Meara P (2005) Small-world network approach to identify key residues in protein-protein interaction. Proteins 58(3): 672–682
45. King OD (2004) Comment on "Subgraphs in random networks". Phys Rev E Stat Nonlin Soft Matter Phys 70(5 Pt 2):058101, author reply 058102
46. Itzkovitz S, Milo R, Kashtan N, Ziv G, Alon U (2003) Subgraphs in random networks. Phys Rev E Stat Nonlin Soft Matter Phys 68(2 Pt 2):026127
47. Shen-Orr SS, Milo R, Mangan S, Alon U (2002) Network motifs in the transcriptional regulation network of *Escherichia coli*. Nat Genet 31(1):64–68
48. Ashburner M et al (2000) Gene ontology: tool for the unification of biology. The Gene Ontology Consortium. Nat Genet 25(1):25–29
49. da Huang W et al (2007) The DAVID Gene Functional Classification Tool: a novel biological module-centric algorithm to functionally analyze large gene lists. Genome Biol 8(9):R183
50. Hu Z et al (2007) Towards zoomable multidimensional maps of the cell (translated from eng). Nat Biotechnol 25(5):547–554 (in eng)
51. Lee TI et al (2002) Transcriptional regulatory networks in *Saccharomyces cerevisiae*. Science 298(5594):799–804
52. Milo R et al (2002) Network motifs: simple building blocks of complex networks. Science 298(5594):824–827
53. Endy D, Brent R (2001) Modelling cellular behaviour. Nature 409(6818):391–395
54. Stamm S (2002) Signals and their transduction pathways regulating alternative splicing: a new dimension of the human genome. Hum Mol Genet 11(20):2409–2416
55. Boulos MN (2003) The use of interactive graphical maps for browsing medical/health Internet information resources. Int J Health Geogr 2(1):1
56. Green ML, Karp PD (2006) The outcomes of pathway database computations depend on pathway ontology. Nucleic Acids Res 34(13): 3687–3697
57. Fraser AG, Marcotte EM (2004) A probabilistic view of gene function. Nat Genet 36(6): 559–564
58. Guimera R, Nunes Amaral LA (2005) Functional cartography of complex metabolic networks. Nature 433(7028):895–900
59. Ihmels J et al (2002) Revealing modular organization in the yeast transcriptional network. Nat Genet 31(4):370–377
60. Bar-Joseph Z et al (2003) Computational discovery of gene modules and regulatory networks. Nat Biotechnol 21(11):1337–1342

61. Wu J, Hu Z, DeLisi C (2006) Gene annotation and network inference by phylogenetic profiling. BMC Bioinformatics 7:80
62. Oltvai ZN, Barabasi AL (2002) Systems biology. Life's complexity pyramid. Science 298 (5594):763–764
63. Rhee SY, Wood V, Dolinski K, Draghici S (2008) Use and misuse of the gene ontology annotations. Nat Rev Genet 9(7):509–515
64. Reimand J, Tooming L, Peterson H, Adler P, Vilo J (2008) GraphWeb: mining heterogeneous biological networks for gene modules with functional significance. Nucleic Acids Res 36(Web Server issue):W452–W459
65. Zhang M et al (2008) Interactive analysis of systems biology molecular expression data. BMC Syst Biol 2:23
66. Brohee S et al (2008) NeAT: a toolbox for the analysis of biological networks, clusters, classes and pathways. Nucleic Acids Res 36(Web Server issue):W444–W451
67. Alibes A, Canada A, Diaz-Uriarte R (2008) PaLS: filtering common literature, biological terms and pathway information. Nucleic Acids Res 36(Web Server issue):W364–W367
68. Antonov AV, Schmidt T, Wang Y, Mewes HW (2008) ProfCom: a web tool for profiling the complex functionality of gene groups identified from high-throughput data. Nucleic Acids Res 36(Web Server issue):W347–W351
69. Lee T, Desai VG, Velasco C, Reis RJ, Delongchamp RR (2008) Testing for treatment effects on gene ontology. BMC Bioinformatics 9 (Suppl 9):S20
70. Salomonis N et al (2007) GenMAPP 2: new features and resources for pathway analysis. BMC Bioinformatics 8:217
71. Zhu J et al (2007) GO-2D: identifying 2-dimensional cellular-localized functional modules in gene ontology. BMC Genomics 8:30
72. Antonov AV, Tetko IV, Mewes HW (2006) A systematic approach to infer biological relevance and biases of gene network structures. Nucleic Acids Res 34(1):e6
73. Draghici S et al (2003) Onto-Tools, the toolkit of the modern biologist: Onto-Express, Onto-Compare, Onto-Design and Onto-Translate. Nucleic Acids Res 31(13):3775–3781
74. Khatri P, Bhavsar P, Bawa G, Draghici S (2004) Onto-Tools: an ensemble of web-accessible, ontology-based tools for the functional design and interpretation of high-throughput gene expression experiments. Nucleic Acids Res 32 (Web Server issue):W449–W456
75. Khatri P et al (2007) Onto-Tools: new additions and improvements in 2006. Nucleic Acids Res 35(Web Server issue):W206–W211
76. Khatri P, Draghici S (2005) Ontological analysis of gene expression data: current tools, limitations, and open problems. Bioinformatics 21 (18):3587–3595
77. Wang JZ, Du Z, Payattakool R, Yu PS, Chen CF (2007) A new method to measure the semantic similarity of GO terms. Bioinformatics 23(10):1274–1281
78. Maglott D, Ostell J, Pruitt KD, Tatusova T (2007) Entrez gene: gene-centered information at NCBI. Nucleic Acids Res 35(Database issue):D26–D31
79. Benjamini Y, Drai D, Elmer G, Kafkafi N, Golani I (2001) Controlling the false discovery rate in behavior genetics research. Behav Brain Res 125(1–2):279–284
80. Barry WT, Nobel AB, Wright FA (2005) Significance analysis of functional categories in gene expression studies: a structured permutation approach. Bioinformatics 21(9): 1943–1949
81. Mootha VK et al (2003) PGC-1alpha-responsive genes involved in oxidative phosphorylation are coordinately downregulated in human diabetes. Nat Genet 34(3):267–273
82. Subramanian A et al (2005) Gene set enrichment analysis: a knowledge-based approach for interpreting genome-wide expression profiles (translated from eng). Proc Natl Acad Sci USA 102(43):15545–15550 (in eng)
83. Volinia S et al (2004) GOAL: automated gene ontology analysis of expression profiles. Nucleic Acids Res 32(Web Server issue): W492–W499
84. Zhou X, Kao MC, Wong WH (2002) Transitive functional annotation by shortest-path analysis of gene expression data. Proc Natl Acad Sci USA 99(20):12783–12788

Chapter 12

Data Mining in the MetaCyc Family of Pathway Databases

Peter D. Karp, Suzanne Paley, and Tomer Altman

Abstract

Pathway databases collect the bioreactions and molecular interactions that define the processes of life. The MetaCyc family of pathway databases consists of thousands of databases that were derived through computational inference of metabolic pathways from the MetaCyc pathway/genome database (PGDB). In some cases, these DBs underwent subsequent manual curation. Curated pathway DBs are now available for most of the major model organisms. Databases in the MetaCyc family are managed using the Pathway Tools software. This chapter presents methods for performing data mining on the MetaCyc family of pathway DBs. We discuss the major data access mechanisms for the family, which include data files in multiple formats; application programming interfaces (APIs) for the Lisp, Java, and Perl languages; and web services. We present an overview of the Pathway Tools schema, an understanding of which is needed to query the DBs. The chapter also presents several interactive data mining tools within Pathway Tools for performing omics data analysis.

Key words: Metabolic pathways, Pathway databases, Systems biology

1. Introduction

Pathway DBs describe the functional landscape of the cell. They collect the bioreactions and molecular interactions that comprise the processes of life. This chapter presents methods for performing data mining on the MetaCyc family of pathway DBs. Most DBs in the MetaCyc family contain metabolic pathway data, although some contain signaling pathway and regulatory information. Databases in the MetaCyc family are created, updated, and queried using the Pathway Tools software (1).

The MetaCyc family consists of thousands of pathway/genome databases (PGDBs) whose contents were derived through computational inference of metabolic pathways from the MetaCyc PGDB (2), followed in some cases by additional manual curation. Curated pathway DBs are now available for a number of model organisms

Hiroshi Mamitsuka et al. (eds.), *Data Mining for Systems Biology: Methods and Protocols*, Methods in Molecular Biology, vol. 939, DOI 10.1007/978-1-62703-107-3_12, © Springer Science+Business Media New York 2013

Table 1
Other Pathway Tools-based websites

Website	Contents	URL
ApiCyc	Apicomplexan genomes	http://apicyc.apidb.org/PLASMO/server.html
AraCyc	*Arabidopsis thaliana*	http://www.arabidopsis.org/biocyc/index.jsp
BeoCyc	33 BioEnergy genomes	http://cricket.ornl.gov/cgi-bin/beocyc_home.cgi
BioCyc	1,004 microbial Genomes	http://biocyc.org/
CalbiCyc	*Candida albicans*	http://pathway.candidagenome.org/
ChlamyCyc	*Chlamydomonas reinhardtii*	http://chlamycyc.mpimp-golm.mpg.de
EcoCyc	*Escherichia coli*	http://ecocyc.org/
FungiCyc	Multiple fungal genomes	http://fungicyc.broadinstitute.org:1555
DictyCyc	*Dictyostelium discoideum*	http://dictybase.org/Dicty_Info/dictycyc_info.html
LeishCyc	*Leishmania major*	http://leishcyc.bio21.unimelb.edu.au/
MicroScope	542 Genomes	http://www.genoscope.cns.fr/agc/microscope/
MouseCyc	*Mus musculus*	http://mousecyc.jax.org:8000/
PseudoCyc	*Pseudomonas aeruginosa*	http://v2.pseudomonas.com:1555/
RiceCyc	Multiple plant genomes	http://www.gramene.org/pathway/
ScoCyc	Streptomyces genomes	http://scocyc.streptomyces.org.uk:14980/SCO/server.html
ShewCyc	18 Shewanella genomes	http://spruce.ornl.gov:1555/SHEON/organism-summary?object=SHEON
SolCyc	Solanaceae genomes	http://solcyc.sgn.cornell.edu/
SoyCyc	Soybean	http://soybase.org
TBestDB	Taxonomically Broad EST DB	http://pepdbpub.bcm.umontreal.ca/pathway/
TBCyc	*Mycobacterium tuberculosis*	http://tbcyc.tbdb.org/
SGD	*Saccharomyces cerevisiae*	http://www.yeastgenome.org/

including *Escherichia coli*, *Saccharomyces cerevisiae*, *Arabidopsis thaliana*, *Mus musculus*, and *Homo sapiens*. Table 1 lists the locations of these PGDBs.

The methodology we present is as follows:

- Choose the DB or DBs to be used for your data mining project. The MetaCyc family contains thousands of DBs created by many different researchers.

- Choose among the many data access mechanisms that exist for these DBs:
  - Files in BioPAX format.
  - Files in attribute-value format.
  - Perl API.
  - Java API.
  - Lisp API.
  - Web services API.
- Study the Pathway Tools schema to understand what data fields to query.
- Use interactive data mining tools provided by the Pathway Tools software to analyze data of interest.

Pathway Tools has many additional capabilities that are not presented here including tools for identification of dead-end metabolites and metabolic choke points, and tools for inferring pathway hole fillers and operons. For more details, see (1).

2. Materials

For web access to PGDBs, we recommend use of Firefox, Safari, or Chrome. Versions of Internet Explorer released prior to 2011 are not recommended because of their slow speed and other problems.

BioCyc PGDBs are freely and openly available to all users, and are available in a variety of file formats listed at (3).

The Pathway Tools software is available for the Linux, Windows, and Macintosh platforms. It is freely available to academic users and is available to commercial users for a fee. BioCyc and Pathway Tools may be obtained from (4).

3. Methods

3.1. Choose the Database(s) for Your Data Mining Project

The following factors should be considered when choosing a PGDB for a data mining project.

Does a PGDB exist for the organism of interest? The websites listed in Table 1 contain thousands of PGDBs that overlap to some degree. If a PGDB does not exist for an organism of interest, a researcher can create a new PGDB using Pathway Tools.

Does a curated PGDB exist for the organism of interest? Curated DBs are generally significantly more accurate than uncurated DBs (note that most pathway DBs outside the MetaCyc family are uncurated). Curated DBs contain minireviews that summarize the biology of a gene or pathway. They contain citations to the literature, and they contain curated data that cannot be predicted computationally (such as enzyme inhibitors and cofactors). The curation level of a PGDB can be assessed by invoking the command *Tools → Comparative Analysis* and then selecting the organisms report for the PGDB of interest. The resulting report identifies the number of publications cited in the PGDB and the number of genes and pathways with experimental evidence. In some cases a curated PGDB contains a large and unique collection of data. For example, EcoCyc contains extensive curated information on the regulation of *E. coli* genes by a variety of mechanisms including transcriptional regulation, regulation by attenuation, and regulation by small RNAs.

Depending upon how you plan to access the PGDB for data mining, a local copy of the PGDB may be required for use within a local copy of Pathway Tools. Accessing PGDB data using web services does not require having a local copy. Accessing PGDB data using the Java, Perl, and Lisp APIs does require having a local copy, because those APIs are accessed from within the Pathway Tools software using a local Unix socket. Pathway Tools must load the PGDB into its virtual memory.

A local copy of a PGDB can be downloaded via the PGDB registry (5), which is a peer-to-peer DB server. The Pathway Tools command *Tools → Browse PGDB Registry* will list all PGDBs defined within the registry. The user can click on a PGDB to download it locally for use within Pathway Tools. If a known PGDB of interest is not within the registry, contact its authors to request that they deposit it there.

3.2. Choose a Data Access Mechanism

Three different approaches can be used to obtain and operate on PGDB data: (1) download the full data files from the BioCyc website and operate on them locally, (2) use the web services interface to retrieve selected data, or (3) install the Pathway Tools software on a local machine and use one of the programmer APIs.

3.2.1. Data Files

Data files can be downloaded from the BioCyc website by following the instructions at (4). For PGDBs hosted elsewhere, contact the administrators of that site directly. The downloadable archive for a given PGDB contains all relevant files for that PGDB, including BioPAX files, attribute-value files, sequence files, and various tabular files. If the Pathway Tools software is installed locally, these files can also be generated using the **File → Export** command.

BioPAX Files

BioPAX (6) is an XML-based standard format that facilitates the exchange of pathway-related information. A number of software tools can import or use data in BioPAX format. Two versions of the BioPAX format file are provided for each PGDB: a level 2 file, which conforms to an older BioPAX version, and a level 3 file, which uses the most recent BioPAX version (the two versions of BioPAX are incompatible). A single file contains data for all pathways and metabolic and transport reactions in the PGDB—including all metabolites, enzymes, and regulators—relevant to those pathways and reactions. It does not contain information about genes, transcription units, or transcriptional or translational regulation. To obtain BioPAX data for just a single pathway rather than for all pathways, use the web services interface instead.

Attribute-Value Files

Attribute-value files contain data in a format that corresponds closely to the Pathway Tools schema. A different file is provided for each class of data object, such as genes, proteins, or reactions. For each Pathway Tools object (such as a single gene), the file will list its unique identifier and the values of its attributes, with each attribute on a different line. More details can be found at (7). Users may have to parse and integrate data from multiple files to extract the information they want. For example, to determine the reaction catalyzed by an enzyme, search the `proteins.dat` file for the value of the CATALYZES attribute of the enzyme (which will give the id for the catalysis object, known as an *enzymatic reaction*), search for that value in the `enzrxns.dat` file to extract the REACTION attribute, and then search for that reaction in the `reactions.dat` file to retrieve properties of the reaction, such as its reactants and products.

Other Files

The file archive for a PGDB also contains other miscellaneous files. FASTA files contain the sequence for each gene or protein in the PGDB in FASTA format. Tabular data files summarize selected commonly used relationships between objects, such as genes of a pathway or reactions of an enzyme, in tab-delimited table format. This format can be more convenient than having to extract and integrate information from multiple attribute-value files if those are the relationships you are interested in retrieving. The full list of files can be found at (7).

3.2.2. Web Services

Data can be retrieved from the BioCyc website (or from any other website that runs version 14.5 or higher of the Pathway Tools software) in XML format using the web services API. Simple queries can return the data for a single Pathway Tools object (such as a gene or a reaction) given its unique id. More complex queries can be constructed using the BioVelo query language to retrieve data for one or more objects based on their properties. For a detailed description of the web services API, including sample queries, see (8).

Pathway data for a single pathway can be optionally retrieved in BioPAX format. All other requests through the web services API generate data in a format known as **ptools-xml**. This format is closely related to the underlying Pathway Tools object representation, so an understanding of the Pathway Tools schema is necessary in order to interpret the data. The ptools-xml format is further described at (9).

To retrieve data for a single object (such as a gene or reaction), you must supply both the PGDB identifier (a short string that uniquely identifies the organism DB) and the object identifier. Objects can be retrieved in either low or full detail. Low detail includes only selected attributes and relationships. For example, low detail for a protein or RNA includes only its name and synonyms and links to its gene, its components (if a complex) or complexes, and any reactions it catalyzes. Full detail (the default) includes all information associated with that object in the DB, including but not limited to textual summaries, citations to the literature, links to other DBs, reactions in which the object participates, and regulation information.

The URL to retrieve an object from the BioCyc website in ptools-xml format is `http://wgetxml?i/getxml?[ORGID]:[OBJECT-ID]` or `http://wgetxml?i/getxml?id=[ORGID]:[OBJECT-ID]&detail=[low|full]`, where `[ORGID]` and `[OBJECT-ID]` are the PGDB identifier and the object identifier, respectively.

The BioVelo query language (10) allows you to write precise queries to extract a set of objects from one or more PGDBs that satisfy specific criteria. Some example searches might be for all proteins whose name includes some search string, all heteromultimeric complexes in the PGDB, all genes associated with a particular pathway, or all compounds in a particular molecular weight range. Query results can be returned at either low detail (the default), full detail, or no detail. In the no detail case, the query returns unique ids only, without any additional attributes, for objects that satisfy the query.

The URL to issue a BioVelo query to the BioCyc website that returns a list of objects in ptools-xml format is `http://websvc.biocyc.org/xmlquery?[QUERY]` or `http://websvc.biocyc.org/xmlquery?query=[QUERY]&detail=[none|low|full]`, where `[QUERY]` is a properly escaped BioVelo query string that returns a single list of Pathway Tools objects. For example, the request `http://websvc.biocyc.org/xmlquery?[x:x<-ecoli^^pathways]` retrieves data for all pathways in EcoCyc. The request `http://websvc.biocyc.org/xmlqu ery?[x:x<-mtbrv^^Genes,x^left-end-position>100000,x^right-end-position<200000, %23[pp:pp<-x^product,pp^molecular-weight-kd>=50,pp^molecular-weight-kd<=60]>0]` retrieves data for all genes in the *M. tuberculosis* H37Rv PGDB whose map position on the chromosome is between 100,000 and 200,000, and that have a product whose molecular weight is between 50 and 60 kdaltons.

BioVelo can be a tricky language to master, but several example queries are provided at (8). In addition, BioVelo is the engine underlying a number of the search mechanisms available from the

Search menu on the BioCyc website, including the Compound, Genes/Proteins/RNAs, Reactions, Pathways, and Advanced search forms. Each search conducted using these forms also returns the BioVelo query that was evaluated, so some experimentation with these can provide guidance when generating your own BioVelo queries (but bear in mind that these query forms often generate multicolumn tables, whereas the web services API is more limited and supports only queries that return a single column).

3.2.3. Lisp, Perl, and Java APIs

If the Pathway Tools software has been installed on a local machine, programs can be written that query the data directly through the Lisp, Perl, or Java APIs. A wealth of information and relevant links can be found at (11), including the list of API functions, example queries, download links for PerlCyc and JavaCyc, and general information about Lisp and the Lisp debugger. It is strongly recommended to explore these resources before beginning to use any of the APIs, so this discussion will provide just a general overview of the APIs.

There are two layers of API functions. The bottom level is the basic object access protocol, known as the generic frame protocol (GFP) (12), which includes operations for enumerating all objects in a class and accessing the attributes (slot values) of an object. These operations require precise knowledge of the Pathway Tools schema in order to utilize them effectively. On top of that are a number of higher-level functions, known collectively as the Pathway Tools API, that encapsulate some of the biological knowledge and relationships between objects, such as how to query the pathways containing a gene, the genes regulated by a given gene, or whether a protein is an enzyme or a transporter. Your programs will probably make use of functions from both layers.

Lisp API

The Pathway Tools software is written in the Common Lisp language, so when writing a new program starting from scratch (as opposed to interfacing with an existing Perl or Java program), the Lisp API is the most convenient API to use. There is no separate package to install, and no interprocess communication to worry about. Learning enough Lisp to write simple queries is quite straightforward (some good resources for learning Lisp are (13, 14)), and will greatly increase your productivity when working with the data.

To use the Lisp API, start Pathway Tools in Lisp mode. On a Linux or Macintosh machine, this is done by supplying the -lisp command-line argument when starting Pathway Tools. On a Windows machine, a separate desktop icon is provided to start in Lisp mode. In either case, the normal Pathway/Genome Navigator interface will not appear. Instead, you will see a lisp prompt in the console window, which indicates that the software is ready to accept commands. Commands and code snippets can be either typed or pasted directly to the Lisp prompt, or they may be loaded from a separate file. Commands and code are interpreted as they are entered and, in case of errors, can be debugged interactively. Some example queries can be found at (15).

PerlCyc and JavaCyc

The Perl and Java APIs are known as PerlCyc (16) and JavaCyc (17), respectively, and must be downloaded as separate modules. Unix file sockets are used for communication between the process running the PerlCyc or JavaCyc program and the Pathway Tools server, which leads to several important limitations: (1) the Java and Perl APIs work only with a Unix installation of Pathway Tools (Linux or MacOS), not with Windows; (2) the machine running the Java or Perl program must be the same as, or share, a file system with the machine running Pathway Tools; and (3) only one external process may interact with a Pathway Tools process at a time.

To use PerlCyc or JavaCyc, Pathway Tools must have been invoked with the -api command-line argument. When invoked in this fashion, no change is noticeable in the operation of the Pathway Tools program—the Pathway/Genome Navigator appears as usual, and users can interact with it in the normal fashion. The only difference is that simultaneously the program is alert to connections from an external process.

The external Perl or Java program must load the PerlCyc or JavaCyc module and then create a new PerlCyc or JavaCyc object to connect to the desired PGDB. Perl functions and Java methods are available that are analogous to most of the corresponding GFP and Pathway Tools API Lisp functions. When one of these functions is invoked on the PerlCyc or JavaCyc object, the function is converted to a Lisp query and submitted to the Pathway Tools process. The subsequent response is then converted back into the appropriate data type in Perl or Java. Examples showing exactly how to write JavaCyc or PerlCyc queries are included in the appropriate module documentation and in Fig. 1. Note that the JavaCyc interface does not support passing of actual objects between Java and Pathway Tools, merely object identifiers. An additional JavaCyc query would be necessary to retrieve some attribute of a returned object.

3.3. Study the Pathway Tools Schema

PGDB data is stored in an object management system called Ocelot (18). Objects are organized into a class/instance hierarchy. Each class can have its own set of attributes, known as slots. A slot can have one or multiple values and either describes an attribute of the object (such as the name or molecular weight of a protein) or defines a relationship between that object and some other object in the DB (such as the gene for a protein or the complexes of which a protein is a component). Some of the major classes in the Pathway Tools schema are listed in Table 2.

a Lisp

```
(select-organism :org-id 'ECOLI)
(loop for x in (get-class-all-instances '|Regulation-of-Enzyme-Activity|)
  ;; Does the regulator slot contain ATP, and is the mode of regulation
  ;; negative (representing inhibition)?
  when (and (member-slot-value-p x 'regulator 'atp)
            (member-slot-value-p x 'mode "-"))
  ;; The value of the regulated-entity slot is an Enzymatic-Reaction
  ;; object, so we need to access its enzyme slot to retrieve the enzyme.
  ;; The collected values are returned as a list.
  collect (get-slot-value (get-slot-value x 'regulated-entity) 'enzyme))
```

b Perl

```
use perlcyc;
my $cyc = perlcyc->new("ECOLI");
my @regs = $cyc->get_class_all_instances("|Regulation-of-Enzyme-Activity|");
foreach my $reg (@regs) {
  my $bool1 = $cyc->member_slot_value_p($reg, "regulator", "ATP");
  my $bool2 = $cyc->member_slot_value_p($reg, "mode", "-");
  if ($bool1 && $bool2) {
     my $enzrxn = $cyc->get_slot_value($reg, "regulated-entity");
     my $enz = $cyc->get_slot_value($enzrxn, "enzyme");
     ##Print the results on the terminal
     print STDOUT "$enz\n";
  }}
```

c Java

```
Import java.util.*;

public class JavaCycSample {
  public static void main(String[] args) {
    Javacyc cyc = new Javacyc("ECOLI");
    ArrayList regs =
      cyc.getClassAllInstances("|Regulation-of-Enzyme-Activity|");
    for (int i=0; i<regs.size(); i++) {
      String reg = (String)regs.get(i);
      boolean bool1 = cyc.memberSlotValueP(reg, "regulator", "ATP");
      boolean bool2 = cyc.memberSlotValueP(reg, "mode", "-");
      if (bool1 && bool2) {
        String enzrxn = cyc.getSlotValue(reg, "regulated-entity");
        String enzyme = cyc.getSlotValue (enzrxn, "enzyme");
        System.out.println(enz);
      }}}}
```

Fig. 1. Sample code showing how to find all enzymes inhibited by ATP in EcoCyc using the (**a**) Lisp, (**b**) Perl, and (**c**) Java APIs.

3.4. Use Interactive Data Mining Tools

Pathway Tools includes powerful interactive tools for analyzing data, including enrichment/depletion analysis, visualization tools for high-throughput experimental data, and comparative analysis.

3.4.1. Enrichment Analysis

Consider the analysis of a gene-expression experiment in which 200 genes are found to be significantly up- or downregulated. Biologists frequently want to ask whether those 200 genes contain significant numbers of genes involved in one or more biological processes (such as cell division) or in one or more biological pathways. That is, are genes from one or more processes statistically overrepresented in that set of genes? Put another way, is the given set of genes enriched for genes from one or more processes or pathways? Similarly, in analysis of metabolomics data, one might ask whether a set of metabolites observed to have changed between two experiments is enriched with respect to the metabolites in one or more metabolic pathways.

Table 2
Some of the major classes and subclasses in the Pathway Tools schema

Compounds-And-Elements	Compounds Elements
Proteins	Polypeptides Protein-Complexes
RNAs	tRNAs rRNAs Regulatory-RNAs
Genetic-Elements	Chromosomes Plasmids Contigs
DNA-Segments	Genes DNA-Binding-Sites Promoters Terminators Transcription-Units
Protein features	Modified-Residues Active-Sites Signal-Sequences
Reactions	Chemical-Reactions Transport-Reactions Binding-Reactions Redox-Half-Reactions
Pathways	
Enzymatic-Reactions	
Regulation	Regulation-of-Enzyme-Activity Regulation-of-Transcription-Initiation Transcriptional-Attenuation Regulation-of-Translation
Organisms	
Publications	
Evidence	
CCO (Cell Component Ontology)	
Gene-Ontology-Terms	
Databases	

Enrichment analysis is a statistical analysis tool that is able to answer this type of question. Gene or metabolite lists such as those in our examples are usually generated as a result of a high-throughput experiment. High-throughput experiments are noisy, and genes and compounds can participate in multiple biological processes or pathways. So in the context of the above example, it is a mistake to take all the pathways in which at least one gene from a list of 200 genes is involved and assume that they all participate in the phenomenon studied in the gene-expression experiment. Enrichment analysis enables us to statistically distinguish the pathways explaining the phenomenon that underlies the expression experiment from those that contain genes from the list by accident.

Enrichment analysis was initially described (19–21) for lists of genes obtained using microarray experiments and for gene-ontology (GO) terms. We have developed a general framework for enrichment analysis. It is tied in to our object groups functionality to allow facile creation of groups of genes, compounds, or other types of objects and then evaluating whether their enrichment or depletion of GO terms, pathways, or other sets is statistically significant.

3.4.2. Omics Viewers

Pathway Tools Omics Viewers use the Cellular, Regulatory, and Genome Overviews to illustrate the results of high-throughput experiments in a global metabolic and genomic context.

Omics Viewers can show absolute data values (such as the concentration of a metabolite or protein, or the absolute expression level of a gene) and can be used to compare two sets of experimental data by computing a ratio and mapping the ratios onto a color spectrum. Multiple sets of experimental data can be superimposed on the same overview diagram so that users can, for example, combine gene expression and metabolomics in the same figure, or view the results of two different microarray experiments together. When combining multiple datasets, users should be careful to assign color schemes that avoid ambiguity. For example, you might want to use "warm colors" like yellow and red for one dataset and "cool colors" like blue and purple for a different dataset, to allow them to be seen side by side.

Superposition of multiple sets of experimental data on the overviews can also be animated to show, for example, how gene expression levels of enzymes change with time over the course of an experiment. The animation can be exported to HTML so that it can be published online.

After displaying Omics data on one of the overviews, navigating to any pathway display will show the Omics data superimposed on the individual pathway. If a particular reaction step has multiple isozymes, then rather than just choosing one value as is done on the Cellular Overview, all values are shown.

Cellular Overview Omics Viewer

The Cellular Overview diagram is a representation of all metabolic pathways and reactions, signaling pathways, membrane proteins, and transporters defined for the current organism. In this diagram, each icon (e.g., circle, square, ellipse) represents a single metabolite. The shape of the icon encodes the chemical class of the metabolite. Each thick line in the overview diagram represents a single reaction. Neither the icons nor the lines are unique in the sense that a given

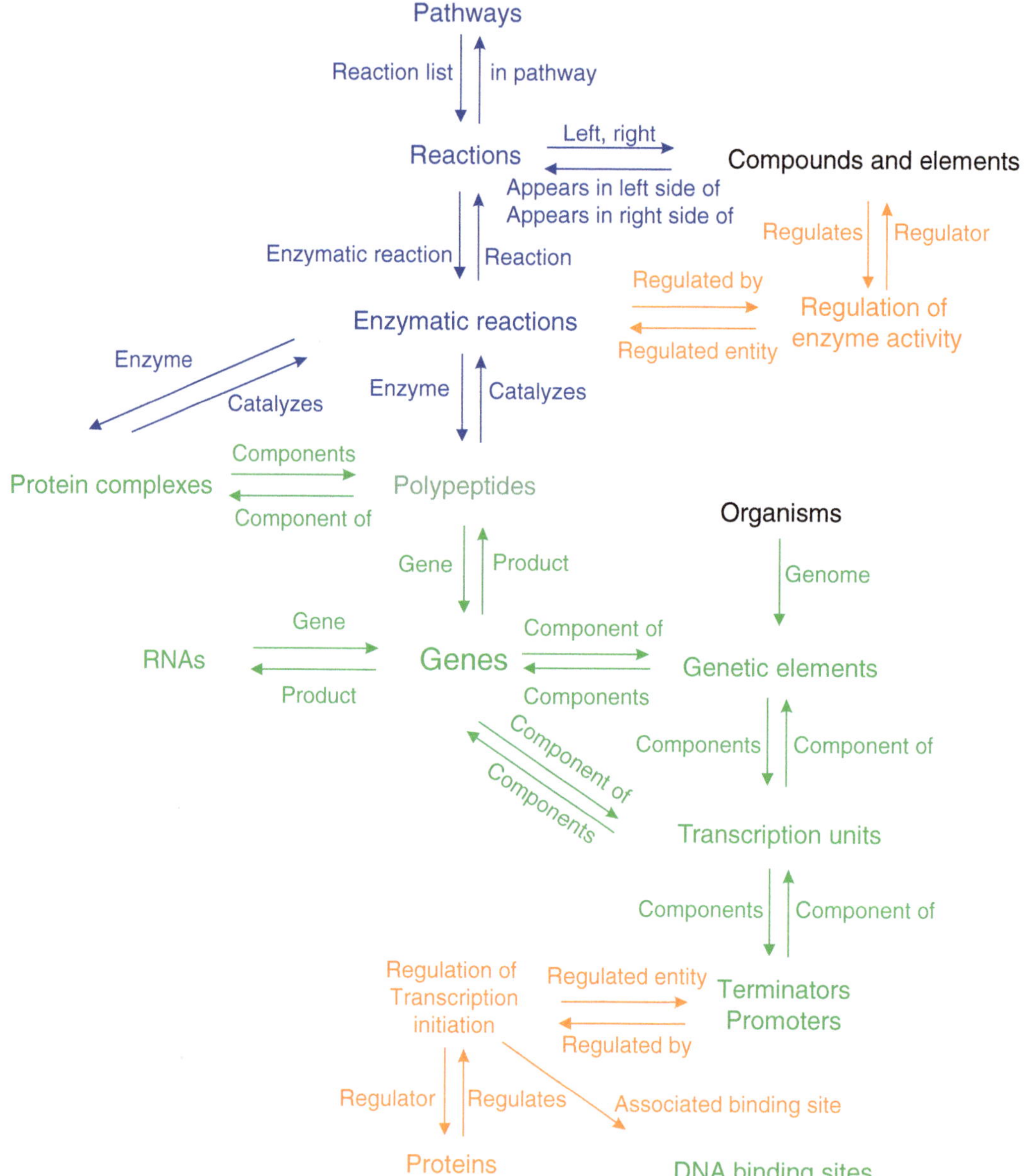

Fig. 2. Some of the major relationships between classes of objects in the Pathway Tools schema. For example, the arrow labeled reaction-list going from Pathways to Reactions indicates that the Pathways class has a slot reaction-list, whose values are members of the Reactions class.

metabolite or a given reaction may occur in more than one position in the diagram. If there are any thin gray reaction lines, these represent reactions for which no enzymes have been identified in the PGDB—in other words, pathway holes.

The Cellular Overview Omics Viewer can be used to illustrate an even wider range of high-throughput experimental results in a global metabolic pathway context (Fig. 2). Genes (in the case of a gene-expression experiment) and proteins (in the case of a proteomics experiment) that are involved in metabolism are mapped to reaction steps in the Cellular Overview, and the range of data values in a given experimental dataset is mapped to a spectrum of colors. Reaction steps in the Cellular Overview are colored according to the corresponding data value. Similarly, for metabolomics experiments, compound nodes are colored according to the data value for the corresponding compound. This facility enables the user to see instantly which pathways are active or inactive under some set of experimental conditions.

The Cellular Omics Viewer can be used for:

Microarray gene-expression data: Reaction lines (and protein icons, where present) are color coded according to the relative or absolute expression level of the gene that codes for the enzyme that catalyzes that reaction step. The Cellular Omics Viewer allows a scientist to interpret the results of gene-expression experiments in a pathway context.

Proteomics data: Reaction lines (and protein icons, where present) are color coded according to the concentration of the enzyme that catalyzes that reaction step.

Proteomics data: Reaction lines (and protein icons, where present) are color coded according to the concentration of the enzyme that catalyzes that reaction step.

Metabolomics data: Compound icons are color coded according to the concentration of the compound.

Reaction flux data: Reaction lines are color coded according to reaction flux values.

Other experimental data: Any experiment, high throughput or otherwise, in which data values are assigned to genes, proteins, reactions, or metabolites can be viewed in a pathway context using the Omics Viewer.

Genome Overview Omics Viewer

The Genome Overview Omics Viewer can map any dataset that focuses on genes (such as gene-expression studies) onto the full genome of the organism, using a spectrum of colors to display the numerical values associated with each gene (Fig. 3).

The Genome Overview shows in one screen all the genes in an organism's genome as well as additional information about their transcription units and products. The Genome Overview has several key differences from the genome browser. Unlike the genome browser, the overview is not to scale nor does it reflect spacing between genes. Conversely, the Genome Overview shows the full genome of an organism at once, even if that genome contains multiple chromosomes or plasmids. Each individual replicon (chromosome, plasmid) is displayed on the page with an appropriate identifying label.

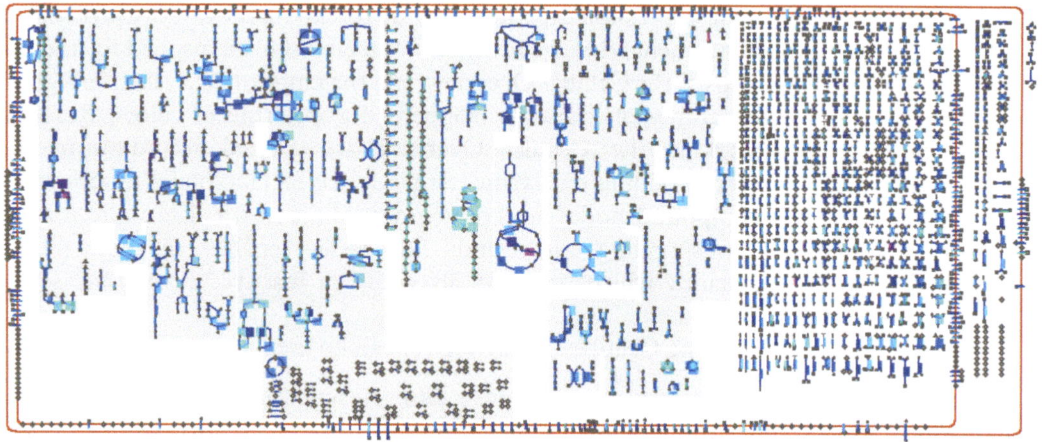

Fig. 3. The Cellular Omics Viewer displays many kinds of high-throughput data.

The Genome Overview uses iconography similar to that of the genome browser, showing the direction of gene transcription with a sloping line on the top (protein-coding gene) or on the bottom (RNA-coding gene) of the gene. Lines underneath genes indicate the extent of transcription units and are particularly useful for identifying multiple promoters. Transcription units are also indicated by shared gene color, although this coloring is replaced if you choose to map expression data onto the overview using the Genome Overview Omics Viewer function.

You can identify a gene in the overview by clicking on it to go to its gene page or by mousing over it to display its gene name, product, and distance from neighboring genes at the bottom of the screen. The Genome Overview Omics Viewer can be used for:

Microarray gene-expression data: Genes are color coded according to the relative or absolute expression level of the gene.

Other experimental data: Any experiment, high throughput or otherwise, in which data values are assigned to genes, can be viewed using the Genome Omics Viewer. One such possible use is the mapping of a set of ESTs that have been assigned to genes onto a sequenced genome, thus offering a view of how much of, and which parts of, the genome are covered by that EST set.

Regulatory Overview Omics Viewer

The Regulatory Overview Omics Viewer, like the Genome Overview Omics Viewer, can display gene- or protein-oriented data. It utilizes the Regulatory Overview to display experimental data in the context of a PGDB's regulatory network.

The Regulatory Overview enables the user to visually analyze the regulatory relationships between genes for a specific organism. These relationships are based on the regulatory data available in the PGDB of the organism. Currently, the relationships are based on transcriptional regulatory data (future versions may cover other types of regulation). The Regulatory Overview is represented as a network with nodes and arrows (i.e., arcs). Each node represents a gene of a specific organism. An arrow extends from gene A to gene B if and only if A regulates B.

The Regulatory Overview initially shows the transcriptional regulatory network for the currently selected organism, in one window, without any arrow relationships shown. Otherwise, their great number would clutter the overview. These arrows can be selectively added by using highlighting commands. Each node icon in the diagram, such as a plus sign or circle, depicts one gene. Not all genes in the genome are shown in the diagram. The genes shown are regulators (transcription factors and sigma factors) and all other genes for which the PGDB encodes regulatory information for the gene.

Two network layouts are available: three nested ellipses and top to bottom rows. Three nested ellipses is the default layout when displaying the entire overview for the first time. Top to bottom rows is the default layout when redisplaying only the highlighted genes.

For the three nested ellipses layout, the genes are partitioned into three groups, each group being laid out on a separate ellipse. The two inner ellipses contain all the genes that regulate at least one gene. The innermost ellipse contains the genes that regulate the most. Typically, about 15% of the regulators are in the innermost ellipse. The outermost ellipse contains genes that are regulated but that do not regulate.

The outermost ellipse is further partitioned into groups of genes such that all genes within one group are regulated by the same set of genes. We call these groups *multi-regulons*. Some groups are shown as a line or as a line leading to a triangle, perpendicular to the outermost ellipse. Note that although all genes within a multi-regulon respond to the same set of regulator genes, different genes in the group may be controlled in different ways. For example, consider a multi-regulon, comprising genes A and B, that are regulated by genes X and Y. Genes X and Y might both activate the transcription of A, but they might both inhibit transcription of B.

It is possible to display a regulatory subnetwork of a specific organism by doing a series of highlighting operations and then using the command Redisplay Highlighted Genes Only. This command will create a new, smaller layout of the regulatory network containing only the genes that are highlighted. Genes that do not regulate, or are not regulated by any highlighted genes, are not included in the subnetwork. Further operations can be done on this subnetwork as for the complete overview.

The Regulatory Overview Omics Viewer illustrates the results of high-throughput experiments in the context of gene regulation. Genes that are involved in regulation are mapped to their gene icon in the Regulatory Overview diagram, and the range of data values in a given experimental dataset is mapped to a spectrum of colors.

Fig. 4. The Genome Omics Viewer displays gene expression results.

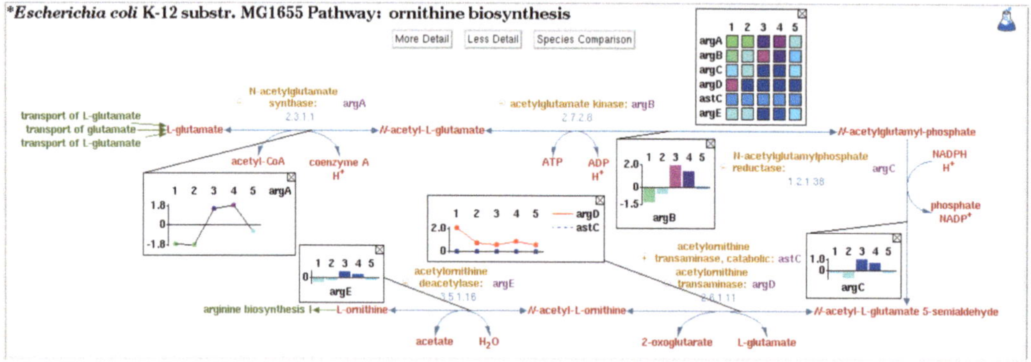

Fig. 5. A pathway display showing the three different styles of omics popups.

This facility enables the user to see instantly which genes are active or inactive under some set of experimental conditions.

The Regulatory Overview Omics Viewer is very similar to the Cellular Overview Omics Viewer. The data file can utilize the names or DB identifiers of genes, proteins, reactions, EC numbers, or compounds in the first column. The data file supports having a mix of objects specified on separate lines (e.g., you can specify metabolomics data in one set of lines in the data file and microarray data in a different set of lines in the same file).

Omics Graphing

When an experimental dataset includes multiple individual experiments, such as in a time-series experiment, the Omics Viewer animation offers one view of the data. However, it can also be useful to visualize the data from all time points simultaneously. Thus, for a given object (such as a gene or a metabolite) or set of objects (such as all the genes or metabolites in a pathway), the software can generate a popup overlay depicting the omics data for all time points (Fig. 4). The data can be displayed either as a heat map, a bar graph, or a plot. Any number of pop-ups can be displayed simultaneously, and the user can drag them with the mouse to reposition them as desired.

Figure 2 illustrates some of the relationships between different classes of objects. More detailed information about classes and their slots can be found at (22–25) or in the Pathway Tools User's Guide.

Acknowledgements

The projects described were supported by award numbers GM077678, GM080746, and GM75742 from the National Institute of General Medical Sciences, respectively. The content is solely the responsibility of the authors and does not necessarily represent the official views of the National Institute of General Medical Sciences or the National Institutes of Health.

References

1. Karp PD, Paley SM, Krummenacker M, Latendresse M, Dale JM, Lee T, Kaipa P, Gilham F, Spaulding A, Popescu L, Altman T, Paulsen I, Keseler IM, Caspi R (2010) Pathway Tools version 13.0: Integrated software for pathway/genome informatics and systems biology. Briefings in Bioinformatics 11:40–79. doi: 10.1093/bib/bbp043
2. Caspi R, Altman T, Dale JM, Dreher K, Fulcher CA, Gilham F, Kaipa P, Karthikeyan AS, Kothari A, Krummenacker M, Latendresse M, Mueller LA, Paley S, Popescu L, Pujar A, Shearer AG, Zhang P, Karp PD (2010) The MetaCyc database of metabolic pathways and enzymes and the BioCyc collection of pathway/genome databases. Nuc Acids Res 38: D473–9. advanced access doi: 10.1093/nar/gkp875
3. Pathway Tools File Formats. http://bioinformatics.ai.sri.com/ptools/flatfile-format.html
4. BioCyc Downloads. http://biocyc.org/download.shtml
5. Pathway Tools PGDB Registry. http://BioCyc.org/registry.html
6. Demir E et al (2010) The BioPAX community standard for pathway data sharing. Nat Biotechnol 28(12):935–942
7. Pathway Tools Data-File Formats. http://bioinformatics.ai.sri.com/ptools/flatfile-format.html
8. Pathway Tools Web Services. http://biocyc.org/PToolsWebsiteHowto.shtml\#webServices
9. Guide to ptools-xml. http://biocyc.org/ptools-xml-guide.shtml
10. The BioVelo Query Language. http://biocyc.org/bioveloLanguage.html
11. Querying Pathway/Genome Databases. http://brg.ai.sri.com/ptools/ptools-resources.html
12. Karp PD, Myers K, Gruber T (1995) The generic frame protocol. In: Proceedings of the 1995 international joint conference on artificial intelligence, pp 768–774
13. Seibel P (2005) Practical common lisp. APress, USA
14. Graham P (1995) ANSI common lisp. Prentice Hall, New Jersey
15. Example Pathway Tools queries in Common Lisp. http://bioinformatics.ai.sri.com/ptools/examples.lisp
16. PerlCyc API to Pathway Tools. http://solgenomics.net/downloads/perlcyc.pl

17. JavaCyc API to Pathway Tools. http://solgenomics.net/downloads/index.pl
18. Karp PD, Chaudhri VK, Paley SM (1999) A collaborative environment for authoring large knowledge bases. J Intell Inform Syst 13:155–94
19. Grossmann S, Bauer S, Robinson PN, Vingron M (2007) Improved detection of overrepresentation of gene-ontology annotations with parent child analysis. Bioinformatics 23 (22):3024–3031
20. Khatri P, Draghici S, Ostermeier GC, Krawetz SA (2002) Profiling gene expression using onto-express. Genomics 79(2):266–270
21. Rivals I, Personnaz L, Taing L, Potier M-C (2007) Enrichment or depletion of a GO category within a class of genes: which test? Bioinformatics 23:401–407
22. Guide to the Pathway Tools Schema. http://biocyc.org/schema.shtml
23. Karp PD (2000) An ontology for biological function based on molecular interactions. Bioinformatics 16(3):269–285
24. Karp PD, Paley SM (1994) Representations of metabolic knowledge: pathways. In: Altman R, Brutlag D, Karp P, Lathrop R, Searls D (eds) Proceedings of second international conference on intelligent systems for molecular biology. AAAI Press, Menlo Park, CA, pp 203–211
25. Karp PD, Riley M (1993) Representations of metabolic knowledge. In: Hunter L, Searls D, Shavlik J (eds) Proceedings of first international conference on intelligent systems for molecular biology. AAAI Press, Menlo Park, CA, pp 207–215

Chapter 13

Gene Set/Pathway Enrichment Analysis

Jui-Hung Hung

Abstract

Thanks for the dramatic reduction of the costs of high-throughput techniques in modern biotechnology, searching for differentially expressed genes is already a common procedure in identifying biomarkers or signatures of phenotypic states such as diseases or compound treatments. However, in most of the cases, especially in complex diseases, even given a list of biomarkers, the underlying biological mechanisms are still obscure to us. In other words, rather than knowing what genes are involved, we are more interested in discovering the common, collective roles of all these genes. Based on the assumption that genes involved in the same biological processes, functions, or localizations present correlated behaviors in terms of expression levels, signal intensities, allele occurrences, and so on, we can therefore apply statistical tests to find perturbed pathways. Gene Set/Pathway enrichment analysis is one of such techniques; a step-by-step instruction is described in this chapter.

Key words: Gene set enrichment analysis, Pathway enrichment analysis, GO term analysis, Overrepresentation analysis, Biomarker

1. Introduction

Gene Set/Pathway enrichment analysis can identify statistically significant gene sets that represent functions, mechanisms, processes, etc. from genes that are either differentially expressed (by microarray probing or RNA-seq techniques) or having strong binding signals of a transcription factor (by ChIP techniques) or of any collection that we believe to share properties. To demonstrate the idea of a gene set enrichment analysis (GSEA), we begin with going to Gene Expression Omnibus (GEO) (1) and accessing microarray experiment GDS3716 (2) (see Fig. 1) as our example. In these experiments, there are 42 human breast epithelium microarrays of either normal or cancerous samples. In Fig. 1a, b, we see some of the putative oncogenes, KPNA3, VPS4B, and others, exhibiting

Hiroshi Mamitsuka et al. (eds.), *Data Mining for Systems Biology: Methods and Protocols*, Methods in Molecular Biology, vol. 939, DOI 10.1007/978-1-62703-107-3_13, © Springer Science+Business Media New York 2013

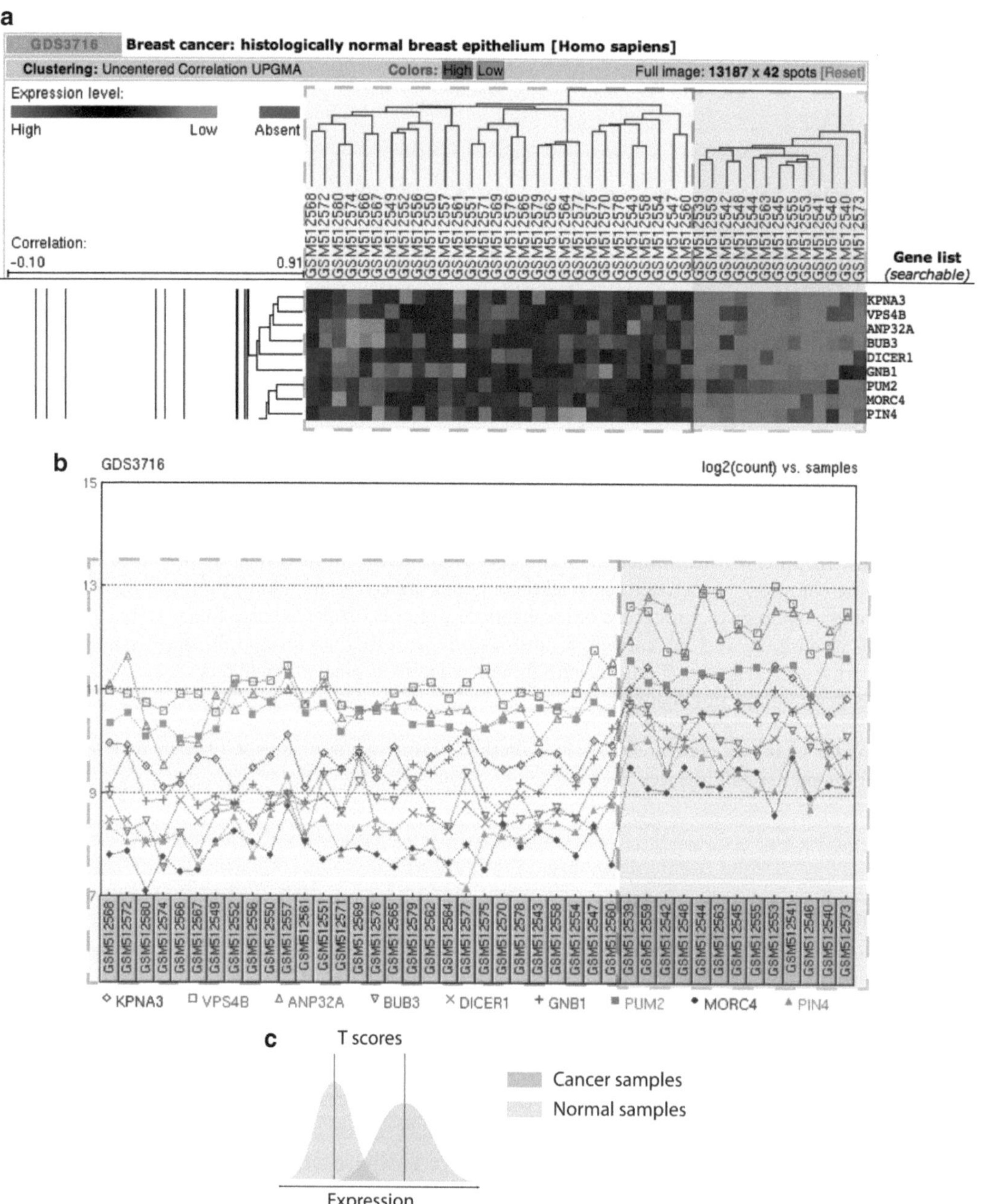

Fig. 1. (**a**) A snapshot of the heatmap of the expression profiles of GEO GDS3716. Heatmap encompassed by the dash line rectangle on the left represents expression of normal samples, and the dash line rectangle on the right for cancer samples. (**b**) Using line plot to present the same data. It is obvious that the expression levels of these genes are higher in cancer than in normal samples. (**c**) Based on the Student *T* test, the difference of the mean expression levels between two groups (i.e., normal and cancer samples) should follow a *T* distribution, and a *T* statistic and corresponding *p*-value can be estimated.

a

Entrez Gene ID (filtered)
853271
853339
853900
856519
856614

b

Entrez Gene ID (complete)	Quan. Anno. (T test score)
850359	0.016
855037	0.010
853271	3.064
853339	15.024
...	...
853867	0.187
853900	4.521
856519	9.802
856614	3.056

c

Gene ID	Normal 1	Normal 2	Normal 3	Normal 4	Normal 5	Cancer 1	Cancer 2	Cancer 3	Cancer 4	Cancer 5	T score	P-value
850359	452	679	461	746	304	241	247	229	312	302	1.660	0.128
855037	609	657	410	571	355	286	165	244	268	223	1.20	0.26
853271	1522	1350	1307	1473	1020	572	377	808	559	530	4.493	0.001
...	...	...	...	...	...	...	...	...	...	...	...	...
853867	813	877	1391	1336	766	458	645	513	639	586	3.008	0.013
853900	9716	10840	14542	15935	7817	4143	5831	6134	5887	5549	3.583	0.005
856614	5557	4480	9779	11528	3385	928	1735	1522	2837	1193	3.249	0.009

Fig. 2. Typical gene lists. (**a**) Gene identities of interests only. Filtered list with only genes that have *T* test scores >3 marked in *grey*. (**b**) All genomic genes with the quantitative annotations without filtering. (**c**) Like **b**, however accompanied by the raw data.

higher expression level in cancer samples than normal. Systematically examining each gene with a statistical test (Fig. 1c) can give each gene a score that represents the level of differential expression. Applying a stringent cutoff on the statistical scores leaves a list of differentially expressed (DE) genes.

A list of genes of interested and a label-to-gene mapping file (see Subheading 2.3) is the minimum requirement for performing a Gene Set/Pathway enrichment analysis. The typical types of gene lists are shown in Fig. 2. The simplest gene list (type 1) is just a list of genes (Fig. 1a), which specifies only the identifications of genes of interests, for example, differentially expressed genes with Student's *T* test *p*-values <0.01.

Type 2 gene list is a complete list of all possible candidates (e.g., all known genes), with a quantification field that indicates the magnitude (e.g., the level of DE, or the Student's *T* test statistic, see Fig. 1b). Keeping a complete gene list is required for the gene-shuffling background (GSB) simulation of the corresponding enrichment analysis (see below).

Type 3 (Fig. 1c) is like Type 2, but requires the original experimental raw data as the input as well (e.g., the expression profile of each gene across all samples). Type 3 input is used for phenotype-shuffling background (PSB) simulation for statistical inference making (see below). In general, type 2 (GSB) and type 3 (PSB) simulation helps increase the sensitivity and specificity of the enrichment analyses than that of type 1.

Respective to three types of gene lists, there are basically two types of Gene Set/Pathway enrichment analysis that can be performed: Overrepresentation analysis (ORA; for Type 1 input, see Fig. 1a) and Quantitative Enrichment Analysis (QEA; for Type 2 and 3 input, see Fig. 1b, c).

1.1. Over Representation Analysis

The concept of ORA (3) stems from the assumption that if, from our gene list, genes annotated to a known pathway significantly outnumber the expected amount of background genes (i.e., from a collection of randomly genome-wide selected, equal-sized gene lists) annotated to the same pathway, then it alludes that our gene list is overrepresented with genes in the pathway. It is a typical approach for finding enriched Gene Ontology (GO) terms (4) of a gene list without additional information given, and sometimes known as GO term enrichment analysis. A limitation of this approach is at the qualitative input of the collection of genes, which disregards magnitude information; therefore all genes in the collection are assumed to contribute equally to the identification of a correlated gene set.

1.2. Quantitative Enrichment Analysis

Since ORA assumes that every gene is independent and equally weighted, which is not concordant with biological facts, this assumption is a target for QEA to improve. QEA hypothesizes that genes with higher magnitude give higher contribution than genes with lower magnitude, and weighs each gene according to its magnitude; by doing so it increases the chances of finding gene sets with fewer but higher weighted genes (see Note 1).

2. Materials

First, collect a list of genes that we want to perform Gene set/ Pathway enrichment on, for example: a list of genes that have a specific transcription factor binding at their promoter region by ChiP (5) experiment with or without the intensities of the binding profiles; a list of genes that possess risk alleles of a heritable disease by genome-wide association study (GWAS) (6) with or without the relative risk levels; and a list of genes that differentially expressed between normal and cancer samples with or without the level of their correlations to the phenotypes.

KEGG Pathway	Entrez Gene ID							
KEGG_GLYCOLYSIS_GLUCONEOGENESIS	55902	2645	5232	5230	5160	5161	55276	7167
KEGG_CITRATE_CYCLE_TCA_CYCLE	3420	5106	1743	5162	5105	5160	5161	642502
KEGG_PENTOSE_PHOSPHATE_PATHWAY	6120	22934	55276	5634	8789	5213	5211	6888
KEGG_PENTOSE_AND_GLUCURONATE_INTERCONVERSIONS	54575	6120	54576	54577	54490	54579	51084	7358
KEGG_FRUCTOSE_AND_MANNOSE_METABOLISM	4351	5373	5372	8789	5210	2762	5211	9107
KEGG_GALACTOSE_METABOLISM	2645	2584	2720	2582	55276	5213	3906	5211
KEGG_ASCORBATE_AND_ALDARATE_METABOLISM	54575	54576	54577	54578	217	54490	55586	54579
KEGG_FATTY_ACID_METABOLISM	1374	126129	219	35	217	33	1579	34
KEGG_STEROID_BIOSYNTHESIS	6646	4047	6713	10682	1717	1594	1718	51478

Fig. 3. Label-to-gene mapping file. KEGG pathway to Entrez Gene ID from Molecular Signatures Database 3.0.

2.1. Collecting a List of Genes with or Without Quantitative Annotation

As mentioned previously, ORA needs only gene IDs as input (i.e., the type 1 gene list). In the field of biotechnology, this list is usually based on a series of experiments and in practice it usually requires a cutoff to retain only the most plausible ones. For example, setting a cutoff of binding profiles of 1,000 reads/per 1,000 base pair to retain only genes with higher ChIP signals than the signal magnitudes is no longer needed for the analysis. In contrast, in QEA, all the genes that are being tested in the ChiP experiment should be included and are free from any prerequisite cutoff and the quantitative annotations of all genome-wide genes should be kept (i.e., the Type 2 or 3 gene lists). Although type 2 and type 3 are similar, they are different in the shuffling simulation for generating the statistics backgrounds, and require different quantitative annotations (see Fig. 1b, c).

In addition, it is important to know that each gene is required to have a universal identification. The ambiguity of gene naming systems is unexpectedly high. It is recommended to prepare the gene list using Entrez Gene ID (7) to assure correct annotations that is going to be made in Subheading 2.3.

2.2. Preparing a Label-to-Gene Mapping File

We need to have a catalogue that annotates each gene with labels indicating gene functions, biological properties, or any attribute of our interests. There are several widely used public databases that provide annotations such as Gene Ontology (4), KEGG (8), Biocarta (9), and Molecular Signatures Database (10). A category annotation file should have each gene well documented and tabulated (see Fig. 3). In ORA, it is also required to know how many genes in the genome are assigned to each label for estimating statistical inference (see Subheading 3.1).

2.3. Topology or Interaction Information (Optional)

If the annotations are from biological pathways, network modules, or topological atlas that consist of interactions between genes, for example, KEGG pathways, this topology/interaction information (see Fig. 4) can be also fed into some QEA approaches that consider correlation of neighboring genes to increase sensitivity and specificity of the enriched pathway discovery (see Subheading 3.3.1).

a

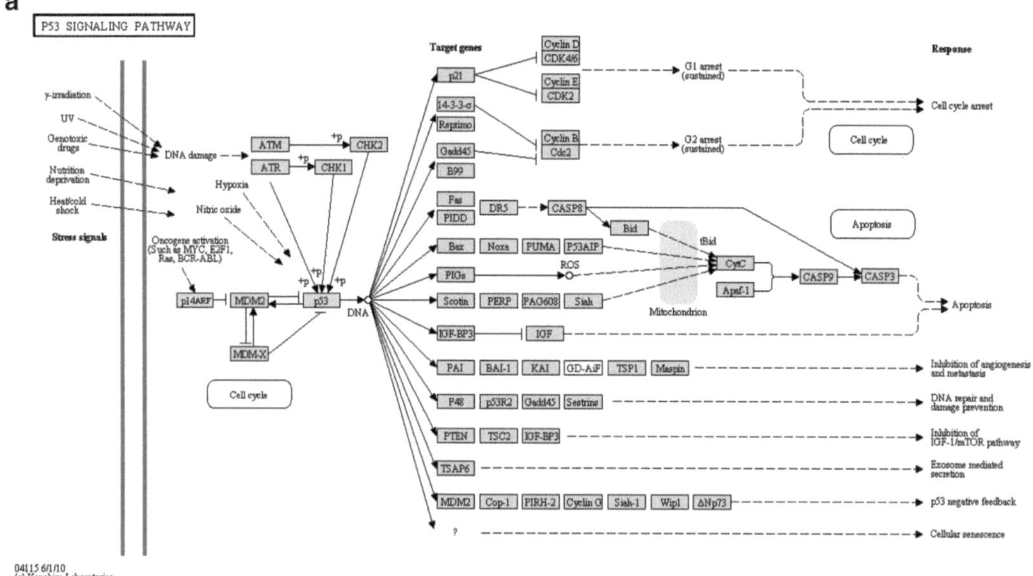

b

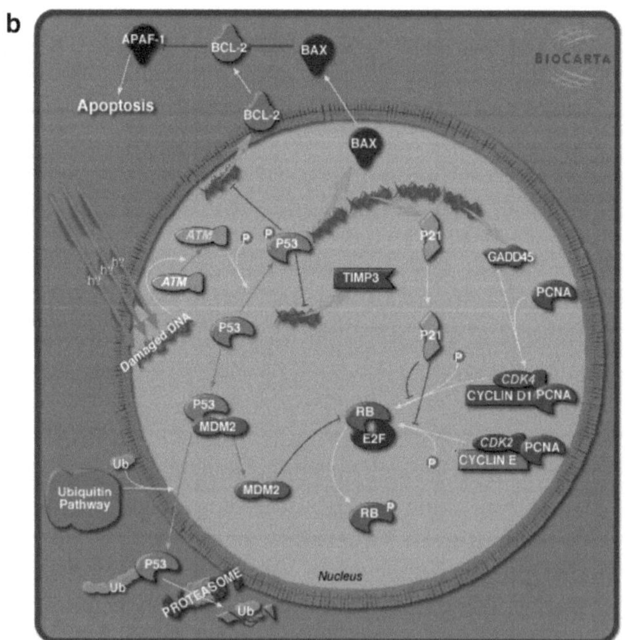

Fig. 4. Topological information of p53 signaling pathway (**a**) From KEGG. (**b**) From Biocarta.

3. Methods

3.1. ORA Approach

A typical ORA (3) starts from tallying the number of genes that are annotated with the predefined gene sets, then uses the Fisher exact tests to verify if the annotations are over represented among input gene sets compared to whole genome (multiple test correction is

	In Gene list	Not in Gene list	total
Annotated with label	*k*	*m-k*	*m*
Not annotated with label	*n-k*	*N+k-n-m*	*N-m*
total	*n*	*N-n*	*N*

Fig. 5. A 2 × 2 contingency table.

needed if more than one gene sets are tested, see below), and then reports gene sets that pass a significance threshold.

Steps:

1. Collect the type 1 gene list and the label-to-gene mapping table.
2. Select a label (e.g., a pathway) to start with.
3. Tally the following 4 numbers: m, N, k, and n, where m is the total number of genes genome-widely annotated with this label, N is the total number of genome-wide genes, k is the number of genes annotated with this label in the gene list, and n is the total number of genes in the list. Constructing a 2-by-2 contingency table as in Fig. 5 might be helpful.
4. Perform a Fisher exact test with the 4 numbers gotten in step 3 as follows:

$$f(k; N, m, n) = \frac{\binom{m}{k}\binom{N-m}{n-k}}{\binom{N}{n}}.$$

The f value is the probability that this random event could happen under the hypergeometric distribution. To get the p-value, sum up all the probability of more extreme cases, as follows:

$$p = \sum_{l=k}^{n} f(l; N, m, n).$$

5. Go to step 2 for another label of interest until all labels are tested.
6. Correct the p-value for multiple testing according to the False Discovery Rate (FDR, see Note 3) correction (11, 12) in the following three steps: (1) All regions are ranked by their p-values in the increasing order; (2) proceeding down the rank, the q-value is computed as $k \times p_k/M$, where k is the rank, p_k is the p-value of the region at rank k, and M is the total number of regions in the genome; (3) only labels with q-values less than a prespecified FDR are retained (e.g., FDR $= 0.01$, or 1% of predicted labels are false positives).

3.2. QEA Approach

One of the most widely used QEA methods, GSEA (10), ranks genes according to their differential expression and then uses a modified Kolmogorov–Smirnov statistic (weighted K-S test, or in short, WKS) as a basis for determining whether genes from a prespecified set (e.g., KEGG pathways or GO terms) are overrepresented among the top or bottom of the list, correcting for false discovery when multiple sets are tested.

There are assorted QEA approaches, and are considered to be a more powerful type of enrichment analysis than ORA. However, QEA requires more input and raises more concerns, and unlike ORA, QEA is still a vivid, unsettled topic in bioinformatics, and one of the major differences between QEA variants is how the background distributions are generated. They can be grouped into three types: Phenotype/sample Shuffling Background (PSB), Gene name Shuffling Background (GSB), and analytical Background (AB). Each of them is applicable to specific experiment setups, and its limitation, feasibility, and assumption are crucial for interpretation of the result.

3.2.1. PSB

This type of background is used for statistics-based gene list (see Subheading 2.1) according to the comparison between phenotypes of multiple samples such as microarray experiments, GWAS, and so on. To make the steps easier to follow, we take microarray expression experiments as our example. The following steps are excerpted from the PathWay Enrichment Analysis (PWEA) (13), which takes advantage of topology information to increase sensitivity and specificity of the enriched pathway found. Another good thing of using PWEA is that when topology information is missing, PWEA can still work as GSEA, one of the most widely used QEA.

1. Prepare original materials (i.e., expression profiles) and the label-to-gene mapping file. Organize the normalized expression levels into an expression matrix with the corresponding phenotypes like in Fig. 2c.
2. Calculate the gene-level statistic (e.g., the Student's *T* statistic) for each gene based on the phenotypes (e.g., normal or cancer). Note that the gene-level statistics can be any kind of test that is suitable for the application and the raw data. If possible, gene-level statistic can be further transformed (see Note 5).
3. Select a label (e.g., a pathway) to start with.
4. (Optional, can be skipped if topological information is missing) For a topological pathway K, compute a topological impact factor (TIF) score for each gene in P_K. TIF is defined as the average of the *mutual influence*, Ψ, with all other reachable genes in the pathway. Ψ_{ij} is used to evaluate the influence between ith gene and the jth gene in P_K, according to both

the absolute value of the correlation of their expression patterns and their topological distances. Ψ_{ij} is defined as

$$\Psi_{ij} = e^{-f_{ij}},$$

where $f_{ij} = d_{ij}/|c_{ij}|$, d_{ij} is the topological shortest distance between gene i and gene j in the pathway calculated using the *Floyd–Warshall algorithm* (with $d_{ii} = 0$), and c_{ij} is the *Pearson correlation coefficient* between gene i and gene j based on their expression profiles over both normal and diseased tissues. The TIF for a gene i is defined by the geometric mean of all influence functions Ψ_{ij} in a given pathway that involve gene i and satisfy $\Psi_{ij} > a$:

$$\text{Tif}_i = \exp\left(-\frac{1}{N} \sum_{\substack{j=1 \\ j \neq i}}^{n} f_{ij} \Theta(f_{ij} \ln \alpha) \right),$$

where

$$\Theta(f_{ij} + \ln \alpha) = \begin{cases} 1 & f_{ij} \leq -\ln \alpha \\ 0 & f_{ij} > -\ln \alpha \end{cases}$$

and

$$N = \sum_{\substack{j=1 \\ j \neq i}}^{n} \Theta(fi_j + \ln \alpha).$$

The significance threshold, α, is used to control the contribution that gene j makes to TIF_i. Note that shorter distances make an exponentially greater contribution to the mutual influence (and TIF) than do longer distances. The parameter α (suggested to be 0.05) is used to control the sensitivity and selectivity of the TIF.

5. Calculate the gene set statistics according to the WKS test (note that there are many other substitutions than WKS test, such as median, mean, chi-square test, and so on). First, rank all genes by their absolute *t score* (denoted as r) by t-test with the weights of TIF scores (if step 4 is skipped, take all TIF $= 0$), which is $r^{1+\text{TIF}}$. The t-test is performed on each gene to compare the expression levels between normal and disease samples. The cumulative distribution functions (CDFs) of P_k and Not P_k at position i in the rank can be written as

$\text{CDF}_{P_k}(i) = \sum_{j \leq i} \frac{r_j^{1+\text{TIF}}}{N_k}$, where $N_k = \sum_j r_j^{1+\text{TIF}}$ and j is the index of all genes belonging to P_k.

$\mathrm{CDF}_{\mathrm{Not}P_k}(i) = \sum_{j \le i} \frac{1}{N_{\mathrm{Not}k}}$, where $N_{\mathrm{Not}\ k}$ is the number of genes belonging to Not P_k.

6. Generate the background distribution of the gene set statistics by shuffling the phenotypes. The statistical significance for rejection of the null hypothesis is determined by comparing the maximum deviation (MD) of two CDFs to other n times of phenotype shuffles. Each randomly generated gene set for which its maximum deviation is higher than the original data will be counted and after n iterations, the *p-value* is computed. n is set at a large number, usually above 1,000.
7. Go back to step 3 until all pathways are tested.
8. Multiple test correction as described in Subheading 3.3.1, step 6, in ORA (see Note 4).

3.2.2. GSB

This type of background is like ORA and no original material is required; however, instead of knowing the number of genes annotated to each specific function label in ORA, GSB needs the entire list of genes with the quantitative annotations (see Note 2). Please note that if the data is applicable to both PSB and GSB, then PSB is always the better choice (14).

1. Collect the genome-wide gene list with the quantitative annotations (type 2 input) for testing and the label-to-gene mapping file.
2. Select a label (e.g., a pathway) to start with.
3. Calculate the gene set statistics, such as the WKS test in Subheading 3.3.1, step 5, or simply a mean test as follows:

   $M = \frac{1}{|P|} \sum_{i \in P, i=1}^{|P|} p_i$, where P is the set of genes with the label and p_i is the quantitative annotation of gene i, representing the importance of gene i in the pathway.
4. Generate the background distribution of the gene set statistics by shuffling the gene name. The statistical significance for rejection of the null hypothesis is determined by comparing the M to other n times of gene name shuffles. Each randomly generated gene set for which its M is higher than the original data will be counted and after n iterations, the *p-value* is computed. n is set at a large number, usually above 1,000.
5. Go to step 2 for another label of interest until all labels are tested.
6. Correct the p-value for multiple testing according to the FDR approach. Specifically, $\mathrm{FDR} = \frac{p \times m}{k}$, where m is the total number of labels tested and k is the rank of the label with p-value p.
7. Multiple test correction as described in Subheading 3.3.1, step 6, in ORA.

3.2.3. AB

This type of background requires no simulation for background but goes with very strong assumptions, and is suggested to be used only when PBS and GBS are not applicable. AB is only suggested in the case where quantitative annotations are from T tests of microarray experiments. For other case, users might need to check if the genome-wide quantitative annotations fit the standard normal distribution for using the analytical background shown below (see Note 6).

1. Collect the genome-wide gene list with the quantitative annotations (Type 2 input) for testing and the label-to-gene mapping file.
2. Select a label (e.g., a pathway) to start with.
3. Calculate the gene set statistics; only several types of statistics can be applied to using analytical backgrounds for estimating significance. One of them is the z test:

$$z = \sum_{i \in P, i=1}^{|P|} p_i,$$

   where p_i can only be the untransformed T test statistic from microarray analysis.
4. By assuming that mean statistic follows standard normal distribution, we can directly get the p-value from the following equation or the Z distribution table:

   p - value $= 1 - \frac{1}{2}\left[\text{erf}\left(\frac{z}{\sqrt{2}}\right) + 1\right]$, where erf is the Gauss error function.

   Note that this is a very strong assumption and is sometimes error-prone; please see Note 6 for more discussion.
5. Go to step 2 for another label of interest until all labels are tested.
6. Multiple test correction as described in Subheading 3.3.1, step 6, in ORA.

4. Notes

1. The key difference between ORA and QEA is that GSB needs weight of each gene and ORA assumes that all genes contribute equally. However, both of them assume no correlation between genes by declaring that every gene is independent to each other, which increases the chance of reporting false positive. Using ORA and QEA does not always give similar results. It is getting more and more recognized that QEA with correct

background simulation gives more reliable results than ORA. However, if quantitative annotations are not available, ORA is the only choice; otherwise, use QEA. We discuss both methods below.

2. FDR correction is required when multiple tests are carried out. An event can be rare if you sample only a few time; however, once you sample frequent enough, it is almost guaranteed that a rare case will appear. In other words, *P*-value for a single test becomes more and more inaccurate when the number of test cases increases.
3. In PSB, usually a gene-level statistic (Student's *T* statistic) will be a signed (±) number and the sign indicates overexpression or depletion of a gene in one of the phenotypes. In many gene set statistic, such as WKS test, the sign values can be cancelled out and reduced the chance of finding enriched pathway having half of genes up-regulated and half of genes down-regulated; taking absolute value (16) of the gene-level statistic can increase the sensitivity of finding such pathway.
4. Actually, when a shuffling simulation is generated, one can do FDR correction directly from the collection of all gene set statistic either from observation or simulation, which is taken as the empirical distribution of gene set statistic (15). For example, among *all* the tests, we want to estimate the FDR of a gene set statistics X, as we know that there are totally 50 shuffled cases having gene set statistics bigger than X from the real observation, and 10 other statistics bigger than *X* from the simulation, and then the FDR of *X* is $100\% \times 10/50 = 20\%$. However, it is difficult to assure that the distribution of gene set statistics is correctly generated for calculating FDR, since the distribution of *X* from different gene set can be different (determined by the size of a gene set or other properties), and is not suitable for a universal use of FDR estimation. Therefore, to simplify the complexity of it, we used the Benjamini–Hochberg procedure all throughout our steps in this chapter.
5. By shuffling phenotypes (PSB), rather than gene name (GSB), the correlation between genes is preserved in the simulation; therefore, PSB gives inference based on a more realistic background distribution and should be used whenever applicable.
6. AB is based on the observation that the Student T statistic distributes as a standard normal distribution in many microarray experiments. By assuming this is true for our data, it becomes very easy to get the significance as described in Subheading 3.3.3, steps 3 and 4. Although convenient, however, it has been suggested to use AB with cautions since it causes over-significant *p*-values and unexpected false positive (17).

Acknowledgments

I would like to thank everyone who supported me during the production of this chapter. Special thanks to the editor and proofreader. A final thank you is due to Charles DeLisi, who never gets tired of my stupidity.

References

1. Barrett T, Troup DB et al (2009) NCBI GEO: archive for high-throughput functional genomic data. Nucleic Acids Res 37(Database issue):D885–D890
2. Graham K, de las Morenas A et al (2010) Gene expression in histologically normal epithelium from breast cancer patients and from cancer-free prophylactic mastectomy patients shares a similar profile. Br J Cancer 102(8):1284–1293
3. Cho RJ, Huang M et al (2001) Transcriptional regulation and function during the human cell cycle. Nat Genet 27(1):48–54
4. Ashburner M, Ball CA et al (2000) Gene ontology: tool for the unification of biology. The Gene Ontology Consortium. Nat Genet 25(1):25–29
5. Collas P (2010) The current state of chromatin immunoprecipitation. Mol Biotechnol 45 (1):87–100
6. Manolio TA (2010) Genomewide association studies and assessment of the risk of disease. N Engl J Med 363(2):166–176
7. Maglott D, Ostell J et al (2007) Entrez gene: gene-centered information at NCBI. Nucleic Acids Res 35(Database issue):D26–D31
8. Kanehisa M, Goto S et al (2010) KEGG for representation and analysis of molecular networks involving diseases and drugs. Nucleic Acids Res 38(Database issue):D355–D360
9. Charting pathways of life. http://www.biocarta.com.
10. Subramanian A, Tamayo P et al (2005) Gene set enrichment analysis: a knowledge-based approach for interpreting genome-wide expression profiles. Proc Natl Acad Sci USA 102(43): 15545–15550
11. Draghici S, Khatri P et al (2003) Global functional profiling of gene expression. Genomics 81(2):98–104
12. Benjamini Y, Drai D et al (2001) Controlling the false discovery rate in behavior genetics research. Behav Brain Res 125(1–2):279–284
13. Hung JH, Whitfield TW et al (2010) Identification of functional modules that correlate with phenotypic difference: the influence of network topology. Genome Biol 11(2):R23
14. Tian L, Greenberg SA et al (2005) Discovering statistically significant pathways in expression profiling studies. Proc Natl Acad Sci USA 102 (38):13544–13549
15. Irizarry RA, Yun Zhou CW, Speed TP. Gene set enrichment analysis made simple. Johns Hopkins University, Dept. of Biostatistics Working Papers, Working Paper 185; 2009.
16. Noble WS (2009) How does multiple testing correction work? Nat Biotechnol 27(12): 1135–1137
17. Saxena V, Orgill D et al (2006) Absolute enrichment: gene set enrichment analysis for homeostatic systems. Nucleic Acids Res 34 (22):e151

Chapter 14

Construction of Functional Linkage Gene Networks by Data Integration

Bolan Linghu, Eric A. Franzosa, and Yu Xia

Abstract

Networks of functional associations between genes have recently been successfully used for gene function and disease-related research. A typical approach for constructing such functional linkage gene networks (FLNs) is based on the integration of diverse high-throughput functional genomics datasets. Data integration is a nontrivial task due to the heterogeneous nature of the different data sources and their variable accuracy and completeness. The presence of correlations between data sources also adds another layer of complexity to the integration process. In this chapter we discuss an approach for constructing a human FLN from data integration and a subsequent application of the FLN to novel disease gene discovery. Similar approaches can be applied to nonhuman species and other discovery tasks.

Key words: Gene networks, Functional association, Data integration, Data heterogeneity, Disease gene prediction

1. Introduction

In the post-genomic era it remains a significant challenge to fully understand the cellular functions of genes, including how the dysfunction of one or more genes causes a disease. Genes or proteins cooperate with each other in specific functional modules for particular cellular tasks (1, 2). Such functional modules can be represented as discrete biological processes or pathways. Genes or proteins can have multiple types of functional association. For instance, certain proteins physically interact with each other to form a protein complex and function as a single unit (3); certain transcription factors regulate a group of target genes in order to coordinate particular biological processes, such as the cell cycle (4); certain genes with similar sequences encode members of a protein family such that different members can be used under different

Hiroshi Mamitsuka et al. (eds.), *Data Mining for Systems Biology: Methods and Protocols*, Methods in Molecular Biology, vol. 939, DOI 10.1007/978-1-62703-107-3_14, © Springer Science+Business Media New York 2013

cellular conditions (5). On the other hand, when one or more genes involved in a particular biological process becomes dysfunctional, the normal status of that biological process might be perturbed, leading the organism to show abnormal physiological phenotypes referred to as a disease (6). Therefore, for gene function and disease-related research, it is very important to consider individual genes as functionally related components within a coherent biological system.

Recent network-based approaches have demonstrated great success in representing functional relationships among genes for the purpose of understanding gene function and human disease (7–19). In these networks, nodes represent genes, and edges represent functional associations between linked genes. These networks are referred to as functional linkage gene networks (FLNs). Typically, to construct such an FLN, multiple types of functional genomics data are desired since different data sources reveal different aspects of functional association between genes. For instance, high-throughput yeast two-hybrid (Y2H) data can identify the physical interactions between proteins, and microarray experiments can identify the transcription-level correlations between genes (e.g., co-expression). Data integration is a nontrivial task due to the heterogeneous nature of the different data sources and their variable accuracy and completeness. The presence of correlations between data sources also adds another layer of complexity to the integration process. To address these challenges, we present in this chapter an approach for constructing a human FLN from data integration and an application of the FLN to novel disease gene discovery. Similar approaches can be applied to nonhuman species and other discovery tasks.

2. Methods

2.1. Collecting Diverse Functional Genomics Data

The first step in constructing the integrated FLN is to collect the results of diverse functional genomics experiments. For instance, Y2H and mass spectrometry experiments can reveal protein–protein binary physical and co-complex interactions (20, 21); microarray experiments can reveal co-expression relationships between genes at the transcriptional level (22–25); using sequence data, computational approaches such as the phylogenetic profile method can predict gene pairs with similar function based on their correlated occurrence patterns across a set of taxa (23). Since each data source usually reveals a specific dimension of functional association between genes, the challenge is to integrate different data sources in order to evaluate the functional associations between genes in a comprehensive way.

Collecting individual data sources can be a tedious task given that different data sources are often found in diverse locations (see Note 1). Additionally, due to the unique properties of each data source, a source-specific scoring scheme is often required. Generally, such scores are either binary or continuous. For instance, for scoring protein–protein interaction (PPI), a binary score of 1 or 0 can be used to denote the presence or absence of the interaction. For microarray expression data, the Pearson correlation coefficient between two genes' expression levels across a set of experimental conditions can serve as the score. The major data sources and their scoring schemes are summarized in Note 1.

The diverse gene ID systems used in different data sources also add complexity to the data integration process. A typical solution is to use common IDs to refer to genes across all data sources. In practice, different gene identifiers, transcript identifiers, and protein identifiers can be mapped onto their corresponding Entrez gene IDs (see Note 2). This also unifies the levels of functional associations inferred from different data sources to the gene level, as opposed to the protein level or transcript level.

The currently available functional genomics data for human is far from complete. To increase coverage, we can further integrate functional genomics data from other model organisms based on the assumption that gene–gene functional associations are conserved across species (11). Specifically, we can retrieve the functionally associated gene pairs in other model organisms—such as yeast, worm, fly, mouse, and rat—and then map them to human through gene orthology relationships (see Note 1).

Finally, once all individual data sources have been retrieved and scored, the next step is to assemble them as a feature matrix in preparation for data integration (Fig. 1). Each row of the feature matrix represents a gene pair and each column of the matrix represents a particular data source or feature. The value in each cell of the matrix represents the score measuring the functional association for the corresponding gene pair in terms of the corresponding data source. In order to include all the data sources in this matrix, the gene pair rows are obtained by taking the union of the gene pairs present in any individual data source. Since different data sources have different coverage, there exist some gene pairs that have scores in some data sources but not in others; i.e., there are missing values in some cells of the feature matrix. To deal with these missing values, a "missing" label can be assigned in place of the actual value.

2.2. Data Integration Using the Naïve Bayes Approach

After the feature matrix encoding the diverse data sources has been built, the next step is to perform data integration in order to build an integrated FLN. In the FLN, the nodes represent genes, and edges (links) represent the degree of functional association between linked genes after summarizing the evidence provided by the individual data sources. One major challenge at this step results from

Gene pairs	Protein-protein interaction	Expression correlation	Protein domain sharing	Correlated phylogenetic profile	Gene fusion	......
A - B						
A - C	Score each evidence and fill in the matrix					
...						

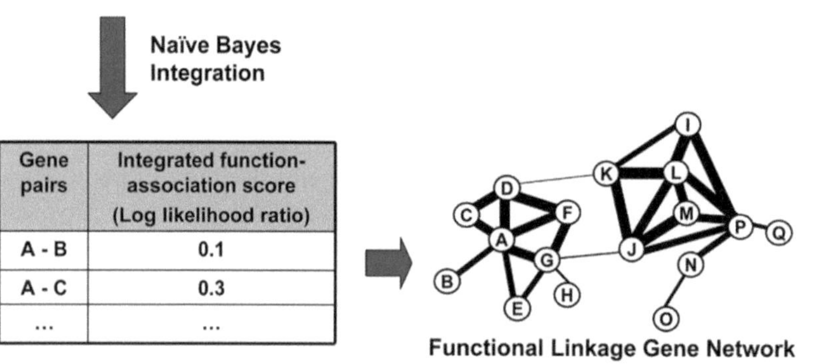

Fig. 1. Construction of the FLN by data integration. Diverse functional genomics data are collected and assembled into a feature matrix with each data source representing a particular feature. In the matrix, each row denotes a gene pair and each column denotes a feature. For each gene pair, each feature is scored individually based on its feature-specific scoring scheme (see Note 1) and the scores are filled in the matrix. A naïve Bayes classifier is then employed to sum up pieces of evidence from individual sources into an integrated functional association score, the total LLR. The gene pairs are then assembled into a weighted FLN with the genes as the nodes and the integrated LLR as the linkage weight for linked genes.

the heterogeneity of the diverse data sources. The different data sources usually vary in their reliability and coverage. As a result, data integration should not be a simple union or intersection of individual data sources. Instead, data integration must take into account the differences among individual sources in terms of their false positive and false negative error rates, as well as their predictive power for functional association. A rigorous solution to this problem requires methods based on (supervised) machine learning, such as Bayesian approaches, support vector machines (SVM), and logistic regression(7, 26–28).

The machine learning-based approaches typically first define a benchmark dataset (gold standard) in order to calibrate and unify the different data sources. To construct such a benchmark dataset, the notion of a true "functional association" between genes must first be defined in an intuitive and consistent manner. One typical way to define functional association is the sharing of Gene Ontology (GO) biological process (29). Specifically, this benchmark dataset is composed of both gold standard positives (GSP; gene pairs sharing the same biological process terms in GO) and gold standard negatives (GSN; gene pairs with both members annotated in GO but not sharing any GO biological process terms; see Note 3). A training set is then constructed based on this benchmark dataset, and each individual data source is calibrated according to its ability to predict the biological process-sharing relationship between genes; in other words, each data source is weighted such that stronger evidence (more predictive power) is given higher weight, and weaker evidence (less predictive power) is given lower weight. Data integration is then performed by summing the pieces of evidence from different data sources using the calibrated weights. Such an integration approach achieves optimal performance in terms of coverage and accuracy because (i) all data sources are taken into account and (ii) gene pairs supported by high-confidence pieces of evidence are given higher weights.

Below we have selected the naïve Bayes method to illustrate machine learning-based data integration. There are two primary reasons that we choose naïve Bayes as the integration approach. First, it allows for the direct integration of heterogeneous data sources in an easily interpretable model. Second, it calculates the probability that two genes have a functional association (share a GO biological process) given their input features. Naïve Bayes classifiers have been previously used to successfully combine heterogeneous data in human (7, 26, 30, 31). The input to the classifier is the feature matrix assembled from various data sources (see Note 1). For each gene pair, the naïve Bayes method generates a total log likelihood ratio (LLR) score denoting how much the current evidence encoded in the feature matrix supports the prediction that the gene pair participates in the same biological process. The associated gene pairs are then assembled into a functional linkage network with individual genes as nodes and the total LLR score as the linkage weight (Fig. 1).

In the Bayesian framework, the posterior log odds that a gene pair participates in the same biological process given the available evidence can be expressed as the sum of two terms:

$$\log\frac{P(I|f_1,\ldots,f_n)}{P(\sim I|f_1,\ldots,f_n)} = \log\frac{P(I)}{P(\sim I)} + \log\frac{P(f_1,\ldots,f_n\mid I)}{P(f_1,\ldots,f_n\mid \sim I)}. \quad (1)$$

The first term is the prior log odds for functional association; this term is the same for all gene pairs. The second term is the integrated LLR, which measures how much the predicted functional association for this gene pair is supported by the integrated evidence:

$$\mathrm{LLR}(f_1,\ldots,f_n) = \log\frac{P(f_1,\ldots,f_n\mid I)}{P(f_1,\ldots,f_n\mid \sim I)}, \quad (2)$$

where I is a binary variable representing the existence of a functional association (GSP pairs), $\sim I$ represents the absence of a functional association (GSN pairs), and f_1 through f_n are the genomic features.

In naïve Bayes, we assume that features are conditionally independent. As a result, $\mathrm{LLR}(f_1,\ldots,f_n)$ can be rewritten as:

$$\mathrm{LLR}\,(f_1,...,f_n) = \mathrm{LLR}(f_1) + \ldots + \mathrm{LLR}(f_n), \tag{3}$$

where LLR(f_i) represents the individual LLR score for feature i:

$$\mathrm{LLR}\,(f_i) = \log\frac{P(f_i|I)}{P(f_i|\sim I)}. \tag{4}$$

The integrated LLR score (left-hand side of Eq. 3) represents the functional association strength of a gene pair after summing up the evidence from individual data sources (individual LLR scores in Eq. 4). The LLR scores for the individual data sources are calculated by calibrating the corresponding raw scores represented in the input feature matrix against the common training dataset defined above. Specifically, all continuous features are converted into categorical features by binning the values into nonoverlapping intervals, with LLR scores calculated for each individual bin:

$$\mathrm{LLR}(f_i \in \mathrm{bin}_{ij}) = \log\frac{P(f_i \in \mathrm{bin}_{ij}|I)}{P(f_i \in \mathrm{bin}_{ij}|\ \sim I)}, \tag{5}$$

where I and ~I denote belonging to the GSP set and the GSN set, respectively; f_i denotes the value of feature i; and bin_{ij} denotes the j-th bin for feature i. Gene pairs with missing values are put into a separate bin for LLR calculation.

As a result of this procedure, the heterogeneous measurements from different data sources are normalized into the individual LLR scores in Eq. 4, which can then be combined to form the total LLR score representing the functional association strength.

Another major challenge for data integration comes from the correlations between data sources. This is especially important for the naïve Bayes method since naïve Bayes assumes conditional independence among features, although in practice the naïve Bayes classifier can still be applied even when this assumption is not strictly satisfied (26, 30, 31). To reduce instances in which this assumption is violated, we can combine related datasets or features into single features before final integration. Specifically, for correlated data sources, an individual LLR is calculated independently for each individual source, and then the maximum LLR among these individual LLRs is used as the final LLR for each gene pair (27, 30).

2.3. FLN Quality Assessment

After naïve Bayes integration, each gene pair is assigned an LLR score for sharing a GO biological process. The associated gene pairs are then assembled into a functional linkage network with individual genes as nodes and the LLR score as the linkage weight (Fig. 1). The quality of the FLN can be evaluated by comparing the performance of the integrated data versus the performance of individual data sources. Typically, performance is measured by plotting linkage precision versus linkage sensitivity at various linkage weight cutoffs using the gold standard dataset defined above. Given a prespecified linkage weight cutoff, "linkage precision" is defined as the fraction of the linked gold standard gene pairs that belong to the GSP set, and "linkage sensitivity" is defined as the fraction of the GSP pairs that are linked. Generally a K-fold cross-validation approach is used for the assessment. Specifically, the gold standard dataset is randomly partitioned into K equal segments; $K-1$ segments serve as the training set and the remaining segment serves as the test set. This procedure is repeated K times such that each segment serves as the test set exactly once. Next, all K test sets are combined and the performance is assessed by plotting linkage precision versus linkage sensitivity. In a successful data integration procedure, the integrated FLN should have higher linkage precision at the same linkage sensitivity level compared with each individual data source. Finally the whole gold standard dataset is used as a training set to generate one final FLN, which is then used in downstream analysis.

2.4. Use FLN for Novel Disease Gene Discovery

One major FLN-based application is novel disease gene discovery. The key assumption behind the utility of an FLN in disease research, which is supported by diverse empirical evidence, is that genes underlying the same or related diseases tend to be functionally related. Because an FLN represents functional associations among genes, genes associated with the same or related diseases are expected to be located close to each other ("in the same neighborhood") of the network (7, 9, 10, 32, 33). Based on this assumption, FLNs have been successfully used to predict new disease genes in recent studies. Given a particular disease of interest, the FLN-based approaches typically start with the identification of known disease genes in the network as "seeds," followed by exploration of the network neighborhoods of these seeds, and finally prioritization of new candidate disease genes based on how closely connected they are to the seeds (Fig. 2).

The seed disease genes can be obtained from the OMIM database, a compendium of human disease genes and phenotypes

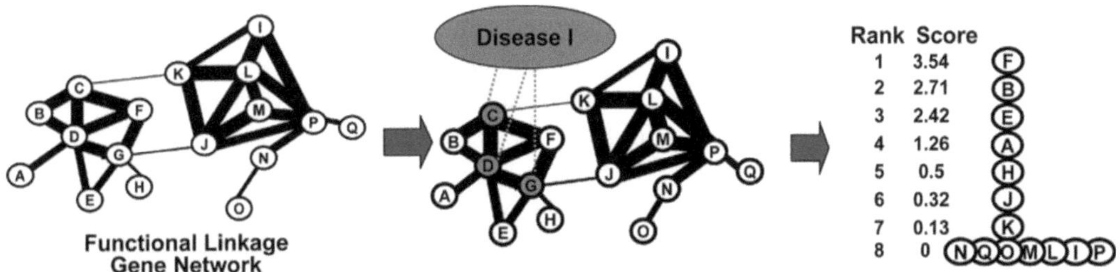

Fig. 2. Prediction of novel disease genes using the FLN. To predict novel genes associated with a particular disease, we first label known genes associated with the disease as seeds, and then rank order the neighboring genes based on their connectivity to seed genes as measured by the neighborhood weighting rule (Eq. 6).

(34) (see Note 4). To obtain reliable test statistics, diseases with at least five seed genes are included in the analysis. Once the seed genes are identified, an appropriate decision rule can then be employed to rank candidate genes based on the strength of their association with the seed genes.

The choice of the optimal decision rule is determined by the nature of the FLN. In our case, the FLN is weighted and the linkage weight is derived from extensive data integration. As a result, the FLN is very dense: for example, in our previous work, we constructed an FLN containing over 21,000 genes and over 22,000,000 edges, with each gene having 2,000 neighbors on average (7). Since links in this FLN are weighted, the high density of the network allows one to directly measure the strength of the functional associations between one gene and thousands of other genes. Taking these features into account, a simple yet effective "neighborhood weighting rule" can be used to rank candidate genes for a particular disease by focusing on the immediate neighborhood of the seed genes (7). In particular, candidate genes are rank-ordered by the sum of the weights of their functional links to the seed genes in the network. Given a particular disease and its seed gene set, we quantify the association of each candidate gene i with the disease using the following disease association score, S_i:

$$S_i = \sum_{j \in \text{seeds}} w_{ij}, \qquad (6)$$

where w_{ij} is the linkage weight connecting gene i and seed j (7, 11, 13). This score is 0 for genes that are not connected to any seed. This approach successfully ranked ~12,000 genes for each disease in our previous work (7). A general justification for the neighborhood weighting rule is given in Note 5.

2.5. FLN Visualization

FLNs can be conveniently visualized by software tools designed for exploring and analyzing biological networks. One such software tool is VisANT, a Web-based open-source platform for the visualization and analysis of different types of biomolecular networks (35–37). Users can upload and explore their own FLNs or the FLNs stored in the VisANT database, which includes integrated FLNs and FLNs derived from single types of data sources. VisANT enables users to interactively query genes of interest in an FLN, explore their network neighborhoods, and perform topological analyses or calculate network degree for selected nodes. Users can also filter an FLN with different linkage weight thresholds and visualize the weights by edge color or edge thickness. As described earlier, genes work in functional groups for specific cellular tasks, and genes underlying the same diseases tend to (i) be functionally associated and (ii) belong to the same functional groups. VisANT introduces meta-nodes—a special type of node that contains associated sub-nodes—to represent these functional groups, which include protein complexes, molecular pathways, and sets of genes associated with a given disease. As a result, the hierarchical structure of the FLN can be visualized: users see not only the functional associations between individual genes in a low-level map but also functional modules composed of gene groups in a high-level map. The disease gene prediction function is also available in VisANT, allowing the user to specify the seed disease genes in the FLN and to rank novel candidate disease genes in the neighborhood of the seed genes using the neighborhood weighting rule.

3. Notes

1. The major data sources for human FLN construction and their scoring schemes are listed as below.
   1.1. Curated PPIs. Curated human PPIs can be downloaded from HPRD, BIND, BioGRID, IntAct, MIPS, DIP, and MINT (38–44). To differentiate these data from high-throughput PPIs, large-scale Y2H data and mass spectrometry data are excluded. A binary score is used to denote the presence or absence of an interaction. To minimize redundancy, these seven datasets are combined as one feature as described in Subheading 2.2.
   1.2. High-throughput Y2H. Y2H data can be downloaded from Rual et al.'s datasets (21). A binary score is also used to denote the presence or absence of an interaction.
   1.3. Large-scale mass spectrometry. The mass spectrometry data can be downloaded from Ewing et al. (20). A binary score is used to denote whether two proteins occur in the same protein complex. The high-throughput Y2H experiment and the large-scale mass spectrometry are complementary techniques revealing different aspects of PPI, and so they are treated as two separate features.
   1.4. PPI inferred from domain–domain interaction. Physical PPIs involve interactions between protein domains. Therefore, new PPIs can be predicted by identifying pairs of protein domains enriched in known interacting protein pairs (30, 31). Two studies have identified enriched domain pairs (with confidence scores) by mining the HPRD database (30, 31). The domain pairs from both studies can be downloaded, and protein pairs containing these domain pairs can be retrieved using the InterPro database (45). Each retrieved protein pair is given the same confidence score as the corresponding

domain pair. Since both studies identified the interacting domain pairs by mining the HPRD database, the two resulting datasets are combined as one feature to minimize redundancy, as described in Subheading 2.2. Protein pairs already included in the curated PPI dataset are excluded here as their interactions are known.

1.5. Text mining. Text mining is also an important source for retrieving functionally associated gene pairs. It involves searching for co-occurrence of gene names in PubMed abstracts (23). Text mining data can be downloaded from the String database with each gene pair associated with a corresponding text mining score (23).

1.6. Correlated gene expression. Correlated gene expression can be inferred based on experiments from the GEO database (http://www.ncbi.nlm.nih.gov/geo/) (46). The Pearson correlation coefficient for each data set can be used as a measure of functional association. To reduce the effects of correlations between datasets, multiple datasets can be combined as one feature, as described in Subheading 2.2.

1.7. Phylogenetic profile. The presence and absence of a gene across a set of genomes can be represented by a binary string called a phylogenetic profile (23, 47, 48). Genes with sufficiently similar profiles tend to be functionally related (23, 47, 48). Gene pairs with correlated phylogenetic profiles and their associated correlation scores can be downloaded from the String database (23, 47, 48).

1.8. Gene neighbor. If two genes are found to be chromosomal neighbors in several different genomes, a functional linkage can be inferred between the proteins they encode (23, 47, 48). Gene pairs identified by the gene neighbor method and their associated interaction scores can be downloaded from the String database (23, 47, 48).

1.9. Gene fusion. Some protein pairs with related functions in one species are observed as a single fused protein in other species (23, 47, 48). Gene pairs involved in such fusion events and the associated confidence scores can be downloaded from the String database (23, 47, 48).

1.10. Functional associations inferred from model organisms (yeast, worm, fly, and mouse/rat). It has been suggested that gene–gene functional associations tend to be conserved across species (11). Such conserved functional associations are defined as "associalogs" (11). We can retrieve functionally associated gene pairs from yeast (49), worm (11), fly, mouse, and rat (23) and then map them onto their human associalogs using the

INPARANOID database (50). For each model organism, the integrated functional association score from species-specific data sources is already available (11, 23, 49), and can serve as the functional association score for the corresponding human associalogs. Associalogs from each model organism encode one single feature. We previously evaluated the correlation of associalogs mapped from pairs of model organisms. A strong correlation between mouse and rat was found, and so the associalogs from mouse and rat can be combined as one feature, as described in Subheading 2.2.

1.11. Sharing molecular function terms in Gene Ontology. Two genes sharing the same GO molecular function term (51) are more likely to belong to the same biological process (be functionally linked) (8). Furthermore, genes sharing more specific molecular function terms should be more likely to belong to the same biological process than genes sharing more general terms. We can download human molecular function annotations from Entrez Gene (ftp://ftp.ncbi.nih.gov/gene/DATA/gene2go.gz). The functional association between two genes with one or more shared GO terms is measured as the number of genes sharing the most specific of those GO terms (i.e., the GO term shared by the smallest number of genes) (30). A lower score corresponds to a higher degree of functional association.

1.12. Sharing cellular component GO terms. Cellular localization information can be described by GO cellular component terms (51). Genes sharing cellular localizations tend to be functionally associated. We can download human cellular component annotations from Entrez Gene (ftp://ftp.ncbi.nih.gov/gene/DATA/gene2go.gz). The scoring scheme is the same as that described in "Sharing molecular-function GO terms."

1.13. Protein domain sharing. Proteins containing the same protein domains tend to have similar functions (52, 53). To retrieve protein domain information, we use InterPro database, which is a database of protein families, domains, repeats, and sites (45). We refer to these protein families, domains, repeats, and sites as InterPro terms. As is the case for GO terms, there exist hierarchical organizations among these InterPro terms, and hence we can use the same scoring scheme described in "Sharing molecular-function GO terms."

2. Different gene, transcript, or protein identifiers from individual data sources are all mapped to Entrez gene IDs using the cross-reference files from Entrez Gene (ftp://ftp.ncbi.nih.gov/gene/DATA/gene_info.gz), the International Protein Index (IPI) database (ftp://ftp.ebi.ac.uk/pub/databases/IPI/current/ipi.genes.HUMAN.xrefs.gz), or Biomart (http://uswest.ensembl.org/biomart/index.html). The reason we provide three cross-reference sources is that they are not identical to each other and using the union of the three can increase coverage.

3. We define gene pairs sharing the same Gene Ontology (GO) biological process term as being functionally associated. Since GO terms are organized hierarchically according to different levels of specificity, we must choose a specificity level that constitutes the appropriate degree of functional association. We count the total number of genes annotated with a GO term as a measure of that term's specificity; low counts correspond to high-specificity GO terms and high counts correspond to low-specificity terms (54–56). To select GO terms with the appropriate level of specificity, we use the definition of an "informative GO term" given by Zhou et al. (54, 56). Specifically, informative GO terms need to satisfy two criteria: (i) at least 175 genes are annotated with the term, and (ii) none of its descendant terms are used to annotate over 175 genes. The files for mapping genes to GO terms can be downloaded from Entrez Gene (ftp://ftp.ncbi.nih.gov/gene/DATA/gene2go.gz), and we only consider annotations which are not predictions themselves, i.e., with non-IEA evidence codes (57). We define gene pairs sharing one or more informative biological process GO terms as the *gold standard positive* (*GSP*) set. Our *gold standard negative* (*GSN*) set is defined as the collection of gene pairs that are annotated with GO informative terms but that do not share common GO biological process terms except for the root term.

4. The seed disease genes are obtained from the Morbid Map (ftp://ftp.ncbi.nih.gov/repository/OMIM/morbidmap) in the Online Mendelian Inheritance in Man (OMIM) database, a compendium of human disease genes and phenotypes (34). To obtain highly reliable seeds, we select only entries with the "(3)" tag, which indicates "mapping of a wild-type gene combined with demonstration of a mutation in that gene in association with the disorder" (6, 34). Next, the subtypes of a single disease are merged into one unique disorder entry according to the merging files provided by Gol et al. (Supporting Information Table 1 in Goh et al.'s paper (6)). For example, "Alport syndrome, 301050 (3)" and "Alport syndrome, autosomal recessive, 203780 (3)" are merged as "Alport syndrome." Each disease is assigned a unique disease ID.

5. Here we provide a general justification for the neighborhood weighting rule for FLN-based ranking of disease candidate genes. Consider the following binary classification problem: for each human gene we have n feature variables $x = (x_1, x_2, ..., x_n)$ representing various properties of the gene and a class variable y (1 or -1) representing whether or not the gene is implicated in a particular disease. In addition, we have a training set $T = \{(x^{(i)}, y^{(i)}); i = 1, ..., m\}$ consisting of m genes known to be either involved or not involved in the disease. The goal is to build a linear classifier which integrates gene features to accurately predict if a gene is implicated in the disease:

$$\hat{y}(x) = b + w^T x \tag{7}$$

Although we focus on linear classifiers here, this argument is very general and can be applied to nonlinear classifiers as well. The performance of the classifier on the i-th gene in the training set can be characterized by the margin t_i:

$$t_i = y^{(i)}\hat{y}^{(i)} = y^{(i)}(b + w^T x^{(i)}) \tag{8}$$

The margin t_i is positive when the predicted class $\text{sgn}(\hat{y}^{(i)})$ agrees with the actual class $y^{(i)}$, and negative otherwise. In other words, a positive margin means a correct prediction, and a negative margin means an incorrect prediction.

Our task is to train the classifier, i.e., tune the weight vector w and intercept b so as to minimize the objective function (J), which includes a loss term measuring the total prediction error on the training set T, plus a second regularization term which penalizes complex models:

$$\min_{w,b} J(w, b) = \min_{w,b} \sum_{i=1}^{m} l(t_i) + \lambda \|w\|^2. \tag{9}$$

Here, $l(t)$ is a predefined loss function which converts the margin t into a loss for each gene in the training set. The first term on the right-hand side of Eq. 9 is the total loss over the entire training set. The second term is a commonly used L2 regularization term that penalizes large weights. λ is a nonnegative constant that quantifies the relative importance of the regularization and loss terms.

The weight vector w can be decomposed into two orthogonal component vectors in the following way:

$$w = w_1 + w_2, \quad w_1 = \sum_{j=1}^{m} \beta_j x^{(j)}, \quad w_2^T x^{(j)} = 0, \quad w_2^T w_1 = 0. \tag{10}$$

The loss term in Eq. 9 now becomes:

$$\sum_{i=1}^{m} l(t_i) = \sum_{i=1}^{m} l(y^{(i)}(b + w^T x^{(i)})) = \sum_{i=1}^{m} l(y^{(i)}(b + w_1^T x^{(i)})). \quad (11)$$

The L2 regularization term in Eq. 9 becomes:

$$\|w\|^2 = \|w_1\|^2 + \|w_2\|^2 \geq \|w_1\|^2. \quad (12)$$

So the following inequality is always true:

$$\sum_{i=1}^{m} l(y^{(i)}(b + w_1^T x^{(i)})) + \lambda\|w_1\|^2 \leq \sum_{i=1}^{m} l(y^{(i)}(b + w^T x^{(i)})) + \lambda\|w\|^2. \quad (13)$$

It follows that the optimal solution to Eq. 9 must have the following form:

$$w = w_1 = \sum_{j=1}^{m} \beta_j x^{(j)} = \sum_{j=1}^{m} \alpha_j y^{(j)} x^{(j)}. \quad (14)$$

Equation 9 is then equivalent to:

$$\min_{\alpha,b} J = \min_{\alpha,b} \sum_{i=1}^{m} l(by^{(i)} + \sum_{j=1}^{m} \alpha_j y^{(i)} y^{(j)} K(x^{(i)}, x^{(j)})) + \lambda \sum_{i=1}^{m}\sum_{j=1}^{m} \alpha_i \alpha_j y^{(i)} y^{(j)} K(x^{(i)}, x^{(j)}), \quad (15)$$

where $K(x, y) = x^T y$ (called the kernel) measures the "intrinsic similarity" between two genes x and y in terms of the dot product of their corresponding feature vectors.

To find the optimal solution to Eq. 15, we set partial derivatives of the objective function to be zero:

$$\frac{\partial J}{\partial b} = \sum_{i=1}^{m} l'(t_i) y^{(i)} = 0,$$

$$\frac{\partial J}{\partial \alpha_j} = \sum_{i=1}^{m} (l'(t_i) + 2\lambda\alpha_i) y^{(i)} y^{(j)} K(x^{(i)}, x^{(j)}) = 0. \quad (16)$$

Solving Eq. 16 leads to:

$$\alpha_i = -l'(t_i)/(2\lambda)$$
$$\sum_{i=1}^{m} \alpha_i y^{(i)} = 0 \quad (17)$$

The loss function $l(t)$ is generally a monotonically decreasing function of the margin t, i.e., the better the prediction, the smaller the loss. Thus: $\alpha_i = -l'(t_i)/(2\lambda) \geq 0$. As a result, α parameters are always nonnegative.

Equation 7 can now be rewritten as:

$$\hat{y}(x) = b + w^T x = b + \sum_{j=1}^{m} \alpha_j y^{(j)} x^{(j)^T}$$

$$x = b + \sum_{j=1}^{m} \alpha_j y^{(j)} K(x^{(j)}, x). \quad (18)$$

It is clear from Eq. 18 that the overall prediction for the new gene x is based on the "weighted majority vote" from all genes in the training set. The contribution of each gene $x^{(j)}$ in the training set to the overall prediction of the new gene x depends on two factors: $K(x^{(j)}, x)$, the "intrinsic similarity" between genes $x^{(j)}$ and x; and α_j, the "informativeness" of gene $x^{(j)}$ in the prediction task at hand. If the total weighted votes from the genes in the positive training set outweigh the total weighted votes from the genes in the negative training set, then the overall prediction will be positive; otherwise it will be negative. Our neighborhood weighting rule (Eq. 6) can be viewed as an approximate extension of this weighted majority vote rule for the complex task of integrated FLN-based disease gene prioritization.

References

1. Ravasz E, Somera AL, Mongru DA, Oltvai ZN, Barabasi AL (2002) Hierarchical organization of modularity in metabolic networks. Science 297:1551–1555
2. Hartwell LH, Hopfield JJ, Leibler S, Murray AW (1999) From molecular to modular cell biology. Nature 402:C47–C52
3. Stelzl U, Worm U, Lalowski M, Haenig C, Brembeck FH, Goehler H, Stroedicke M, Zenkner M, Schoenherr A, Koeppen S, Timm J, Mintzlaff S, Abraham C, Bock N, Kietzmann S, Goedde A, Toksoz E, Droege A, Krobitsch S, Korn B, Birchmeier W, Lehrach H, Wanker EE (2005) A human protein-protein interaction network: a resource for annotating the proteome. Cell 122:957–968
4. Chen KC, Csikasz-Nagy A, Gyorffy B, Val J, Novak B, Tyson JJ (2000) Kinetic analysis of a molecular model of the budding yeast cell cycle. Mol Biol Cell 11:369–391
5. Iwabe N, Kuma K, Miyata T (1996) Evolution of gene families and relationship with organismal evolution: rapid divergence of tissue-specific genes in the early evolution of chordates. Mol Biol Evol 13:483–493
6. Goh KI, Cusick ME, Valle D, Childs B, Vidal M, Barabasi AL (2007) The human disease network. Proc Natl Acad Sci USA 104: 8685–8690
7. Linghu B, Snitkin ES, Hu Z, Xia Y, Delisi C (2009) Genome-wide prioritization of disease genes and identification of disease-disease associations from an integrated human functional linkage network. Genome Biol 10:R91
8. Franke L, Bakel H, Fokkens L, de Jong ED, Egmont-Petersen M, Wijmenga C (2006) Reconstruction of a functional human gene network, with an application for prioritizing positional candidate genes. Am J Hum Genet 78:1011–1025
9. Kohler S, Bauer S, Horn D, Robinson PN (2008) Walking the interactome for prioritization of candidate disease genes. Am J Hum Genet 82:949–958
10. Lage K, Karlberg EO, Storling ZM, Olason PI, Pedersen AG, Rigina O, Hinsby AM, Tumer Z, Pociot F, Tommerup N, Moreau Y, Brunak S (2007) A human phenome-interactome network of protein complexes implicated in genetic disorders. Nat Biotechnol 25:309–316

11. Lee I, Lehner B, Crombie C, Wong W, Fraser AG, Marcotte EM (2008) A single gene network accurately predicts phenotypic effects of gene perturbation in *Caenorhabditis elegans*. Nat Genet 40:181–188
12. Linghu B, Snitkin ES, Holloway DT, Gustafson AM, Xia Y, DeLisi C (2008) High-precision high-coverage functional inference from integrated data sources. BMC Bioinformatics 9:119
13. McGary KL, Lee I, Marcotte EM (2007) Broad network-based predictability of *Saccharomyces cerevisiae* gene loss-of-function phenotypes. Genome Biol 8:R258
14. Oti M, Brunner HG (2007) The modular nature of genetic diseases. Clin Genet 71:1–11
15. Oti M, Snel B, Huynen MA, Brunner HG (2006) Predicting disease genes using protein-protein interactions. J Med Genet 43: 691–698
16. Schadt EE (2009) Molecular networks as sensors and drivers of common human diseases. Nature 461:218–223
17. Huttenhower C, Haley EM, Hibbs MA, Dumeaux V, Barrett DR, Coller HA, Troyanskaya OG (2009) Exploring the human genome with functional maps. Genome Res 19: 1093–1106
18. Ahmed A, Xing EP (2009) Recovering time-varying networks of dependencies in social and biological studies. Proc Natl Acad Sci USA 106:11878–11883
19. Linghu B, Delisi C (2010) Phenotypic connections in surprising places. Genome Biol 11:116
20. Ewing RM, Chu P, Elisma F, Li H, Taylor P, Climie S, McBroom-Cerajewski L, Robinson MD, O'Connor L, Li M, Taylor R, Dharsee M, Ho Y, Heilbut A, Moore L, Zhang S, Ornatsky O, Bukhman YV, Ethier M, Sheng Y, Vasilescu J, Abu-Farha M, Lambert JP, Duewel HS, Stewart II, Kuehl B, Hogue K, Colwill K, Gladwish K, Muskat B, Kinach R, Adams SL, Moran MF, Morin GB, Topaloglou T, Figeys D (2007) Large-scale mapping of human protein-protein interactions by mass spectrometry. Mol Syst Biol 3:89
21. Rual JF, Venkatesan K, Hao T, Hirozane-Kishikawa T, Dricot A, Li N, Berriz GF, Gibbons FD, Dreze M, Ayivi-Guedehoussou N, Klitgord N, Simon C, Boxem M, Milstein S, Rosenberg J, Goldberg DS, Zhang LV, Wong SL, Franklin G, Li S, Albala JS, Lim J, Fraughton C, Llamosas E, Cevik S, Bex C, Lamesch P, Sikorski RS, Vandenhaute J, Zoghbi HY, Smolyar A, Bosak S, Sequerra R, Doucette-Stamm L, Cusick ME, Hill DE, Roth FP, Vidal M (2005) Towards a proteome-scale map of the human protein-protein interaction network. Nature 437:1173–1178
22. Lee HK, Hsu AK, Sajdak J, Qin J, Pavlidis P (2004) Coexpression analysis of human genes across many microarray data sets. Genome Res 14:1085–1094
23. von Mering C, Jensen LJ, Kuhn M, Chaffron S, Doerks T, Kruger B, Snel B, Bork P (2007) STRING 7—recent developments in the integration and prediction of protein interactions. Nucleic Acids Res 35:D358–D362
24. Jensen LJ, Lagarde J, von Mering C, Bork P (2004) ArrayProspector: a web resource of functional associations inferred from microarray expression data. Nucleic Acids Res 32: W445–W448
25. Griffith OL, Pleasance ED, Fulton DL, Oveisi M, Ester M, Siddiqui AS, Jones SJ (2005) Assessment and integration of publicly available SAGE, cDNA microarray, and oligonucleotide microarray expression data for global coexpression analyses. Genomics 86:476–488
26. Calvo S, Jain M, Xie X, Sheth SA, Chang B, Goldberger OA, Spinazzola A, Zeviani M, Carr SA, Mootha VK (2006) Systematic identification of human mitochondrial disease genes through integrative genomics. Nat Genet 38:576–582
27. Zhong W, Sternberg PW (2006) Genome-wide prediction of *C. elegans* genetic interactions. Science 311:1481–1484
28. Franzosa E, Linghu B, Xia Y (2009) Computational reconstruction of protein-protein interaction networks: algorithms and issues. Methods Mol Biol 541:89–100
29. Berardini TZ, Li D, Huala E, Bridges S, Burgess S, McCarthy F et al (2010) The gene ontology in 2010: extensions and refinements. Nucleic Acids Res 38:D331–D335
30. Rhodes DR, Tomlins SA, Varambally S, Mahavisno V, Barrette T, Kalyana-Sundaram S, Ghosh D, Pandey A, Chinnaiyan AM (2005) Probabilistic model of the human protein-protein interaction network. Nat Biotechnol 23:951–959
31. Scott MS, Barton GJ (2007) Probabilistic prediction and ranking of human protein-protein interactions. BMC Bioinformatics 8:239
32. Janga SC, Tzakos A (2009) Structure and organization of drug-target networks: insights from genomic approaches for drug discovery. Mol Biosyst 5(12):1536–1548
33. Wu X, Jiang R, Zhang MQ, Li S (2008) Network-based global inference of human disease genes. Mol Syst Biol 4:189
34. Hamosh A, Scott AF, Amberger JS, Bocchini CA, McKusick VA (2005) Online Mendelian

Inheritance in Man (OMIM), a knowledgebase of human genes and genetic disorders. Nucleic Acids Res 33:D514–D517

35. Hu Z, Hung JH, Wang Y, Chang YC, Huang CL, Huyck M, DeLisi C (2009) VisANT 3.5: multi-scale network visualization, analysis and inference based on the gene ontology. Nucleic Acids Res 37:W115–W121
36. Hu Z, Snitkin ES, DeLisi C (2008) VisANT: an integrative framework for networks in systems biology. Brief Bioinform 9:317–325
37. Hu Z, Mellor J, Wu J, Kanehisa M, Stuart JM, DeLisi C (2007) Towards zoomable multidimensional maps of the cell. Nat Biotechnol 25:547–554
38. Mishra GR, Suresh M, Kumaran K, Kannabiran N, Suresh S, Bala P, Shivakumar K, Anuradha N, Reddy R, Raghavan TM, Menon S, Hanumanthu G, Gupta M, Upendran S, Gupta S, Mahesh M, Jacob B, Mathew P, Chatterjee P, Arun KS, Sharma S, Chandrika KN, Deshpande N, Palvankar K, Raghavnath R, Krishnakanth R, Karathia H, Rekha B, Nayak R, Vishnupriya G, Kumar HG, Nagini M, Kumar GS, Jose R, Deepthi P, Mohan SS, Gandhi TK, Harsha HC, Deshpande KS, Sarker M, Prasad TS, Pandey A (2006) Human protein reference database—2006 update. Nucleic Acids Res 34: D411–D414
39. Bader GD, Betel D, Hogue CW (2003) BIND: the Biomolecular Interaction Network Database. Nucleic Acids Res 31:248–250
40. Breitkreutz BJ, Stark C, Reguly T, Boucher L, Breitkreutz A, Livstone M, Oughtred R, Lackner DH, Bahler J, Wood V, Dolinski K, Tyers M (2008) The BioGRID Interaction Database: 2008 update. Nucleic Acids Res 36: D637–D640
41. Kerrien S, Alam-Faruque Y, Aranda B, Bancarz I, Bridge A, Derow C, Dimmer E, Feuermann M, Friedrichsen A, Huntley R, Kohler C, Khadake J, Leroy C, Liban A, Lieftink C, Montecchi-Palazzi L, Orchard S, Risse J, Robbe K, Roechert B, Thorneycroft D, Zhang Y, Apweiler R, Hermjakob H (2007) IntAct—open source resource for molecular interaction data. Nucleic Acids Res 35: D561–D565
42. Mewes HW, Dietmann S, Frishman D, Gregory R, Mannhaupt G, Mayer KF, Munsterkotter M, Ruepp A, Spannagl M, Stumpflen V, Rattei T (2008) MIPS: analysis and annotation of genome information in 2007. Nucleic Acids Res 36:D196–D201
43. Salwinski L, Miller CS, Smith AJ, Pettit FK, Bowie JU, Eisenberg D (2004) The Database of Interacting Proteins: 2004 update. Nucleic Acids Res 32:D449–D451
44. Chatr-Aryamontri A, Zanzoni A, Ceol A, Cesareni G (2008) Searching the protein interaction space through the MINT database. Methods Mol Biol 484:305–317
45. Mulder NJ, Apweiler R, Attwood TK, Bairoch A, Bateman A, Binns D, Bradley P, Bork P, Bucher P, Cerutti L, Copley R, Courcelle E, Das U, Durbin R, Fleischmann W, Gough J, Haft D, Harte N, Hulo N, Kahn D, Kanapin A, Krestyaninova M, Lonsdale D, Lopez R, Letunic I, Madera M, Maslen J, McDowall J, Mitchell A, Nikolskaya AN, Orchard S, Pagni M, Ponting CP, Quevillon E, Selengut J, Sigrist CJ, Silventoinen V, Studholme DJ, Vaughan R, Wu CH (2005) InterPro, progress and status in 2005. Nucleic Acids Res 33:D201–D205
46. Barrett T, Troup DB, Wilhite SE, Ledoux P, Rudnev D, Evangelista C, Kim IF, Soboleva A, Tomashevsky M, Marshall KA, Phillippy KH, Sherman PM, Muertter RN, Edgar R (2009) NCBI GEO: archive for high-throughput functional genomic data. Nucleic Acids Res 37:D885–D890
47. von Mering C, Jensen LJ, Snel B, Hooper SD, Krupp M, Foglierini M, Jouffre N, Huynen MA, Bork P (2005) STRING: known and predicted protein-protein associations, integrated and transferred across organisms. Nucleic Acids Res 33:D433–D437
48. von Mering C, Huynen M, Jaeggi D, Schmidt S, Bork P, Snel B (2003) STRING: a database of predicted functional associations between proteins. Nucleic Acids Res 31:258–261
49. Lee I, Li Z, Marcotte EM (2007) An improved, bias-reduced probabilistic functional gene network of baker's yeast, *Saccharomyces cerevisiae*. PLoS One 2:e988
50. Berglund AC, Sjolund E, Ostlund G, Sonnhammer EL (2008) InParanoid 6: eukaryotic ortholog clusters with inparalogs. Nucleic Acids Res 36:D263–D266
51. (2007) The Gene Ontology project in 2008. Nucleic Acids Res.
52. Adie EA, Adams RR, Evans KL, Porteous DJ, Pickard BS (2006) SUSPECTS: enabling fast and effective prioritization of positional candidates. Bioinformatics 22:773–774
53. Aerts S, Lambrechts D, Maity S, Van Loo P, Coessens B, De Smet F, Tranchevent LC, De Moor B, Marynen P, Hassan B, Carmeliet P, Moreau Y (2006) Gene prioritization through genomic data fusion. Nat Biotechnol 24:537–544
54. Zhou X, Kao MC, Wong WH (2002) Transitive functional annotation by shortest-path analysis of gene expression data. Proc Natl Acad Sci USA 99:12783–12788

55. Hughes TR, Roth FP (2008) A race through the maze of genomic evidence. Genome Biol 9 (Suppl 1):S1
56. Huang Y, Li H, Hu H, Yan X, Waterman MS, Huang H, Zhou XJ (2007) Systematic discovery of functional modules and context-specific functional annotation of human genome. Bioinformatics 23:i222–i229
57. Harris MA, Clark J, Ireland A, Lomax J, Ashburner M, Foulger R, Eilbeck K, Lewis S, Marshall B, Mungall C, Richter J, Rubin GM, Blake JA, Bult C, Dolan M, Drabkin H, Eppig JT, Hill DP, Ni L, Ringwald M, Balakrishnan R, Cherry JM, Christie KR, Costanzo MC, Dwight SS, Engel S, Fisk DG, Hirschman JE, Hong EL, Nash RS, Sethuraman A, Theesfeld CL, Botstein D, Dolinski K, Feierbach B, Berardini T, Mundodi S, Rhee SY, Apweiler R, Barrell D, Camon E, Dimmer E, Lee V, Chisholm R, Gaudet P, Kibbe W, Kishore R, Schwarz EM, Sternberg P, Gwinn M, Hannick L, Wortman J, Berriman M, Wood V, de la Cruz N, Tonellato P, Jaiswal P, Seigfried T, White R (2004) The Gene Ontology (GO) database and informatics resource. Nucleic Acids Res 32:D258–D261

Chapter 15

Genome-Wide Association Studies

Tun-Hsiang Yang, Mark Kon, and Charles DeLisi

Abstract

A host of data on genetic variation from the Human Genome and International HapMap projects, and advances in high-throughput genotyping technologies, have made genome-wide association (GWA) studies technically feasible. GWA studies help in the discovery and quantification of the genetic components of disease risks, many of which have not been unveiled before and have opened a new avenue to understanding disease, treatment, and prevention.

This chapter presents an overview of GWA, an important tool for discovering regions of the genome that harbor common genetic variants to confer susceptibility for various diseases or health outcomes in the post-Human Genome Project era. A tutorial on how to conduct a GWA study and some practical challenges specifically related to the GWA design is presented, followed by a detailed GWA case study involving the identification of loci associated with glioma as an example and an illustration of current technologies.

Key words: Genome-wide association studies, Genetic variation markers, Genotyping quality control, Linkage disequilibrium

1. Introduction

A significant scientific breakthrough in genomic research has been made in the first decade of the new millennium. The draft completion of the Human Genome Project in 2001 is a major milestone in human genomics and biomedical sciences (1, 2). It mapped the three billion nucleotide bases that make up the human genetic code, providing the foundation for studying genetic variations in the human genome, and showed that the DNA sequences of any two people are about 99.9% identical. The International HapMap Project (3) (http://www.hapmap.org/) which was completed in 2005 is another scientific landmark in the genomic research. It provides a catalog of common genetic variants, predominantly single nucleotide polymorphisms (SNPs), occurring in humans within and across populations in the world, and identifies chromosomal regions where genetic variants are shared. It further deepens on our understanding

Hiroshi Mamitsuka et al. (eds.), *Data Mining for Systems Biology: Methods and Protocols*, Methods in Molecular Biology, vol. 939, DOI 10.1007/978-1-62703-107-3_15,

of the genetic architecture of the human genome. The linkage disequilibrium (LD) map of the human genome provided by the HapMap project creates a valuable and useful genome-wide database of patterns of human genetic variation and also promotes the breakthrough in large-scale and high-throughput genotyping technological developments.

During the past 5 years, the accumulating knowledge about the correlation structure and frequency of common variants in the human genome combined with rapid advances in array technology have made GWA studies technically feasible. The number of published GWA studies at $p \leq 5 \times 10^{-8}$ has doubled within a year from June 2009 to June 2010 ($N = 439$ through 6/2009, and $N = 904$ for 165 traits through 6/2010 (NHGRI GWA Catalog, www.genome.gov/GWAStudies). In contrast to hypothesis-driven candidate-gene association studies, which largely rely on the understanding of known and suspected pathology in a given trait, GWA studies systematically investigate genetic variation across the genome without the constraints of a priori hypotheses and allow for the possibility of discovering associations in previously unsuspected pathways or in chromosomal regions of as yet undetermined function. This approach provides a comprehensive and unbiased examination of the common genetic basis of various complex traits. GWA studies have expanded our understanding of the complexity and diversity of genetic variations in the human genome, and have led to pivotal discoveries of new genetic loci for a host of common human disorders, including cancer, type 2 diabetes mellitus, and autoimmune diseases (4).

2. How Are Genome-Wide Association Studies Conducted?

As in other genetic association studies (such as candidate gene studies), genome-wide association compares the allele/genotype frequencies between groups that in principle differ in a single well-defined phenotype; e.g., with and without a particular disease, looking for markers that are statistically significant correlates of phenotype.

2.1. Association Study Designs

The principal goal is to minimize systematic bias and maximize power. Two fundamentally different designs are used: population-based designs that collect unrelated individuals (such as case–control or cohort studies) and family-based designs that use families (such as trio or pedigree studies), but case–control studies are most typically used in GWA studies. For common diseases, population-based studies generally have higher statistical power; in addition, in late-onset diseases/disorders such as Alzheimer's disease, parents and siblings may not be available. On the other hand, although family-based design is generally more time- and resource-consuming, it is robust against population stratification and population admixture, and significant findings always imply both linkage and association.

2.2. DNA Sample Collection and Genotyping Technology

After appropriate samples are recruited, DNA is drawn from each participant, usually by either blood draw or buccal swab. Each person's complete set of DNA is then purified from the blood or buccal cells, placed on tiny chips and scanned on automated laboratory machines. The genotyping machines quickly survey each participant's genome for a dense set of strategically selected markers of genetic variation, including either SNPs or copy number polymorphisms (CNPs), or both.

The popular commercially available genotyping arrays for GWA studies include Illumina arrays (such as Human Hap550, Human Hap650, and Infinium HD BeadChips) and Affymetrix arrays (such as Genome-wide Human SNP Array 5.0, SNP Array 6.0, and Human Mapping 500 K Array Set). In terms of the design in general, Affymetrix chips use the "random" design, in which the SNPs on the platform are randomly selected from the genome, without specific reference to the LD patterns. In contrast, Illumina chips use the "tagging" design, where SNPs are explicitly chosen to serve as surrogates for common variants in the HapMap data. The current genotyping platforms can accommodate up to one million or even more markers per chip per person.

2.3. Genotyping Quality Control

A battery of genotyping quality control procedures should be performed and checked after genotyping is completed, including marker completion rate, marker concordance, deviations from Hardy-Weinberg Equilibrium (HWE), sample completion rate, minor allele frequency (MAF), heterozygosity, gender concordance, duplicate sample detection, relatedness check, self-reported ethnicity concordance, and Mendelian consistency for markers and samples (if it is family-based study). In the population-based design, unexpected population structure can cause potential bias due to population stratification when there is confounding due to correlated differences in both allele frequencies and disease risks across unobserved subpopulations. GWA studies therefore typically adjust for multiple random, unlinked markers as a surrogate for genetic variation across subpopulation using EIGENSTRAT software (5). A principal component-based analysis detects and correct for population stratification and false positive results from ethnic mixtures. (http://genepath.med.harvard.edu/~reich/EIGENSTRAT.htm). In addition, linkage agglomerative clustering based on pairwise identity-by-state (IBS) distance followed by multidimensional scaling (MDS) implemented in the PLINK toolset (6) (http://pngu.mgh.harvard.edu/~purcell/plink/) can be used to identify clusters of samples with more homogeneous genetic backgrounds for subsequent association tests.

A Quantile–Quantile (QQ) plot that compares observed order statistics of p-values against the expected order statistics of p-values under the null hypothesis is also useful for visualizing and to summarizing both systematic bias and evidence for association. Early departure from the expected p-values usually suggests systematic bias, whereas late departures suggest true association signals.

2.4. Statistical Analysis

Because the purpose of the GWA studies is to analyze associations between thousands and millions of genetic markers at the genome-wide level and a disease or trait of interest without an a priori hypothesis, the initial association analysis examines marker-disease associations on a marker-by-marker basis from those who pass the quality control filtering. Depending on the assumption of genetic mode of inheritance, researchers may choose either allelic test, dominant, recessive, or codominant genotypic test, or trend test. Association analysis can be performed by several existing programming packages, such as PLINK (6) and EIGENSTRAT (5). PLINK is a free, open-source specifically developed for GWA studies that allows large-scale analyses in a computationally efficient manner for both population-based and family-based designs. EIGENSTRAT uses principal component analysis to model ancestry differences between cases and controls. The resulting correction minimizes spurious associations while maximizing power to detect true associations.

Because a large number of tests are conducted in GWA studies, stringent significance thresholds are essential to rule out false positive results. Several hundred thousand tests require a threshold of $p = 10^{-7}$ to control experiment-wide type I error for all common variants and $p = 5 \times 10^{-8}$ for all variants (7–9). A comprehensive analysis beyond single-marker analyses in the GWA setting is not yet feasible because it can introduce a large number of additional tests. For example, a combinatorial scan for all two-way interaction on one million SNPs is barely feasible. A restriction to only a small subset of the data based on a specific rationale or hypothesis is more desirable, unless solutions on high-dimension data reduction and optimization are developed. SNPs or markers that are identified from the GWAS results can be further assigned to pathway analysis or enrichment analysis which could potentially be very useful for prioritizing genes and pathways within a biological context, which can be done with computational tools and pathway databases (10).

2.5. Validation and Replication

If certain genetic markers are found to be significantly more frequent or less frequent in cases than in controls, the variations are said to be "associated" with the disease. The associated genetic variations can serve as powerful pointers to the region of the human genome where the causal locus resides. However, the significantly associated variants themselves may not always directly cause the disease. In fact, in most cases they may just "tag along" with the actual causal variants due to the LD correlation structure in the human genome. Deeper analysis of the associated regions by sequencing is the best way to identify (a set of possible) causal variant(s), and filtering the list of highly associated variants using biologic annotation, including sequence context or known function (eQTLs), or conducting further in vitro experiments for functionality.

In order to represent credible genotype–phenotype associations observed in a GWA study, replication of the results is especially critical. That means finding the same marker or a marker in perfect or high LD with the prior marker (11).

With the increasing number consortia of multiple GWA studies, meta-analysis of multiple genome-wide studies conducted by different investigative groups, in different populations, using different genotyping technologies and different study designs) becomes an emerging approach of replication of GWA studies in the context of gene discovery (11), as illustrated below. Meta-analysis can increase the sample size effectively by combining different studies, which is especially powerful and useful in genetic association research, particularly when most of the common genetic variants contributing to complex diseases have only small to modest effects (odds ratio <1.5). While a single GWA may not have sufficient statistical power to detect small effects, secondary analyses using a meta-analysis framework that pools information across studies provides an inexpensive and efficient way to accumulate evidence, which can also provide additional power for discovery of new associations by combing association signals across GWA studies, even when the original raw data are unavailable. For example, additional genetic loci with BMI (12) and lipid traits (13, 14) have recently been discovered by meta-analysis.

3. An Example: Glioma Genome-Wide Association Study

A case study involving the identification of loci which are associated with glioma using the GWA approach is presented here as a working example. We follow closely the presentation in (15).

3.1. Study Samples

To identify risk variants for glioma, we conducted a principal component-adjusted genome-wide association study. 226 glioma patients were collected from The Cancer Genome Atlas (TCGA) SNP data (16). The TCGA data portal contains clinical information, genomic characterization data of the tumor genomes and provides a platform for researchers to search, download, and analyze data sets generated by TCGA. Genotypes were determined using the Illumina Human Hap550 Array. We eliminated all samples for which more than 5% of the SNPs are missing, and

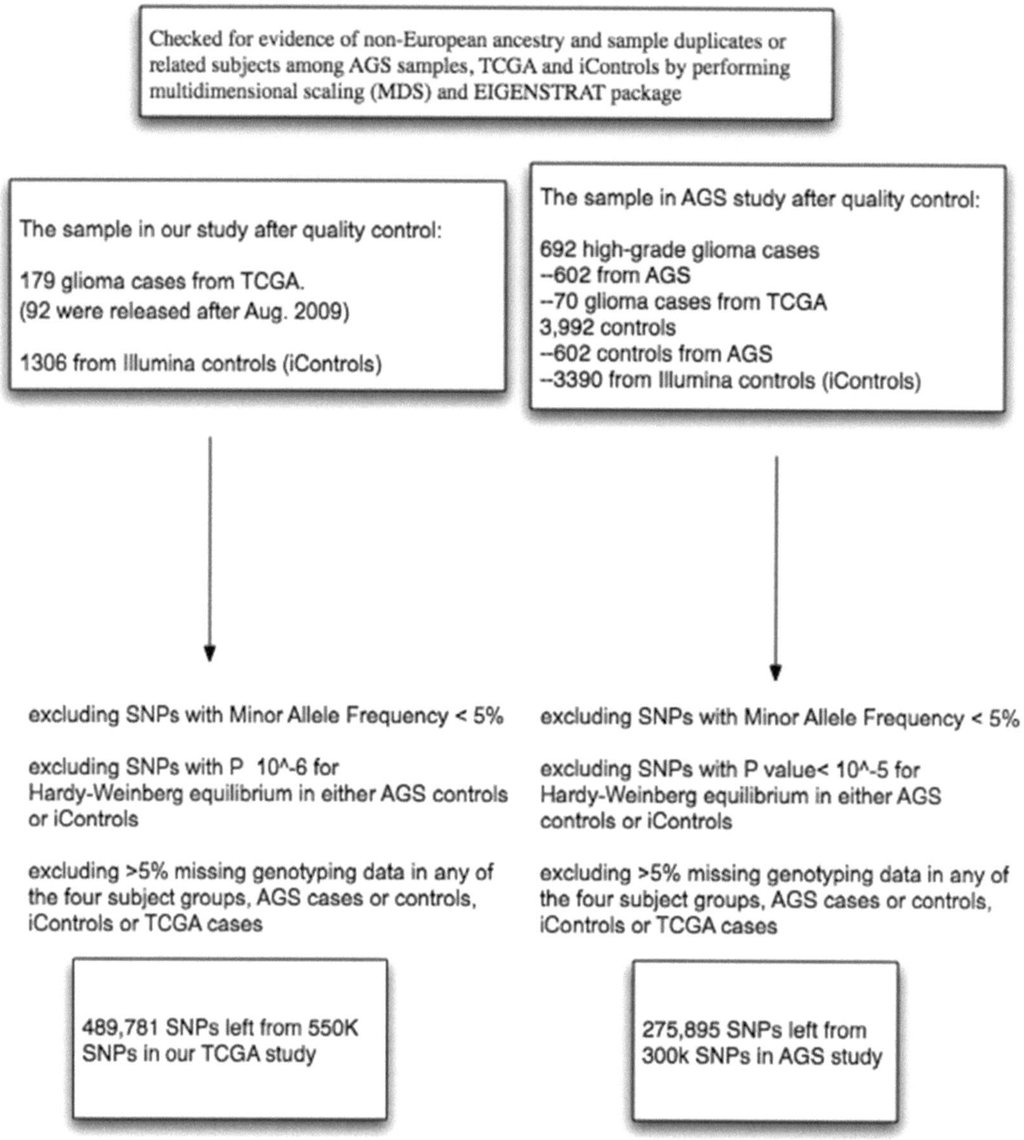

Checked for evidence of non-European ancestry and sample duplicates or related subjects among AGS samples, TCGA and iControls by performing multidimensional scaling (MDS) and EIGENSTRAT package

The sample in our study after quality control:

179 glioma cases from TCGA.
(92 were released after Aug. 2009)

1306 from Illumina controls (iControls)

excluding SNPs with Minor Allele Frequency < 5%

excluding SNPs with P 10^-6 for Hardy-Weinberg equilibrium in either AGS controls or iControls

excluding >5% missing genotyping data in any of the four subject groups, AGS cases or controls, iControls or TCGA cases

489,781 SNPs left from 550K SNPs in our TCGA study

The sample in AGS study after quality control:

692 high-grade glioma cases
--602 from AGS
--70 glioma cases from TCGA
3,992 controls
--602 controls from AGS
--3390 from Illumina controls (iControls)

excluding SNPs with Minor Allele Frequency < 5%

excluding SNPs with P value< 10^-5 for Hardy-Weinberg equilibrium in either AGS controls or iControls

excluding >5% missing genotyping data in any of the four subject groups, AGS cases or controls, iControls or TCGA cases

275,895 SNPs left from 300k SNPs in AGS study

Fig. 1. Subjects and single-SNP exclusion schema for genome-wide association studies.

eliminated all SNPs that (1) are determined in fewer than 95% of the samples, (2) have MAF less than 5%, or (3) have a HWE p-value of less than 10^{-6}. The procedure is outlined in Fig. 1.

In order to adjust the potential confounding effects by ethnicity-specific SNP frequencies, we further confined our study sample to European-Americans only, which is the group from which the majority of samples were obtained. Glioma patient samples were identified by a two-step screening: (a) self-reporting of ancestry, and (b) computationally assisted stratification. The latter was carried out using the EIGENSTRAT package (5). After screening, 179 TCGA samples remained. Of these, we used for confirmation only the 92 that were released after August 2009, since the majority of the earlier samples have been already included in the Adult Glioma Study (AGS) (17).

The comparison group included normal European-American blood samples ($n = 1366$), which was downloaded from the Illumina iControlDB (iControls), an online data repository of genotype and phenotype data from individuals that can be used as controls in association studies. After applying the quality control procedures described above, 1,306 control samples remained.

3.2. Association Analysis

Association analysis was performed with the EIGENSTRAT package, under the null hypothesis of "no association between the glioblastoma multiform (GBM) SNP genotype and the control SNP genotype" based on an additive inheritance model. To set the significance threshold for p we required that the probability of 1 or more false positives be less than 0.05; in particular, $1 - e^{-Np} \leq 0.05$ or $p \leq 0.93 \times 10^{-7} \approx 10^{-7}$ with N, the total number of SNPs examined, taken as 550,000. At this level, few if any SNPs will be detected for typical glioma population sizes. The alternative is to accept a less stringent p-value, and to eliminate false discoveries by seeking confirmation in an independent study.

3.3. Meta-Analysis

Various versions of meta-analysis can be used to combine p-values from two independent studies. Because the AGS and TCGA datasets differ widely in the number of samples, we assigned them different weights (18). In particular

$$p_{12} = P\left(Z > \frac{W_1 Z_1 + W_2 Z_2}{\sqrt{W_1^2 + W_2^2}}\right); \; Z \sim \text{Normal}(0, 1) \qquad (1)$$

where the weights (W_i) are proportional to the square root of the "total number of individuals," $Z_i = F^{-1}(1 - p_i)$, and F^{-1} (.) is an inverse standard normal CDF. The false discovery rate (FDR) is estimated as the fused probability multiplied by the total number of SNPs, which is 300,000.

The procedure for calculating fused p-values begins with lists of SNPs that have p-values of less than 0.001 in each population. We walk down this list, calculating a combined p value Eq. 1 for each pair, and accept all SNPs for which the FDR is less than 0.05 (or equivalently $p_{12} = 0.05/300{,}000 = 1.7 \times 10^{-7}$; see Table 1). For our results, when p exceeds 0.001 in either population, p_{12} no longer meets the required threshold, and the walk stops.[1]

[1] As a practical matter, the walk can be stopped at more stringent p values without changing the main conclusions. In particular stopping AGS at $p = 10^{-5}$, and TCGA at 10^{-3}, while requiring that p_{12} pass the genomic significance level (1.7×10^{-7}), loses only 2 SNPs (rs12021720 and rs2810424), neither of which adds new genomic regions.

Table 1
Concordant SNPs recovered from TCGA and AGS data and associated genes

SNP	chr	gene	Location	left_gene	right_gene	Genes in LD with SNP (R^2)	AGS (P1)	TCGA (P2)	FDR
rs2736100[a]	**5**	**TERT**	5p15.33	**SLC6A18**	**CLPTM1L**	NA	**5.30E-13**	**2.66E-04**	**7.38E-09**
rs1412829[b,a]	**9**	**CDKN2A/2B**	9p21.3	LOC100130239	LOC729983	**CDKN2A**(1.0); **CDKN2B**(1.0); C9orf53(0.87)	**3.40E-08**	**3.26E-03**	**1.27E-03**
rs2157719	9	CDKN2A/2B	9p21.3	LOC100130239	LOC729983	**CDKN2A**(0.97); **CDKN2B**(0.97); C9orf53(0.93); RP11-145E5.4(0.97); LOC100130239(0.97)	6.10E-08	8.00E-03	5.40E-03
rs1063192[a]	9	CDKN2A/2B	9p21.3	**CDKN2A**	LOC100130239	**CDKN2A**(0.97); **CDKN2B**(0.97); C9orf53(0.93); RP11-145E5.4(0.97); LOC100130239(0.97)	9.20E-08	8.31E-03	7.95E-03
rs4977756[a]	9	CDKN2A/2B	9p21.3	LOC100130239	LOC729983	**CDKN2A**(0.97); **CDKN2B**(0.97); C9orf53(0.93); RP11-145E5.4(0.97); LOC100130239(0.97)	4.20E-07	1.12E-02	3.90E-02
rs7530361	**1**	**SLC35A3**	1p21.2	LOC730081	HIAT1	SLC35A3(1.0); CCDC76(0.95); HIAT1(1.0); LRRC39(0.95); **SASS6**(1.0); BRI3P1(0.95); LOC730081(1.0)	**6.50E-07**	**2.19E-06**	**4.29E-05**

rs501700	**1**	**HIAT1**	1p21.2	SLC35A3	**SASS6**	DBT(0.95); RTCD1(0.89); SLC35A3(1.0); CCDC76(0.95); HIAT1(1.0); LRRC39(0.94); **SASS6**(1.0); BRI3P1(0.94); LOC730081(1.0)	**7.10E-07**	**5.99E-06**	**9.72E-05**
rs1920116	3	LRRC31	3q26.2	LRRIQ4	KRT18P43	MYNN(0.89); LRRC31(1); ARPM1(0.85); KRT18P43(1) LRRC34(1)	1.40E-06	2.88E-03	2.81E-02
rs506044	1	SASS6	1p21.2	**SASS6**	LRRC39	DBT(1.0); RTCD1(0.89); SLC35A3(0.95); CCDC76(1.0); HIAT1(1.0); LRRC39(1.0); **SASS6**(1.0); BRI3P1(0.95); LOC730081(0.94)	2.10E-06	2.45E-06	1.57E-04
rs640030	1	SASS6	1p21.2	HIAT1	CCDC76	DBT(1.0); RTCD1(0.89); SLC35A3(0.95); CCDC76(1.0); HIAT1(1.0); LRRC39(1.0); **SASS6**(1.0); BRI3P1(0.95); LOC730081(0.94)	2.40E-06	2.57E-06	1.86E-04

(continued)

Table 1 (continued)

SNP	chr	gene	Location	left_gene	right_gene	Genes in LD with SNP (R^2)	AGS (P1)	TCGA (P2)	FDR
rs687513	1	SASS6	1p21.2	**SASS6**	LRRC39	DBT(0.95); RTCD1(0.90); SLC35A3(0.94); CCDC76(1.0); HIAT1(1.0); LRRC39(1.0); **SASS6**(1.0); BRI3P1(0.95); LOC730081(0.90)	2.90E-06	3.91E-06	3.03E-04
rs3779505	7	**ITGB8**	7p15.3	**MACC1**	LOC100130234	**ITGB8**(1.0)	3.00E-06	5.67E-04	1.35E-02

Concordant SNPs (FDR < 0.05) recovered from TCGA (n = 97) and AGS data (n = 692), and associated genes

[a]Reported by Shete et al.

[b]Reported by Wrensch et al. and validated on Mayo Clinic population

3.4. Genes in Linkage Disequilibrium with SNPs

We use the coefficient of determination, R^2, to identify genes in strong linkage disequilibrium with the SNPs that we identified in the meta-analysis. R^2 is calculated based on the correlation between gene expression level and SNP genotype. Genes with R^2 greater then 0.8 are considered to be in strong LD with the SNP.

3.5. Relative Risk

As indicated below, analysis of TCGA and AGS identifies 12 significant SNPs, 7 of which are new. One of the implications of additional SNPs is that the number of associated genes that can be used to estimate relative glioma risk increases combinatorially. Consequently we can expect higher prognostic reliability for individuals possessing a combination of risk alleles, although at some loss of population coverage. We consider here all combinations of two and three SNPs, while constraining our choices to SNPs that are more than a mega-base pairs (Mb) apart, in order to minimize redundant (disequilibrated) information. Specifically, the 12 SNPs are divided into five groups based on location. Chromosome 1 has 5 SNPs clustered together within 1 Mb, and chromosome 9 has 4 SNPs within 1 Mb around genes *CDKN2A/2B*. The remaining 3 SNPs are located on chromosomes 3, 5, and 7.

If we rule out combinations including any pair of SNPs that are within a single chromosome, we find 50 SNP pairs, and 88 SNP triplets. Statistical analyses were implemented using R (v2.7) and PLINK (v1.07) (6). Combinations with odds ratios greater than three, along with *p*-values, are shown in Table 2, which also shows that SNP combinations from chromosomes 1 and 9 are associated with the highest relative risk.

3.6. Identification of Associated Pathways and Genes

The standard method for identifying altered processes is a pathway enrichment analysis, which can be carried out using a single population (19). In this case pathways would be identified by showing that the number of significant SNPs/genes that occur in a particular pathway is above chance expectation. The procedure that we describe here requires multiple populations. The assignment of a SNP/gene to a particular pathway from a single population meets a significance threshold which is loose enough to allow multiple assignments from that population, but not stringent enough for an acceptable FDR. The FDR is brought down to an acceptable level, as described below, when both populations assign the same gene(s) to the same pathway.

The procedure is as follows: (1) identify SNPs having a *p*-value $<10^{-3}$ in either the populations; (2) identify genes that include these SNPs, and (3) assign the genes thus obtained to KEGG pathways (20). The detailed procedure by which assignments are made is explained elsewhere (15).

Table 2
Pairwise and triplet SNP combinations with odds ratios greater than 3

SNP combinations			Risk allele[a]	Frequency	OR^{eq2}	*p*-value
rs1412829 (1.58)[b]		rs7530361 (1.89)[c]	11	5.45E-02	3.31	3.58E-07
rs1412829 (1.58)[b]		rs501700 (1.90)[c]	11	5.51E-02	3.09	1.95E-06
rs1412829 (1.58)[b]		rs506044 (1.96)[c]	11	5.47E-02	3.23	5.15E-07
rs1412829 (1.58)[b]		rs640030 (1.95)[c]	11	5.42E-02	3.28	4.30E-07
rs1412829 (1.58)[b]		rs687513 (1.93)[c]	11	5.52E-02	3.18	7.32E-07
rs2157719 (1.49)[b]		rs7530361 (1.89)[c]	11	5.51E-02	3.2	6.83E-07
rs2157719 (1.49)[b]		rs506044 (1.96)[c]	11	5.54E-02	3.12	9.64E-07
rs2157719 (1.49)[b]		rs640030 (1.95)[c]	11	5.49E-02	3.16	8.12E-07
rs2157719 (1.49)[b]		rs687513 (1.93)[c]	11	5.59E-02	3.07	1.35E-06
rs1063192 (1.42)[b]		rs7530361 (1.89)[c]	11	5.60E-02	3.12	1.13E-06
rs1063192 (1.42)[b]		rs506044 (1.96)[c]	11	5.63E-02	3.05	1.60E-06
rs1063192 (1.42)[b]		rs640030 (1.95)[c]	11	5.59E-02	3.08	1.35E-06
rs4977756 (1.60)[b]		rs7530361 (1.89)[c]	11	5.35E-02	4.28	3.14E-10
rs4977756 (1.60)[b]		rs501700 (1.90)[c]	11	5.44E-02	4.17	5.57E-10
rs4977756 (1.60)[b]		rs506044 (1.96)[c]	11	5.36E-02	4.18	4.46E-10
rs4977756 (1.60)[b]		rs640030 (1.95)[c]	11	5.31E-02	4.24	3.66E-10
rs4977756 (1.60)[b]		rs687513 (1.93)[c]	11	5.41E-02	4.1	6.86E-10
rs2736100 (0.63)	rs7530361 (1.89)[c]	rs1920116 (0.68)	212	5.01E-02	4.3	5.02E-10
rs11823971 (1.45)	rs1412829 (1.58)[b]	rs7530361 (1.89)[c]	211	5.21E-02	3.04	5.05E-06
rs11823971 (1.45)	rs1412829 (1.58)[b]	rs506044 (1.96)[c]	211	5.26E-02	3.01	4.67E-06

Numbers in parenthesis are single-SNP odds ratios. Last column is the Wald test *p*-value for the odds ratio of the combination. This is an unadjusted *p*-value, with an 0.05 multiple testing adjusted threshold of $p = 0.05/(50 + 88) = 3.6 \times 10^{-4}$. Freq denotes the combined frequency of the given combination in the case and control populations as a whole

[a]Denotes alleles in which significant shifts occur. 11 denotes significant shift in the minor alleles for both SNPs. 212 denotes significant shifts in major, minor major; 211, significant shifts in major, minor, minor

[b]Denotes SNP on chromosome 9 in gene CDKN2A/2B

[c]Denotes SNP on chromosome 1

3.7. Results

1. Significant SNP candidates
   Using TCGA datasets, we validated 4 of the 13 SNPs inferred by Wrensch et al. based on the AGS (Table 1, boldface) (15, 17). SNPs rs7530361 and rs501700, both at 1p21.2, were reported for the first time.

Joint analysis of data, as reported in (15), rather than sequential analysis of two or more populations can increase the power to detect genetic associations (21). In particular using Eq. 1 as described in Methods, we found 12 SNPs (Table 1), confirmed by AGS and TCGA at an FDR <0.05, one of which was previously confirmed by Wrensch et al. (17) and Shete et al. (22). Of the 11 remaining, 4 were reported by Shete et al.; the other 7 are reported for the first time. The 12 SNPs are distributed across five genomic regions: chromosomes 5q15.33, 9q21.3, 1p21.2, 3q26.2 and 7p15.3. Two of these, 5q15.33 and 9q21.3, have been reported in previous studies (17, 22). The 12 candidates are in strong linkage disequilibrium with 25 genes, 8 of which are previously known to be associated with cancer are indicated in boldface in Table 1. An additional SNP of interest is rs12341266 at 9q32, which has an FDR of 0.06 and is in the glioma-associated gene, *RGS3*.

2. Genes identified by conserved pathway analysis
We identified 49 pathways that contain genes associated with loosely defined AGS or TCGA SNPs. Thirty-six of them do not meet the hypergeometric test at a p value of 0.001 (an FDR of 0.05 divided by 49), leaving 13 invariant pathways; i.e., pathways that are relevant to both populations. Each of the 13 pathways has 1 common gene (Table 3) from the two groups. There are 5 such genes—*FHIT*, *GABRG3*, *PRKG1*, *DCC*, and *ITGB8*—each of which occurs in more than one of these pathways.

3. Genes in Strong Linkage Disequilibrium with SNP candidates
The SNP candidates occur within, or are in strong linkage disequilibrium with, 25 genes (Table 1). Eight of which are cancer associated. The latter are *TERT* (17, 23, 24), *SLC6A18* (23), *CLPTM1L* (23, 24), *CDKN2A/2B* (17, 25, 26), *SASS6* (27), *ITGB8* (28), and *MACC1* (29) (Table 1). Five of the genes, *TERT*, *SLC6A18*, *CLPTM1L*, and *CDKN2A/2B*, were previously shown to be associated with glioma by other GWA studies. As explained below, we have predicted by a combination of GWA and pathway analysis, 4 additional genes, which are identified in the literature as cancer related. The detail literature citations and the type of cancers that associated with these genes are discussed in discussion section. We therefore predict 29 glioma-associated genes, 12 of them known by previous studies to be cancer related. It is useful to ask for the probability that as many as 12 cancer related genes in a set of 29 would be found by chance. If we use the fraction of OMIM genes that are cancer related as an estimate of the background frequency of cancer genes in the disease genes population, the probability that 29 genes have 12 cancer-associated genes by chance is 1.4×10^{-6}. The fraction of OMIM genes that are cancer related is 0.1 (750 cancer-associated gene in 7,381 OMIM genes).

Table 3
Pathways that contain significant SNPs ($p<10^{-3}$) inferred from both AGS and TCGA samples

Pathway*	AGS_SNP	Gene	TCGA_SNP	Gene
Purine metabolism (p = 3.50E-04)** Small cell lung cancer (p = 4.35E-04) ** Non-small cell lung cancer (p = 2.6E-04) **	rs7617530	FHIT	rs13059601	FHIT
Neuroactive ligand-receptor interaction (p = 8.00E-04) **	rs1011455 rs4887546 rs1011456	GABRG3 GABRG3 GABRG3	rs12904325	GABRG3
Vascular smooth muscle contraction (p = 3.48E-04) ** Gap junction (p = 1.30E-04) ** Long-term depression (p = 6.95E-04) ** Olfactory transduction (p = 3.47E-04) **	rs4400745 rs4466778	PRKG1 PRKG1	rs1922139	PRKG1
Axon guidance (p = 3.91E-04) ** Pathways in cancer (p = 2.13E-03) Colorectal cancer (p = 8.69E-05) **	rs1145245	DCC	rs11082983 rs11872471 rs12604940	DCC DCC DCC
Focal adhesion (p = 1.95E-03) ECM-receptor interaction (p = 8.69E-04) ** Cell adhesion molecules (CAMs) (p = 1.74E-04) ** Regulation of actin cytoskeleton (p = 1.56E-03) Hypertrophic cardiomyopathy (HCM) (p = 1.22E-03) Arrhythmogenic right ventricular cardiomyopathy (ARVC) (p = 9.12E-04) ** Dilated cardiomyopathy (p = 1.04E-03)	rs3779505 rs2301727 rs3807936 rs2158250	ITGB8 ITGB8 ITGB8 ITGB8	rs3779505	ITGB8

* p = Probability of the gene overlap in two independent populations. Multiple testing adjusted threshold of $p = 0.05/49 = 10^{-3}$
** Pathways with $p < 10^{-3}$

Each of the 8 cancer related genes listed above plays one or more key roles in processes known to be altered during tumor initiation and development (30). For example, MACC1 is a growth pathway regulator influencing angiogenesis and processes related to metastasis (29); CDKN2A is a well studied cell cycle regulator (26) and a known tumor suppressor whose loss results in a diminished ability to regulate growth and predisposition to cancer (25); ITGB8 has been implicated in activities related to metastasis, including adhesion and migration (28); and the telomerase enzyme (TERT) is linked to unlimited replication (17). It is worth noting that CDKN2A/2B are in strong linkage disequilibrium with rs1412829 at 9p21.3, which has now been identified in 3 independent studies and should therefore be considered an unusually high confidence gene marker.

4. Potential Challenges in Genome-Wide Association Studies

While GWA studies open a new avenue of discovering and understanding of the common genetic variation of the human genome in diseases and health, the assessments of the overall evidence deriving from GWA studies remains a complex endeavor. The GWA studies also create some open challenges as the field is still under development and much of the literature remains exploratory.

4.1. From Statistics to Functionality

Although statistically compelling associations have been identified, many association signals identified in GWA studies are not localized to intervals that include a gene, unlike Mendelian human diseases whose genetics is understood that functional rare mutations with large effects act through altering or truncating gene products. However, there is growing evidence that a sizeable proportion, perhaps the majority, of the functional variants that underlie GWA studies exert their effects through gene regulation rather than changing gene products (31). For example, a SNP (rs6983267) in the 8q24 locus implicated in multiple cancer pathogeneses identified in GWA studies is located in a gene desert that is >300 kb away from the most neighboring annotated *MYC* proto-oncogene; recently studies have shown that the region harboring this risk allele is a transcriptional enhancer that interacts with the *MYC* gene (32, 33). How to translating mere statistically association signals to biological relevance of the precise variants that have a causal role in conferring the disease susceptibility remains unclear at present, but more research towards a deeper understanding of the vast regulatory regions within the human genome and functional studies will be the future direction.

4.2. Investigations of Complex Interactions

Given the fact that common complex diseases are multifactorial with each factor contributing a small effect, it is possible that what really counts is not the main effect of the genes but complex gene–gene or gene–environment interactions. How to proceed with the investigations of gene–gene interactions or gene–environment interactions in GWA studies is an important question with no straightforward answers.

4.3. Sufficiency of Common Variants to Account for Genetic Bases of Complex Traits

The current technology for GWAS studies consider common genetic variants, predominantly SNPs, as possible targets for association with a trait or a phenotype, and do not capture information about rare variants. However, not only SNPs, there are also others forms of genetic variations that could account for disease risk. For example, recently, genomic copy number variations (CNVs) have begun investigated in several GWA studies. CNVs are defined as gains or losses of repeats of DNA sequences consisting of between kilo- and mega-base pairs. CNVs have been detected in locations covering about 12% of the human genome (34, 35). As technology and knowledge surrounding CNVs continue to improve, CNVs have become a significantly more mainstream in GWA studies (36, 37). However, in addition to SNPs and CNVs, there are also other types of structural variations and epigenetics in the human genome and it is unclear how much each type of genetic variation contributes to inherited risk and the relative proportion of rare versus common variants. The use of new technologies for assaying DNA sequences can provide important and additional insights about the roles of different types of genetic variants in human disease or health. For example, the 1000 Genomes Project launched in 2008 has used the next-generation sequencing technique to provide a comprehensive resource on human genetic variation with at least 1% across most of the genome and down to 0.5% or lower within genes (38). The 1000 Genomes Project will map not only the SNPs but also will produce a high-resolution map of structural variants, including rearrangements, deletions, or duplications of segments of the human genome.

References

1. Collins FS, Morgan M, Patrinos A (2003) The Human Genome Project: lessons from large-scale biology. Science 300:286–290
2. Roberts L, Davenport RJ, Pennisi E, Marshall E (2001) A history of the Human Genome Project. Science 291:1195
3. Frazer KA, Ballinger DG, Cox DR, Hinds DA, Stuve LL, Gibbs RA, Belmont JW, Boudreau A, Hardenbol P, Leal SM, Pasternak S, Wheeler DA, Willis TD, Yu F, Yang H, Zeng C, Gao Y, Hu H, Hu W, Li C, Lin W, Liu S, Pan H, Tang X, Wang J, Wang W, Yu J, Zhang B, Zhang Q, Zhao H, Zhao H, Zhou J, Gabriel SB, Barry R, Blumenstiel B, Camargo A, Defelice M, Faggart M, Goyette M, Gupta S, Moore J, Nguyen H, Onofrio RC, Parkin M, Roy J, Stahl E, Winchester E, Ziaugra L, Altshuler D, Shen Y, Yao Z, Huang W, Chu X, He Y, Jin L, Liu Y, Shen Y, Sun W, Wang H, Wang Y, Wang Y, Xiong X, Xu L, Waye MM, Tsui SK, Xue H, Wong JT, Galver LM, Fan JB, Gunderson K, Murray SS, Oliphant AR, Chee MS, Montpetit A, Chagnon F, Ferretti V, Leboeuf M, Olivier JF, Phillips MS, Roumy S, Sallee C, Verner A, Hudson TJ, Kwok PY, Cai D, Koboldt DC, Miller RD, Pawlikowska L, Taillon-Miller P, Xiao M, Tsui LC, Mak W, Song YQ, Tam PK, Nakamura Y, Kawaguchi T, Kitamoto T, Morizono T, Nagashima A, Ohnishi Y, Sekine A, Tanaka T, Tsunoda T, Deloukas P, Bird CP, Delgado M, Dermitzakis ET, Gwilliam R, Hunt S, Morrison J, Powell D, Stranger BE, Whittaker P, Bentley DR, Daly MJ, de Bakker PI, Barrett J, Chretien YR, Maller J, McCarroll S, Patterson N, Pe'er I, Price A, Purcell S, Richter DJ, Sabeti P, Saxena R, Schaffner SF, Sham PC, Varilly P, Altshuler D, Stein LD, Krishnan L, Smith AV, Tello-Ruiz MK, Thorisson GA, Chakravarti A, Chen PE, Cutler DJ, Kashuk CS, Lin S, Abecasis GR, Guan W, Li Y, Munro HM, Qin ZS, Thomas DJ, McVean G, Auton A, Bottolo L, Cardin N, Eyheramendy S, Freeman C, Marchini J, Myers S, Spencer C, Stephens M, Donnelly P, Cardon LR, Clarke G, Evans DM, Morris AP, Weir BS, Tsunoda T, Mullikin JC, Sherry ST, Feolo M, Skol A, Zhang H, Zeng C, Zhao H, Matsuda I, Fukushima Y, Macer DR, Suda E, Rotimi CN, Adebamowo CA, Ajayi I, Aniagwu T, Marshall PA, Nkwodimmah C, Royal CD, Leppert MF, Dixon M, Peiffer A, Qiu R, Kent A, Kato K, Niikawa N, Adewole IF, Knoppers BM, Foster MW, Clayton EW, Watkin J, Gibbs RA, Belmont JW, Muzny D, Nazareth L, Sodergren E, Weinstock GM, Wheeler DA, Yakub I, Gabriel SB, Onofrio RC, Richter DJ, Ziaugra L, Birren BW, Daly MJ, Altshuler D, Wilson RK, Fulton LL, Rogers J, Burton J, Carter NP, Clee CM, Griffiths M, Jones MC, McLay K, Plumb RW, Ross MT, Sims SK, Willey DL, Chen Z, Han H, Kang L, Godbout M, Wallenburg JC, L'Archeveque P, Bellemare G, Saeki K, Wang H, An D, Fu H, Li Q, Wang Z, Wang R, Holden AL, Brooks LD, McEwen JE, Guyer MS, Wang VO, Peterson JL, Shi M, Spiegel J, Sung LM, Zacharia LF, Collins FS,

Kennedy K, Jamieson R, Stewart J (2007) A second generation human haplotype map of over 3.1 million SNPs. Nature 449:851–861

4. Hindorff LA, Sethupathy P, Junkins HA, Ramos EM, Mehta JP, Collins FS, Manolio TA (2009) Potential etiologic and functional implications of genome-wide association loci for human diseases and traits. Proc Natl Acad Sci USA 106:9362–9367
5. Price AL, Patterson NJ, Plenge RM, Weinblatt ME, Shadick NA, Reich D (2006) Principal components analysis corrects for stratification in genome-wide association studies. Nat Genet 38:904–909
6. Purcell S, Neale B, Todd-Brown K, Thomas L, Ferreira MA, Bender D, Maller J, Sklar P, de Bakker PI, Daly MJ, Sham PC (2007) PLINK: a tool set for whole-genome association and population-based linkage analyses. Am J Hum Genet 81:559–575
7. Dudbridge F, Gusnanto A (2008) Estimation of significance thresholds for genomewide association scans. Genet Epidemiol 32:227–234
8. Pe'er I, Yelensky R, Altshuler D, Daly MJ (2008) Estimation of the multiple testing burden for genomewide association studies of nearly all common variants. Genet Epidemiol 32:381–385
9. Wakefield J (2007) A Bayesian measure of the probability of false discovery in genetic epidemiology studies. Am J Hum Genet 81:208–227
10. Cantor RM, Lange K, Sinsheimer JS (2010) Prioritizing GWAS results: a review of statistical methods and recommendations for their application. Am J Hum Genet 86:6–22
11. Kraft P, Zeggini E, Ioannidis JP (2009) Replication in genome-wide association studies. Stat Sci 24:561–573
12. Willer CJ, Speliotes EK, Loos RJ, Li S, Lindgren CM, Heid IM, Berndt SI, Elliott AL, Jackson AU, Lamina C, Lettre G, Lim N, Lyon HN, McCarroll SA, Papadakis K, Qi L, Randall JC, Roccasecca RM, Sanna S, Scheet P, Weedon MN, Wheeler E, Zhao JH, Jacobs LC, Prokopenko I, Soranzo N, Tanaka T, Timpson NJ, Almgren P, Bennett A, Bergman RN, Bingham SA, Bonnycastle LL, Brown M, Burtt NP, Chines P, Coin L, Collins FS, Connell JM, Cooper C, Smith GD, Dennison EM, Deodhar P, Elliott P, Erdos MR, Estrada K, Evans DM, Gianniny L, Gieger C, Gillson CJ, Guiducci C, Hackett R, Hadley D, Hall AS, Havulinna AS, Hebebrand J, Hofman A, Isomaa B, Jacobs KB, Johnson T, Jousilahti P, Jovanovic Z, Khaw KT, Kraft P, Kuokkanen M, Kuusisto J, Laitinen J, Lakatta EG, Luan J, Luben RN, Mangino M, McArdle WL, Meitinger T, Mulas A, Munroe PB, Narisu N, Ness AR, Northstone K, O'Rahilly S, Purmann C, Rees MG, Ridderstrale M, Ring SM, Rivadeneira F, Ruokonen A, Sandhu MS, Saramies J, Scott LJ, Scuteri A, Silander K, Sims MA, Song K, Stephens J, Stevens S, Stringham HM, Tung YC, Valle TT, Van Duijn CM, Vimaleswaran KS, Vollenweider P, Waeber G, Wallace C, Watanabe RM, Waterworth DM, Watkins N, Witteman JC, Zeggini E, Zhai G, Zillikens MC, Altshuler D, Caulfield MJ, Chanock SJ, Farooqi IS, Ferrucci L, Guralnik JM, Hattersley AT, Hu FB, Jarvelin MR, Laakso M, Mooser V, Ong KK, Ouwehand WH, Salomaa V, Samani NJ, Spector TD, Tuomi T, Tuomilehto J, Uda M, Uitterlinden AG, Wareham NJ, Deloukas P, Frayling TM, Groop LC, Hayes RB, Hunter DJ, Mohlke KL, Peltonen L, Schlessinger D, Strachan DP, Wichmann HE, McCarthy MI, Boehnke M, Barroso I, Abecasis GR, Hirschhorn JN (2009) Six new loci associated with body mass index highlight a neuronal influence on body weight regulation. Nat Genet 41:25–34
13. Aulchenko YS, Ripatti S, Lindqvist I, Boomsma D, Heid IM, Pramstaller PP, Penninx BW, Janssens AC, Wilson JF, Spector T, Martin NG, Pedersen NL, Kyvik KO, Kaprio J, Hofman A, Freimer NB, Jarvelin MR, Gyllensten U, Campbell H, Rudan I, Johansson A, Marroni F, Hayward C, Vitart V, Jonasson I, Pattaro C, Wright A, Hastie N, Pichler I, Hicks AA, Falchi M, Willemsen G, Hottenga JJ, de Geus EJ, Montgomery GW, Whitfield J, Magnusson P, Saharinen J, Perola M, Silander K, Isaacs A, Sijbrands EJ, Uitterlinden AG, Witteman JC, Oostra BA, Elliott P, Ruokonen A, Sabatti C, Gieger C, Meitinger T, Kronenberg F, Doring A, Wichmann HE, Smit JH, McCarthy MI, van Duijn CM, Peltonen L (2009) Loci influencing lipid levels and coronary heart disease risk in 16 European population cohorts. Nat Genet 41:47–55
14. Kathiresan S, Willer CJ, Peloso GM, Demissie S, Musunuru K, Schadt EE, Kaplan L, Bennett D, Li Y, Tanaka T, Voight BF, Bonnycastle LL, Jackson AU, Crawford G, Surti A, Guiducci C, Burtt NP, Parish S, Clarke R, Zelenika D, Kubalanza KA, Morken MA, Scott LJ, Stringham HM, Galan P, Swift AJ, Kuusisto J, Bergman RN, Sundvall J, Laakso M, Ferrucci L, Scheet P, Sanna S, Uda M, Yang Q, Lunetta KL, Dupuis J, de Bakker PI, O'Donnell CJ, Chambers JC, Kooner JS, Hercberg S, Meneton P, Lakatta EG, Scuteri A, Schlessinger D, Tuomilehto J, Collins FS, Groop L, Altshuler D, Collins R, Lathrop GM, Melander O, Salomaa V, Peltonen L, Orho-Melander M,

Ordovas JM, Boehnke M, Abecasis GR, Mohlke KL, Cupples LA (2009) Common variants at 30 loci contribute to polygenic dyslipidemia. Nat Genet 41:56–65

15. Yang TH, Kon M, Hung JH, Delisi C (2011) Combinations of newly confirmed glioma-associated loci link regions on chromosomes 1 and 9 to increased disease risk. BMC Med Genomics. 4(1):63

16. McLendon R, Friedman A, Bigner D, Van Meir E, Brat DJ, Mastrogianakis GM, Olson JJ, Mikkelsen T, Lehman N, Aldape K, Yung WK, Bogler O, Weinstein JN, Vandenberg S, Berger M, Prados M, Muzny D, Morgan M, Scherer S, Sabo A, Nazareth L, Lewis L, Hall O, Zhu Y, Ren Y, Alvi O, Yao J, Hawes A, Jhangiani S, Fowler G, San Lucas A, Kovar C, Cree A, Dinh H, Santibanez J, Joshi V, Gonzalez-Garay ML, Miller CA, Milosavljevic A, Donehower L, Wheeler DA, Gibbs RA, Cibulskis K, Sougnez C, Fennell T, Mahan S, Wilkinson J, Ziaugra L, Onofrio R, Bloom T, Nicol R, Ardlie K, Baldwin J, Gabriel S, Lander ES, Ding L, Fulton RS, McLellan MD, Wallis J, Larson DE, Shi X, Abbott R, Fulton L, Chen K, Koboldt DC, Wendl MC, Meyer R, Tang Y, Lin L, Osborne JR, Dunford-Shore BH, Miner TL, Delehaunty K, Markovic C, Swift G, Courtney W, Pohl C, Abbott S, Hawkins A, Leong S, Haipek C, Schmidt H, Wiechert M, Vickery T, Scott S, Dooling DJ, Chinwalla A, Weinstock GM, Mardis ER, Wilson RK, Getz G, Winckler W, Verhaak RG, Lawrence MS, O'Kelly M, Robinson J, Alexe G, Beroukhim R, Carter S, Chiang D, Gould J, Gupta S, Korn J, Mermel C, Mesirov J, Monti S, Nguyen H, Parkin M, Reich M, Stransky N, Weir BA, Garraway L, Golub T, Meyerson M, Chin L, Protopopov A, Zhang J, Perna I, Aronson S, Sathiamoorthy N, Ren G, Yao J, Wiedemeyer WR, Kim H, Kong SW, Xiao Y, Kohane IS, Seidman J, Park PJ, Kucherlapati R, Laird PW, Cope L, Herman JG, Weisenberger DJ, Pan F, Van den Berg D, Van Neste L, Yi JM, Schuebel KE, Baylin SB, Absher DM, Li JZ, Southwick A, Brady S, Aggarwal A, Chung T, Sherlock G, Brooks JD, Myers RM, Spellman PT, Purdom E, Jakkula LR, Lapuk AV, Marr H, Dorton S, Choi YG, Han J, Ray A, Wang V, Durinck S, Robinson M, Wang NJ, Vranizan K, Peng V, Van Name E, Fontenay GV, Ngai J, Conboy JG, Parvin B, Feiler HS, Speed TP, Gray JW, Brennan C, Socci ND, Olshen A, Taylor BS, Lash A, Schultz N, Reva B, Antipin Y, Stukalov A, Gross B, Cerami E, Wang WQ, Qin LX, Seshan VE, Villafania L, Cavatore M, Borsu L, Viale A, Gerald W, Sander C, Ladanyi M, Perou CM, Hayes DN, Topal MD, Hoadley KA, Qi Y, Balu S, Shi Y, Wu J, Penny R, Bittner M, Shelton T, Lenkiewicz E, Morris S, Beasley D, Sanders S, Kahn A, Sfeir R, Chen J, Nassau D, Feng L, Hickey E, Barker A, Gerhard DS, Vockley J, Compton C, Vaught J, Fielding P, Ferguson ML, Schaefer C, Zhang J, Madhavan S, Buetow KH, Collins F, Good P, Guyer M, Ozenberger B, Peterson J, Thomson E (2008) Comprehensive genomic characterization defines human glioblastoma genes and core pathways. Nature 455:1061–1068

17. Wrensch M, Jenkins RB, Chang JS, Yeh RF, Xiao Y, Decker PA, Ballman KV, Berger M, Buckner JC, Chang S, Giannini C, Halder C, Kollmeyer TM, Kosel ML, LaChance DH, McCoy L, O'Neill BP, Patoka J, Pico AR, Prados M, Quesenberry C, Rice T, Rynearson AL, Smirnov I, Tihan T, Wiemels J, Yang P, Wiencke JK (2009) Variants in the CDKN2B and RTEL1 regions are associated with high-grade glioma susceptibility. Nat Genet 41:905–908

18. Liptak T (1958) On the combination of independent tests. Magyar Tud Akad Mat Kutato Int Kozl 3:171–197

19. Subramanian A, Tamayo P, Mootha VK, Mukherjee S, Ebert BL, Gillette MA, Paulovich A, Pomeroy SL, Golub TR, Lander ES, Mesirov JP (2005) Gene set enrichment analysis: a knowledge-based approach for interpreting genome-wide expression profiles. Proc Natl Acad Sci USA 102:15545–15550

20. Kanehisa M (2002) The KEGG database. Novartis Found Symp 247:91–101, discussion 101–103, 119–128, 244–152

21. Skol AD, Scott LJ, Abecasis GR, Boehnke M (2006) Joint analysis is more efficient than replication-based analysis for two-stage genome-wide association studies. Nat Genet 38:209–213

22. Shete S, Hosking FJ, Robertson LB, Dobbins SE, Sanson M, Malmer B, Simon M, Marie Y, Boisselier B, Delattre JY, Hoang-Xuan K, El Hallani S, Idbaih A, Zelenika D, Andersson U, Henriksson R, Bergenheim AT, Feychting M, Lonn S, Ahlbom A, Schramm J, Linnebank M, Hemminki K, Kumar R, Hepworth SJ, Price A, Armstrong G, Liu Y, Gu X, Yu R, Lau C, Schoemaker M, Muir K, Swerdlow A, Lathrop M, Bondy M, Houlston RS (2009) Genome-wide association study identifies five susceptibility loci for glioma. Nat Genet 41:899–904

23. Kang JU, Koo SH, Kwon KC, Park JW, Kim JM (2008) Gain at chromosomal region 5p15.33, containing TERT, is the most frequent genetic event in early stages of non-small cell lung cancer. Cancer Genet Cytogenet 182:1–11

24. Stacey SN, Sulem P, Masson G, Gudjonsson SA, Thorleifsson G, Jakobsdottir M, Sigurdsson A,

Gudbjartsson DF, Sigurgeirsson B, Benediktsdottir KR, Thorisdottir K, Ragnarsson R, Scherer D, Hemminki K, Rudnai P, Gurzau E, Koppova K, Botella-Estrada R, Soriano V, Juberias P, Saez B, Gilaberte Y, Fuentelsaz V, Corredera C, Grasa M, Hoiom V, Lindblom A, Bonenkamp JJ, van Rossum MM, Aben KK, de Vries E, Santinami M, Di Mauro MG, Maurichi A, Wendt J, Hochleitner P, Pehamberger H, Gudmundsson J, Magnusdottir DN, Gretarsdottir S, Holm H, Steinthorsdottir V, Frigge ML, Blondal T, Saemundsdottir J, Bjarnason H, Kristjansson K, Bjornsdottir G, Okamoto I, Rivoltini L, Rodolfo M, Kiemeney LA, Hansson J, Nagore E, Mayordomo JI, Kumar R, Karagas MR, Nelson HH, Gulcher JR, Rafnar T, Thorsteinsdottir U, Olafsson JH, Kong A, Stefansson K (2009) New common variants affecting susceptibility to basal cell carcinoma. Nat Genet 41:909–914

25. Bisio A, Nasti S, Jordan JJ, Gargiulo S, Pastorino L, Provenzani A, Quattrone A, Queirolo P, Bianchi-Scarra G, Ghiorzo P, Inga A (2010) Functional analysis of CDKN2A/p16INK4a 5′-UTR variants predisposing to melanoma. Hum Mol Genet 19:1479–1491

26. Sherr CJ (1996) Cancer cell cycles. Science 274:1672–1677

27. Leidel S, Delattre M, Cerutti L, Baumer K, Gonczy P (2005) SAS-6 defines a protein family required for centrosome duplication in C. elegans and in human cells. Nat Cell Biol 7:115–125

28. Culhane AC, Quackenbush J (2009) Confounding effects in “a six-gene signature predicting breast cancer lung metastasis”. Cancer Res 69:7480–7485

29. Boardman LA (2009) Overexpression of MACC1 leads to downstream activation of HGF/MET and potentiates metastasis and recurrence of colorectal cancer. Genome Med 1:36

30. Hanahan D, Weinberg RA (2000) The hallmarks of cancer. Cell 100:57–70

31. Ku CS, Loy EY, Pawitan Y, Chia KS (2010) The pursuit of genome-wide association studies: where are we now? J Hum Genet 55:195–206

32. Pomerantz MM, Ahmadiyeh N, Jia L, Herman P, Verzi MP, Doddapaneni H, Beckwith CA, Chan JA, Hills A, Davis M, Yao K, Kehoe SM, Lenz HJ, Haiman CA, Yan C, Henderson BE, Frenkel B, Barretina J, Bass A, Tabernero J, Baselga J, Regan MM, Manak JR, Shivdasani R, Coetzee GA, Freedman ML (2009) The 8q24 cancer risk variant rs6983267 shows long-range interaction with MYC in colorectal cancer. Nat Genet 41:882–884

33. Tuupanen S, Turunen M, Lehtonen R, Hallikas O, Vanharanta S, Kivioja T, Bjorklund M, Wei G, Yan J, Niittymaki I, Mecklin JP, Jarvinen H, Ristimaki A, Di-Bernardo M, East P, Carvajal-Carmona L, Houlston RS, Tomlinson I, Palin K, Ukkonen E, Karhu A, Taipale J, Aaltonen LA (2009) The common colorectal cancer predisposition SNP rs6983267 at chromosome 8q24 confers potential to enhanced Wnt signaling. Nat Genet 41:885–890

34. Perry GH, Ben-Dor A, Tsalenko A, Sampas N, Rodriguez-Revenga L, Tran CW, Scheffer A, Steinfeld I, Tsang P, Yamada NA, Park HS, Kim JI, Seo JS, Yakhini Z, Laderman S, Bruhn L, Lee C (2008) The fine-scale and complex architecture of human copy-number variation. Am J Hum Genet 82:685–695

35. Redon R, Ishikawa S, Fitch KR, Feuk L, Perry GH, Andrews TD, Fiegler H, Shapero MH, Carson AR, Chen W, Cho EK, Dallaire S, Freeman JL, Gonzalez JR, Gratacos M, Huang J, Kalaitzopoulos D, Komura D, MacDonald JR, Marshall CR, Mei R, Montgomery L, Nishimura K, Okamura K, Shen F, Somerville MJ, Tchinda J, Valsesia A, Woodwark C, Yang F, Zhang J, Zerjal T, Zhang J, Armengol L, Conrad DF, Estivill X, Tyler-Smith C, Carter NP, Aburatani H, Lee C, Jones KW, Scherer SW, Hurles ME (2006) Global variation in copy number in the human genome. Nature 444:444–454

36. Korn JM, Kuruvilla FG, McCarroll SA, Wysoker A, Nemesh J, Cawley S, Hubbell E, Veitch J, Collins PJ, Darvishi K, Lee C, Nizzari MM, Gabriel SB, Purcell S, Daly MJ, Altshuler D (2008) Integrated genotype calling and association analysis of SNPs, common copy number polymorphisms and rare CNVs. Nat Genet 40:1253–1260

37. McCarroll SA, Kuruvilla FG, Korn JM, Cawley S, Nemesh J, Wysoker A, Shapero MH, de Bakker PI, Maller JB, Kirby A, Elliott AL, Parkin M, Hubbell E, Webster T, Mei R, Veitch J, Collins PJ, Handsaker R, Lincoln S, Nizzari M, Blume J, Jones KW, Rava R, Daly MJ, Gabriel SB, Altshuler D (2008) Integrated detection and population-genetic analysis of SNPs and copy number variation. Nat Genet 40:1166–1174

38. Kuehn BM (2008) 1000 Genomes Project promises closer look at variation in human genome. JAMA 300:2715

Chapter 16

Viral Genome Analysis and Knowledge Management

Carla Kuiken, Hyejin Yoon, Werner Abfalterer, Brian Gaschen, Chienchi Lo, and Bette Korber

Abstract

One of the challenges of genetic data analysis is to combine information from sources that are distributed around the world and accessible through a wide array of different methods and interfaces. The HIV database and its footsteps, the hepatitis C virus (HCV) and hemorrhagic fever virus (HFV) databases, have made it their mission to make different data types easily available to their users. This involves a large amount of behind-the-scenes processing, including quality control and analysis of the sequences and their annotation. Gene and protein sequences are distilled from the sequences that are stored in GenBank; to this end, both submitter annotation and script-generated sequences are used. Alignments of both nucleotide and amino acid sequences are generated, manually curated, distilled into an alignment model, and regenerated in an iterative cycle that results in ever better new alignments. Annotation of epidemiological and clinical information is parsed, checked, and added to the database. User interfaces are updated, and new interfaces are added based upon user requests. Vital for its success, the database staff are heavy users of the system, which enables them to fix bugs and find opportunities for improvement. In this chapter we describe some of the infrastructure that keeps these heavily used analysis platforms alive and vital after nearly 25 years of use.

The database/analysis platforms described in this chapter can be accessed at

http://hiv.lanl.gov
http://hcv.lanl.gov
http://hfv.lanl.gov

Key words: RNA virus, Alignment, Database, Ontology, Taxonomy, Annotation

1. Background

The original HIV database project in Los Alamos was devoted to storing HIV sequence data. In its 20 years of existence, it has evolved to include data on immunological (B and T cell) epitopes in HIV and a large set of tools for data manipulation and analysis. Two other sites were added more recently that offer similar genomic data analysis platforms for hepatitis C virus (HCV) and hemorrhagic fever viruses (HFV). Collectively, these repositories are

Hiroshi Mamitsuka et al. (eds.), *Data Mining for Systems Biology: Methods and Protocols*, Methods in Molecular Biology, vol. 939, DOI 10.1007/978-1-62703-107-3_16, © Springer Science+Business Media New York 2013

denoted as the h*v databases. The HFV site especially requires more sophisticated content management. Providing reference information for one (or in the case of HIV, three: HIV-1, HIV-2 and similarly structured SIV, and other SIV) is relatively simple, since the information is static enough to manually update if there is a modification. Making such information available for a large and increasing number of viruses means it has to be parsed from a number of relatively unstructured external sources.

A number of complementary sources of viral taxonomical and gene structure information are available (1). For viral taxonomy, the authoritative body is the International Committee on the Taxonomy of Viruses, ICTV. This committee is an unfunded community effort. It is working hard to automate and structure its work, but lacks resources to make great progress. Another important player is the Uniprot database (2), which provides universal protein annotation. Unfortunately, there are very few back-links from protein sequence to nucleotide sequence. The Gene Ontology (GO) project has only recently turned its attention to viral gene information, and the effort is not yet mature enough to be able to solve many of these problems. At the same time, this information is vital for the functionality of HFV database.

"Annotation" in a genomic context means identifying which genes and coding regions are present in a genome, and assigning functions to these regions. One cause of the problematic state of viral annotation is a confusion of these two roles. Many regions are labeled "hypothetical protein," because at the time the sequence was submitted to GenBank, the encoded protein had not been positively identified.

Annotation is often done by analogy: a previously annotated genome is aligned to the new genomes, and similar genes are identified and named. New genomes that do not significantly resemble any annotated ones have to be annotated de novo. This is usually done by blindly translating the all sequences to amino acids (six-way translation, three reading frames and two orientations), and seeing which translated regions are long enough to potentially encode a protein. Next, the translations are compared to each other to find groups of similar potential proteins in the new sequence data or in stored proteins or profiles. Lab assays are then usually needed to determine the function of the newly identified proteins.

Future functionality for the viral databases will likely include analysis of next-generation viral or metagenomic sequences, short sequence fragments (200–300 nucleotides) that are random pieces of genomes. Identifying which viruses are contained in a sample means identifying the most similar viral sequences in the database (using the Blast algorithm), and determining which genes are included. If the sample viruses are sufficiently similar to existing viruses, again the identification can be done by analogy.

A trivial first step in annotating the sequences consists of parsing submitted information such as sampling country and city, sampling year, information about host, and sample treatment. Some of this information is provided with fixed labels by GenBank, some of it is harvested by free text searching. Additional information is added through manual curation; among these are patient, patient cluster (for transmission clusters), origin of infection, treatment history, and HLA information.

2. Automated Alignment of Viral Sequences

Broadly, to do any kind of automated sequence comparison, including annotation, the first step is to align the sequences to each other so the similarity is maximized, and the effect of length variation is minimized. This step is trivial for highly conserved genes and genomes, but highly challenging for variable ones. Viruses have tiny genomes; the ones in the h*v databases are usually around 10,000 bases. However, the divergence and length variation within their genomes and genes are very large—within one patient, viruses can be as much as 5% different, and between types of the same viral species the difference goes up to 20%. Consequently, the result of any analysis depends greatly on the quality of the alignment. Existing alignment programs can provide good starting points for an alignment, but they all make mistakes that are obvious to a human and that need to be corrected by hand. We will not dwell here on the alignment problems, but assume there is an optimal starting alignment at the nucleotide level. For the h*v databases, these manually curated alignments are available to the user, and they form the basis of further processing.

Next, all new sequences are aligned to our existing HMM model alignment using Sean Eddy's HMMer program suite (3). The programs HMMA and HMMSW are used consecutively; HMMSW is sometimes able to align sequences for which HMMA fails (*opposite. First, try HMMSW and then next is HMMA*). If the alignment succeeds, the result file is parsed, and information is added to the database about

(a) Gaps added to each sequence to align it to the model

(b) Gaps added to the model to accommodate any inserts in that sequence

Upon retrieval, then, the gaps are added back to either the sequence [in case of (a)] or to all other sequences in the alignment [in case of (b)], and the resulting alignment can be downloaded as a quasi-multiple alignment of all the sequences the user requested.

To identify the location of each sequence relative to the complete genome, and the location of every gene and annotated region

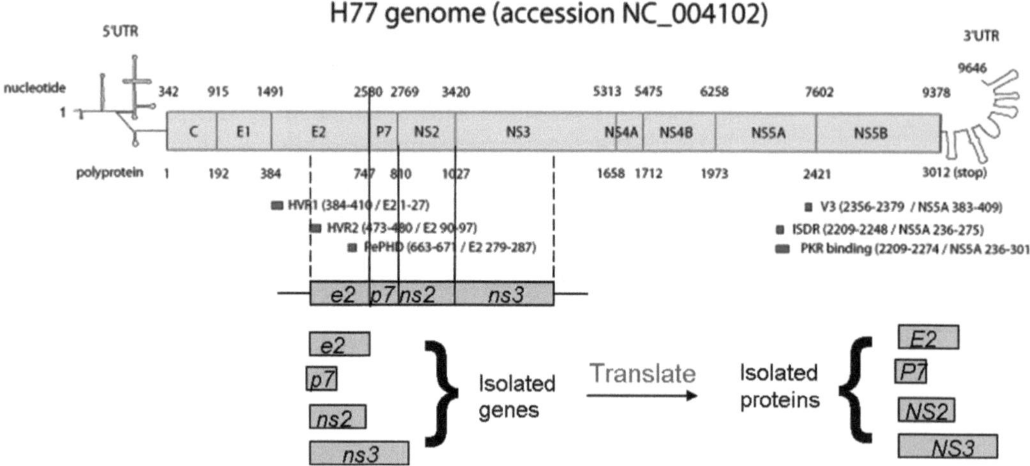

Fig. 1. Schematic of gene cutter algorithm.

in the sequence, the program Gene cutter (4) was developed. It uses the align0 algorithm (5) to align each sequence to the reference sequence for the species (HIV-1, HIV-2 and SIV) and determines the boundaries by analogy to the reference sequences. It can then retrieve individual genes and regions by clipping them out of longer sequences. The end result is a set of codon-aligned and (optionally) translated alignments of all CDSs and proteins that are present in the original alignment (Fig. 1). For the HFV platform, a Blast search was added to identify the most similar reference sequence; this option can be overridden by the user. (*No GeneCutter for HFV. Blast search is used in VirAlign*).

3. Application: The GenBank Submission Tool

While creating a general-purpose sequence submission tool for GenBank is complicated, viral sequences require only a limited subset of the possible options. The Gene cutter tool described above can be expanded fairly easily to create GenBank submission files. This involves finding the genes and CDSs, codon-aligning them to compensate for length variation, assigning gene and protein names and coordinates, and automatically annotating known features in the identified regions based on a reference sequence. In addition, the information needs to be submitted in a format that genBank accepts; we use ASN.1. The sequence submission tool for HIV is already available (http://www.hiv.lanl.gov/content/sequence/GENBANK/genbank.html), and for HCV and HFV the tool is complete and in the final testing phase.

Once again, a similar tool that can handle viruses from multiple families requires more thought. The reference-sequence information is used to inform the framework of the annotation features, but many specific cases cannot be dealt with in a general way. For example, in Reston Ebolavirus a posttranscriptional modification is used to create a CDS for the GP protein. Since this feature is not inherent in the sequence, it is difficult to annotate based purely on sequence data; the annotation will only succeed if the same feature is present in the reference sequence. The tool is designed to catch the standard and trivially predictable features, and otherwise warn users if expected features are missing or malformed, e.g., because of missing start- or premature stop codons.

4. Automated Protein Retrieval and Alignment

We recently started an effort to expand the pre-alignment and region identification to amino acids (Fig. 2). There are many obvious reasons for this. Alignments based on amino acids tend to be

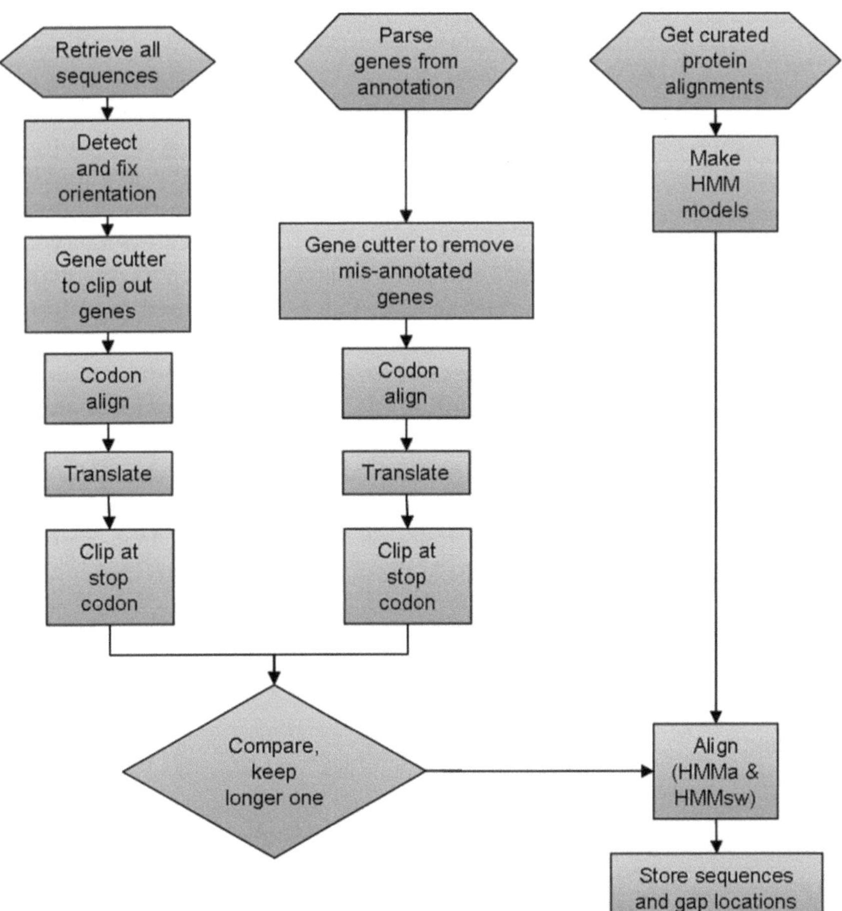

Fig. 2. Flowchart of the gene/protein extraction and alignment process.

better; many users are interested in proteins; translation of the aligned nucleotide sequences is problematic since they are not codon-aligned; neither the Gene cutter nor the user-provided annotation always gets the start location right; codon-alignment for HIV is complicated further by its overlapping reading frames and spliced genes. In the HFV database, several viruses have genes containing a forward-encoding and a reverse-encoding region. All these factors contribute to making this effort a large and complicated undertaking.

First, we harvested all individual genes or coding regions (the terms gene and coding region are used interchangeably here) and gene fragments we could from the available sequences, without regard to completeness. We used both Gene cutter and the user-provided annotation for this. The annotation of HIV sequences dates back about 25 years, and gene and region names have changed over that period, so this required some sophisticated text parsing. In addition, there is currently no spell check for gene names in the GenBank submission tools, so annotation spelling mistakes also had to be incorporated.

The next step was to compare the results from Gene Cutter and from the annotation parsing. For this, sequences obtained from the annotation were processed by Gene cutter to remove obvious mistakes (i.e., a Rev gene annotated as "Tat"). Next, sequences were codon-aligned, translated, and cut off at the stop codon. The process of codon alignment considers the reading frame of the translated protein and adjusts the nucleotide alignment so that the protein alignment stays in frame based on reference gene. For each gene and entry, the longer sequence was retained. The resulting sets of amino acid sequences were aligned to amino acid models that were made for all proteins based on manually curated alignments (available at http://www.hiv.lanl.gov/content/sequence/NEWALIGN/align.html). Next, an adaptation of an in-house code, described previously (4), was used to determine the location of all gaps and store those locations in the database. This information is used s necessary to recreate the alignment when users retrieve an aligned set of sequences. The final steps are currently being taken to make these data accessible to users.

5. Identifying Homologous Genes in Heterogeneous Sequence Sets

The curated RefSeq database provided by NCBI (6) has made it possible to locate and isolate individual genes for each species. This quality is used in the HFV database to allow users to retrieve species-level alignments of individual genes and coding sequences (CDSs), using a method that is exactly analogous to that described in the previous section. However, the next step in an analysis is

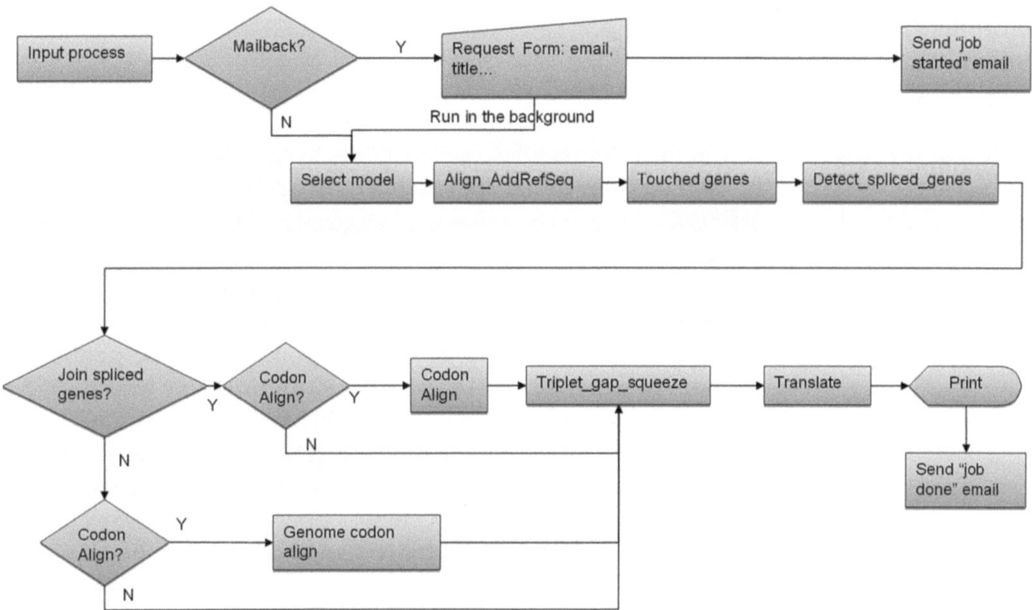

Fig. 3. Flowchart of the VirAlign program algorithm.

usually to compare genes between different species, at least up to the genus level. To achieve this, HMM models were built for each genus represented in the database by aligning all reference genomes within each genus. The resulting model sequences can be used with remarkable success to align sequences at the genus level.

For the HFV, individual genes are isolated using the same clip-by-sequence-homology method, represented in the HFV database by a generalized Gene cutter-like code called VirAlign (7) (Fig. 3). VirAlign uses the predefined reference sequence for a species or the curated HMM model sequence for genus-level alignments; it will also execute a quick Blast search if the species does not have a reference sequence, or is novel. Unfortunately, the alignment is not always good enough to accurately isolate genes and proteins for each species and variant. To produce an annotation-based alignment analogous to the HIV case, the gene names would need to be enumerated or matched using a text search. Our efforts to achieve this so far have not yielded reliable enough results to improve much on the results produced by VirAlign. We are hopeful that efforts such as NCBI's Annotation Workshops, and Uniprot's pilot implementation of gene ontologies will lead to a more consistent gene- and protein naming, and a mapping table to allow old-style, unstructured naming to be used for annotation-based gene identification.

6. Bringing Gene/Protein Identification to Higher Taxonomic Levels

The problem of missing reliable gene identifiers plays a much greater role in the HFV database, where genes from different species have to be aligned. Our initial approach to this problem was to provide an artificial genus-level reference sequence. The individual species sequences are aligned to the species reference sequences, and those in turn are aligned to the genus-level one. Then, if needed, the gene coordinates from species X are mapped onto the genus reference, and for each species the gene corresponding to the genus-level location is retrieved. This method works remarkably well as long as the gene locations are close and well behaved. However, when one protein (for example, RNA-directed RNA polymerase) is translocated, or a segmented genome needs to be compared to a nonsegmented one, it breaks down. It also does not function well when there is no species-level reference sequence yet, or when a virus is novel or unclassified. In that case the most similar (by Blast) reference sequence is substituted for reference-based analyses, but this solution has obvious shortcomings.

A better method would be to incorporate the annotation into the process. The parsing process described above for HIV is manageable when there are only a few synonymous gene names to consider. However, for the greater viral field the situation is much more complex. A quick analysis of gene naming mapping using UniProt's standardized protein names showed that approximately half the annotated gene names in the HFV database could be conclusively mapped into UniProt's framework. The other half comprised large number of different naming variations, such as "West Nile Virus RDRP," "putative RDRP," "hypothetical RDRP," and "viral RNA polymerase (L protein)". To conclusively solve this issue (rather than relying on parsing scripts of ever-increasing complexity) it requires a unified, agreed-upon viral naming system and a set of mapping rules from the current free-text annotation.

Even though viral genomes even in the same genus can be nearly impossible to align due to their extreme variability, when classified by function the protein encoding schemata of many RNA viral orders are fairly simple and remarkably uniform. However, the naming of the genes and the proteins and products they encode is a tangle, and trying to identify CDSs on that basis is a daunting task. We are currently in the process of investigating the possibility of using the functional classification as the basis of a naming system. If this translates down to lower levels, it will be possible to create a synonym list on the basis of the annotated gene names and the actual gene location and (predicted) protein function that allows users to retrieve even highly dissimilar genes and CDSs by selecting a (putative) gene/protein function.

7. The Future

Currently, several efforts are proceeding in parallel. The expansion of the viral gene and protein ontologies is ongoing, and we hope the ontology can begin to be used by the start of 2012. Automated viral annotation pipelines also require some information about the correspondence between coding sequences, posttranscriptional modification, and translation to protein.

Future plans include expanding the "Los Alamos viral genomic database platform" to include further emerging viruses that are human pathogens, and to provide an annotation and analysis facilities for both current and next-generation sequence data. This will include expanded and more sophisticated data manipulation, for example faster and less labor-intensive ways to create reference alignments that cover as much genetic and geographic variation as possible, to identify contaminants and unusual outliers and to classify new viruses. Meanwhile, the creation of new tools for sophisticated analysis of viral and human immune data is ongoing.

References

1. Brister JR, Bao Y, Kuiken C, Lefkowitz EJ, Le Mercier P, Leplae R, Madupu R, Scheuermann RH, Schobel S, Seto D et al (2010) Towards Viral Genome Annotation Standards, Report from the 2010 NCBI Annotation Workshop. Viruses 2:2258–2268
2. Magrane M, Consortium U (2011) UniProt Knowledgebase: a hub of integrated protein data. Database (Oxford), 2011, bar009.
3. Eddy SR (1996) Hidden markov models. Curr Opin Struct Biol 6:361–365
4. Gaschen B, Kuiken C, Korber B, Foley B (2001) Retrieval and on-the-fly alignment of sequence fragments from the HIV database. Bioinformatics 17:415–418
5. Smith TF, Waterman MS (1981) Identification of common molecular subsequences. J Mol Biol 147:195–197
6. Pruitt KD, Tatusova T, Maglott DR (2007) NCBI reference sequences (RefSeq): a curated non-redundant sequence database of genomes, transcripts and proteins. Nucleic Acids Res 35: D61–D65
7. Lo C-C, Yoon H, Gaschen B, Korber B, Kuiken C (2011) Using virus reference information to improve sequence alignment. Databases (in press)

Chapter 17

Molecular Network Analysis of Diseases and Drugs in KEGG

Minoru Kanehisa

Abstract

KEGG (http://www.genome.jp/kegg/) is an integrated database resource for linking genomes or molecular datasets to molecular networks (pathways, etc.) representing higher-level systemic functions of the cell, the organism, and the ecosystem. Major efforts have been undertaken for capturing and representing experimental knowledge as manually drawn KEGG pathway maps and for genome-based generalization of experimental knowledge through the KEGG Orthology (KO) system. Current knowledge on diseases and drugs has also been integrated in the KEGG pathway maps, especially in terms of known disease genes and drug targets. Thus, KEGG can be used as a reference knowledge base for integration and interpretation of large-scale datasets generated by high-throughput experimental technologies, as well for finding their practical values. Here we give an introduction to the KEGG Mapper tools, especially for understanding disease mechanisms and adverse drug interactions.

Key words: KEGG pathway map, BRITE functional hierarchy, Disease gene, Drug target, KEGG Mapper

1. Introduction

Thanks to the continuous development of sequencing and other high-throughput experimental technologies, genome sequences and various types of large-scale molecular datasets can now be determined rapidly and cost-effectively. The experimental technologies and the associated bioinformatics technologies have revolutionalized many areas of biological sciences. However, the benefits of what may be called genomic revolution have been limited to research communities. The promises of new strategies for diagnosis, treatment, and prevention of diseases have been repeatedly made since the Human Genome Project 20 years ago, but the genomic revolution has not yet been brought to society. We believe that this is partly due to the lack of appropriate resources and methods for linking research results to practical values. Here we introduce the KEGG DISEASE and DRUG database resource (1)

Hiroshi Mamitsuka et al. (eds.), *Data Mining for Systems Biology: Methods and Protocols*, Methods in Molecular Biology, vol. 939, DOI 10.1007/978-1-62703-107-3_17,

and present bioinformatics methods for finding potentially useful information in practice and in society from genome sequences and other large-scale data.

KEGG (Kyoto Encyclopedia of Genes and Genomes) is a database resource that integrates genomic, chemical, and systemic functional information (2). In particular, gene catalogs in the completely sequenced genomes are linked to higher-level systemic functions of the cell, the organism, and the ecosystem. Major efforts have been undertaken to manually create a knowledge base for such systemic functions by capturing and summarizing experimental knowledge in computable forms; namely, in the forms of molecular networks called KEGG pathway maps, BRITE functional hierarchies, and KEGG modules. Continuous efforts have also been made to develop and improve the cross-species (ortholog-based) annotation procedure for linking genomes to the molecular networks. The molecular network-based methods in KEGG now include diseases and drugs. We view diseases as perturbed states of the molecular network system that operates the cell and the organism, and drugs as perturbants to the molecular network system. From this perspective KEGG DISEASE, together with KEGG DRUG, has been developed as a computable disease information resource, enabling integrated analysis and interpretation of various large-scale datasets.

2. Materials

2.1. KEGG Object Identifiers

KEGG consists of 13 main databases shown in Table 1. Each database is identified by the database name (such as pathway) or by its abbreviation (such as path). Each entry in a database is identified by an entry name (such as hsa04930 in pathway). Generally, in order to uniquely identify an entry across all the databases, it is necessary to give the combination of the database name and the entry name in the form of db:entry (such as path:hsa04930). However, for the 11 databases that are originally developed by KEGG the database name may be omitted because the entry name, called the KEGG object identifier, consists of a database-dependent prefix and a five-digit number (such as hsa04930). KEGG GENES and KEGG ENZYME are the databases derived from NCBI RefSeq (3) and IUBMB Enzyme Nomenclature (4), respectively, and the identifiers are in the form of db:name. Specifically, the KEGG GENES identifier consists of the three-letter organism code (alias of the T number identifier of KEGG GENOME) and the gene identifier (usually locus_tag or Gene ID in the RefSeq database). These identifiers represent molecular objects, such as genes (K numbers), small molecules (C numbers),

Table 1
KEGG databases and KEGG object identifiers

Database	Content	Prefix or db:entry	Example
KEGG PATHWAY	Pathway maps	map/ko/ec/rn/(org)	hsa04930
KEGG BRITE	Functional hierarchies (ontologies)	br/jp/ko/(org)	ko01003
KEGG MODULE	KEGG modules	M	M00008
KEGG DISEASE	Human diseases	H	H00004
KEGG DRUG	Drugs	D	D01441
KEGG ENVIRON	Crude drugs, etc.	E	E00048
KEGG ORTHOLOGY	KEGG Orthology (KO) groups	K	K04527
KEGG GENOME	KEGG organisms	T	T01001 (hsa)
KEGG GENES	Gene catalogs	org:gene	hsa:3643
KEGG COMPOUND	Small molecules	C	C00031
KEGG GLYCAN	Glycans	G	G00109
KEGG REACTION	Biochemical reactions	R	R00259
KEGG RPAIR	Reactant pairs	RP	RP004458
KEGG RCLASS	Reaction class	RC	RC00046
KEGG ENZYME	Enzyme nomenclature	ec:number	ec:2.7.10.1

org: three-letter organism code; gene: locus_tag or Gene ID

and drugs (D numbers), and higher-level objects, such as pathways (map numbers), ontologies (br numbers), organisms (T numbers), and diseases (H numbers).

2.2. KEGG PATHWAY Database

The KEGG PATHWAY database is a collection of manually drawn KEGG pathway maps representing our knowledge on the molecular interaction and reaction networks for metabolism, genetic information processing, environmental information processing, cellular processes, organismal systems, human diseases, and drug development. Each map is a result of knowledge-intensive work summarizing experimental evidence in published literature. This is similar to writing a review article, but domain-specific knowledge is represented as molecular networks enabling computational processing rather than text description for humans to read and understand.

For example, the pathway map for a human disease, type II diabetes mellitus, can be retrieved by entering hsa04930 in the search box of the KEGG top page (http://www.genome.jp/kegg/). The molecular network is represented as a wiring diagram of boxes and circles, where boxes correspond to genes and proteins

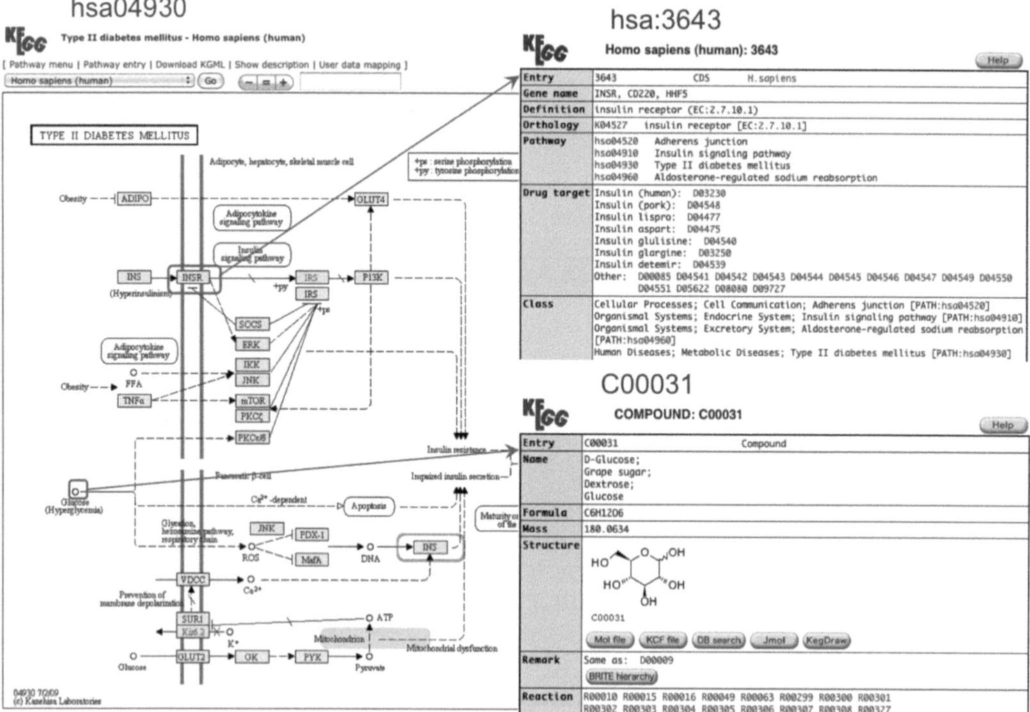

Fig. 1. KEGG pathway map for type II diabetes mellitus (hsa04930) including insulin receptor (hsa:3634) and D-glucose (C00031).

and circles to chemical substances. As shown in Fig. 1, by clicking on a box or a circle in the pathway map, the corresponding molecular object can be retrieved, such as insulin receptor (hsa:3643) in the KEGG GENES database or D-glucose (C00031) in the KEGG COMPOUND database.

While experiments are usually done in specific organisms, the KEGG pathway map is drawn as a generic molecular network that can be extended to other organisms based on the genome information. In fact Fig. 1 was a human version of the diabetes map, and there are versions for mouse (mmu04930), rat (rno04930), and many other organisms, which can be selected from the pulldown menu. The generic version is called the reference pathway map (ko04930), which is linked to the KEGG ORTHOLOGY database as described next.

2.3. KEGG ORTHOLOGY and BRITE Databases

When the KEGG pathway map is manually drawn, the boxes are linked to ortholog groups rather than individual genes in specific organisms. Figure 2 shows an example of the ortholog entry (K04527 for insulin receptor) in the KEGG ORTHOLOGY (KO) database, which contains a list of orthologous genes in all available genomes together with a link to the KEGG pathway node. Thus, having the same K number (KO identifier) represents

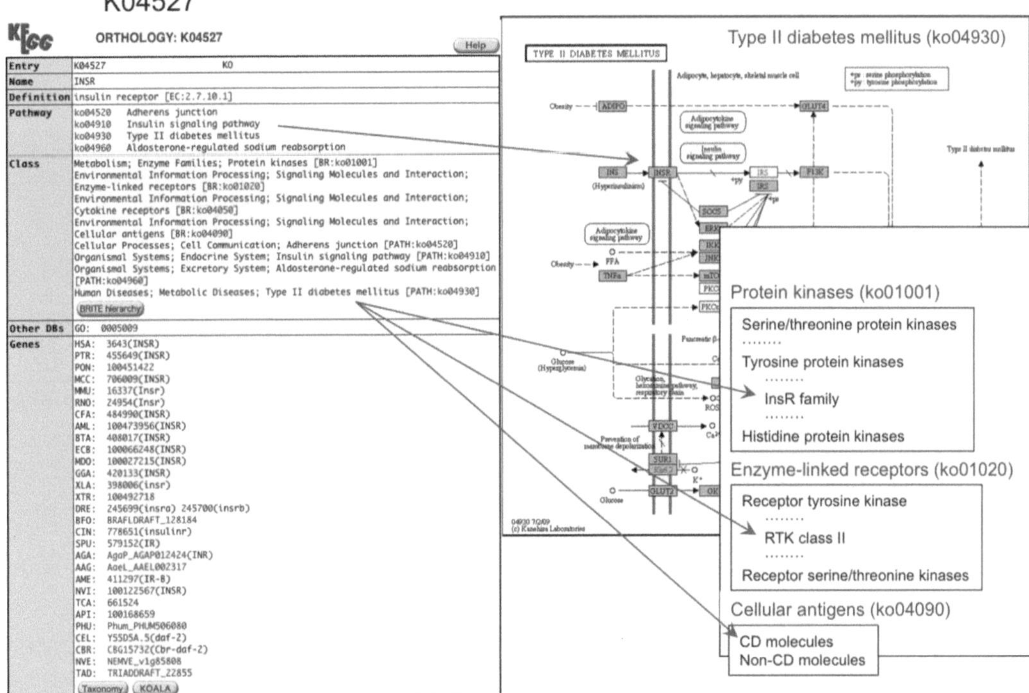

Fig. 2. The KEGG ORTHOLOGY (KO) entry for insulin receptor (K04527) with links to the KEGG pathway map and BRITE functional hierarchies.

functional equivalence in the context of KEGG pathways. The KO entry may also be defined in the context of BRITE functional hierarchies, such as protein family classifications shown in Fig. 2. The KEGG BRITE database contains hierarchical classifications of molecular and higher-level KEGG objects shown in Table 1, and classifications are developed in terms of orthologs (K numbers) for genes and proteins.

The ortholog grouping and membership assignment require continuous efforts. First, the initial grouping and assignment are done manually when the pathway map or the BRITE functional hierarchy is developed. Second, both computational and manual membership assignments (genome annotations) are performed with the KOALA (KEGG Orthology And Links Annotation) tool for all the complete genomes that have become publicly available. Third, in order to automate the KOALA annotation as much as possible, existing ortholog groups are often redefined by splitting, merging, and adding groups. In essence, the KEGG ORTHOLOGY database represents a genome-based generalization of experimental knowledge.

2.4. KEGG DISEASE and DRUG Databases

In KEGG, diseases are viewed as perturbed states of the molecular system, and drugs as perturbants to the molecular system. Different types of diseases, including single-gene (monogenic) diseases,

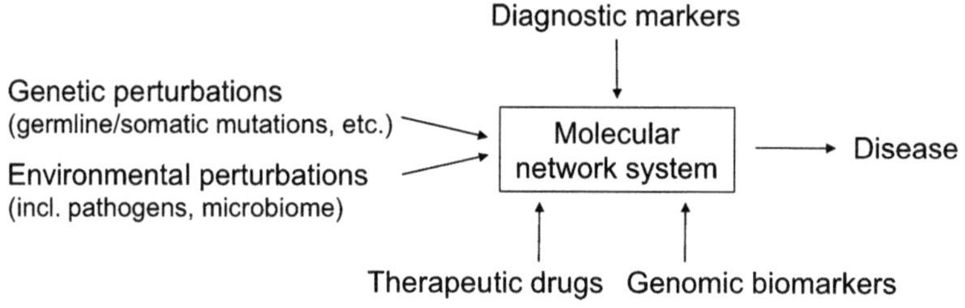

Fig. 3. Molecular network-based view on diseases and drugs.

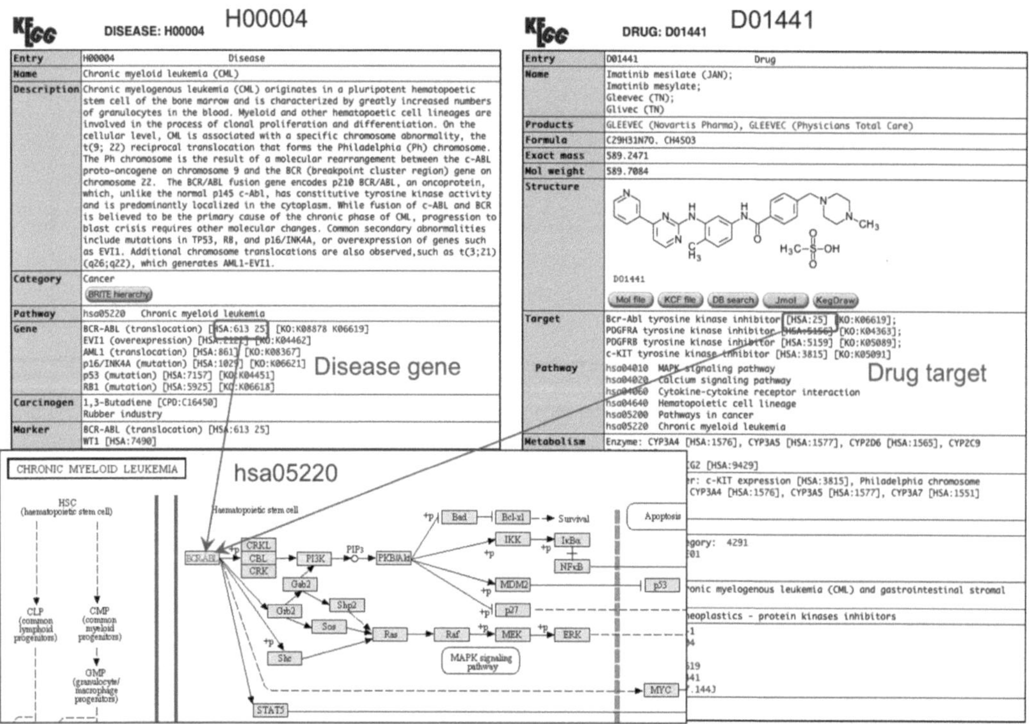

Fig. 4. Chronic myeloid leukemia (H00004) and Gleevec (D01441). The fused gene BCR-ABL is both a disease gene and a drug target as shown in the chronic myeloid leukemia pathway map (hsa05220).

multifactorial diseases, and infectious diseases, are all treated in a unified manner as shown in Fig. 3. Our knowledge on perturbed molecular networks has been captured and represented as disease pathway maps in the KEGG PATHWAY database (see, for example, the disease pathway map of chronic myeloid leukemia hsa05220 in Fig. 4). The KEGG DISEASE database is a collection of disease entries capturing knowledge on genetic and environmental perturbations. Each disease entry is identified by the H number and contains a list of known genetic factors (disease genes), environmental factors, diagnostic markers, and therapeutic drugs (see, for example, the disease entry of chronic myeloid leukemia H00004 in Fig. 4).

The KEGG DRUG database is a comprehensive collection of prescription drugs marketed in Japan, USA, and Europe unified based on the chemical structures and/or the chemical components. Each KEGG DRUG entry is identified by the D number distinguishing the chemical structure of chemicals or the chemical component of mixtures. It is associated with the following information (see, for example, the molecular target drug for chronic myeloid leukemia Gleevec D01441 in Fig. 4): generic names, representative trade names, links to FDA-approved drug labels in DailyMed (http://dailymed.nlm.nih.gov/) and Japanese labels in JAPIC (http://www.japic.or.jp/), target molecules in the context of KEGG pathway maps, drug metabolizing enzymes and transporters, other interacting molecules including genomic biomarkers and CYP inducers/inhibitors, adverse drug–drug interaction data (collected from the Japanese labels) (5), text description of activity and efficacy, drug classification information in BRITE hierarchies such as ATC codes, history of drug development represented as a KEGG DRUG structure map (6), and links to outside databases.

3. Methods

The KEGG resource is accessible either at the GenomeNet Web site http://www.genome.jp/kegg/ or at the KEGG Web site http://www.kegg.jp/. The Web site is hierarchically organized as follows. The first layer is the two top pages of the KEGG resource. The database home (KEGG) page contains links to main databases, selected computational tools, and documents. The KEGG Table of Contents (KEGG2) page contains links to all databases and computational tools. The second layer is the top page of each database, and in the third layer each database entry is found. It is good to remember the color coding of Web pages that correspond to the category of databases: green for systemic information (PATHWAY, BRITE, and MODULE), purple for practical information (DISEASE, DRUG, and ENVIRON), red or brown for genomic information (GENES, GENOME, and ORTHOLOGY), and blue for chemical information (COMPOUND, GLYCAN, REACTION, and ENZYME).

3.1. KEGG Mapper

KEGG PATHWAY and KEGG BRITE are the reference knowledge bases for biological interpretation of molecular datasets, especially large-scale datasets generated by high-throughput experimental technologies. This is accomplished by the process of KEGG mapping, which is to map, for example, a genomic or transcriptomic content of genes to KEGG pathway maps in order to see which parts of pathways are reconstructed from the genome or up/

Table 2
KEGG mapper tools

Tool	Query data	Reference knowledge	Example[a]
Search Pathway	Objects	KEGG PATHWAY database	
Search&Color Pathway	Object-attributes relations	KEGG PATHWAY database	3.1, 3.2
Color Pathway	Object-attributes relations	KEGG PATHWAY map	3.4
Search Brite	Objects	KEGG BRITE database	3.3
Search&Color Brite	Object-attributes relations	KEGG BRITE database	
Join Brite	Object-attributes relations	BRITE functional hierarchy	3.3

[a]Subsection numbers are shown

down regulated in the transcriptome. The KEGG mapping operations are incorporated in the daily database update procedure in KEGG. Especially, organism-specific pathway maps are computationally generated for all available genomes by combining genome annotation data (gene to K number relations) and the reference pathway maps.

The user interface for KEGG mapping is called KEGG Mapper, which currently consists of six tools as shown in Table 2. Query data may be a collection of molecular objects (genes, proteins, small molecules, etc.) or a more ordered set of object-attributes relations; for example, genes annotated with K numbers, genes associated with up/down expression levels, and genes associated with somatic mutation frequency. Target data may be the KEGG PATHWAY database, the KEGG BRITE database, or one entry of these databases. The mapping procedure is considered a set operation between the query data, which can be large-scale data, and the target data of reference knowledge. In this Methods section we show examples of how data and knowledge are computationally processed to obtain new insights and to find new discoveries.

Figure 5 shows an example of using the Search&Color pathway tool in KEGG Mapper. Here we prepare a list of genes involved in chronic myeloid leukemia (CML) and ask KEGG Mapper how this gene list is related to other pathways. The gene list contains a color specification, background color, and optionally followed by foreground color, for the matched objects found (in this case, boxes). The RGB code of #bfffbf is a greenish color used in organism-specific pathways in KEGG, and the foreground color red is used to identify known disease genes in CML. Once the query gene list is ready, enter the data in the search box or upload the data from a file. Select "*Homo sapiens* (human)" in the "Search against" pulldown menu, check the "Use uncolored diagrams" option, and click on "Exec." Then KEGG Mapper gives you

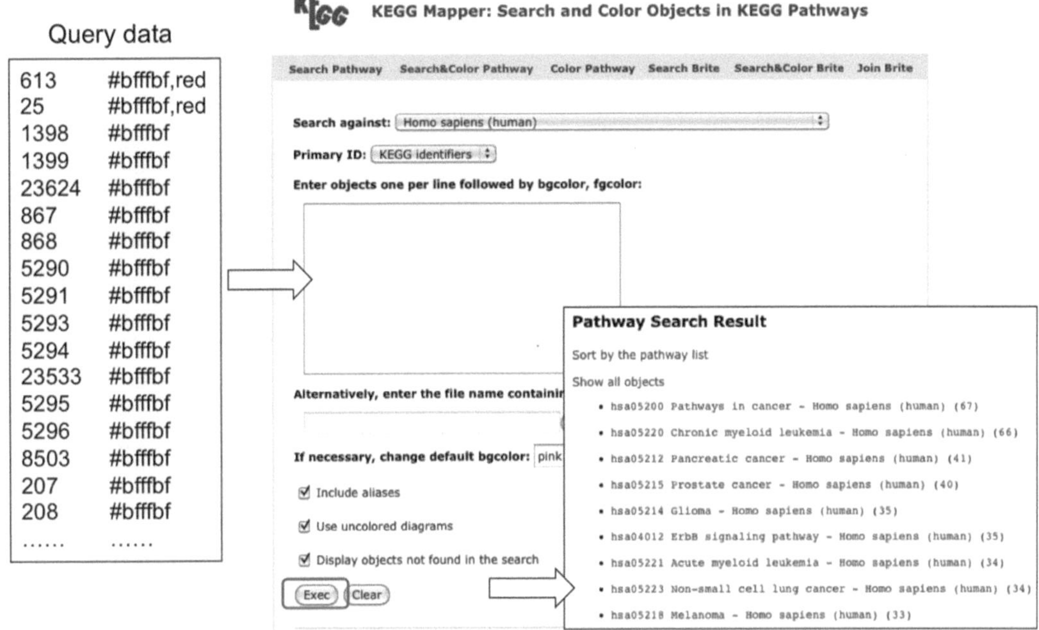

Fig. 5. An example of using the Search & Color Pathway tool in KEGG Mapper. The query data contains a human gene list with color specification.

"Pathway Search Result," a list of KEGG pathway maps that contain the query genes sorted by the number of genes found. The top hit "hsa05200 Pathways in cancer" is a global cancer pathway map, which is shown in Fig. 6. This result shows the CML pathway in relation to various cancer signaling pathways. By examining other pathways in the "Pathway Search Result," commonality of signaling pathways between CML and other cancers may become apparent.

3.2. Disease/Drug Mapping

During the daily update of the KEGG PATHWAY database, manually drawn reference pathways (prefix map) are combined with genome annotation data to generate organism-specific pathways such as for human (prefix hsa). Additionally one special type of pathway maps are generated by incorporating all known disease genes accumulated in KEGG DISEASE and all known drug targets stored in KEGG DRUG. This is called disease/drug mapping, identified by the prefix "hsadd," and displayed as "*Homo sapiens* (human) + Disease/drug" in the organism selection pulldown menu of each pathway map. In addition to greenish coloring of boxes in hsa maps, hsadd maps contain additional coloring. When the gene is known to be associated with a disease, it is marked in pink. When the gene (product) is a known drug target, it is marked in light blue. When the gene is both a disease gene and a drug target, its coloring is split into pink and light blue.

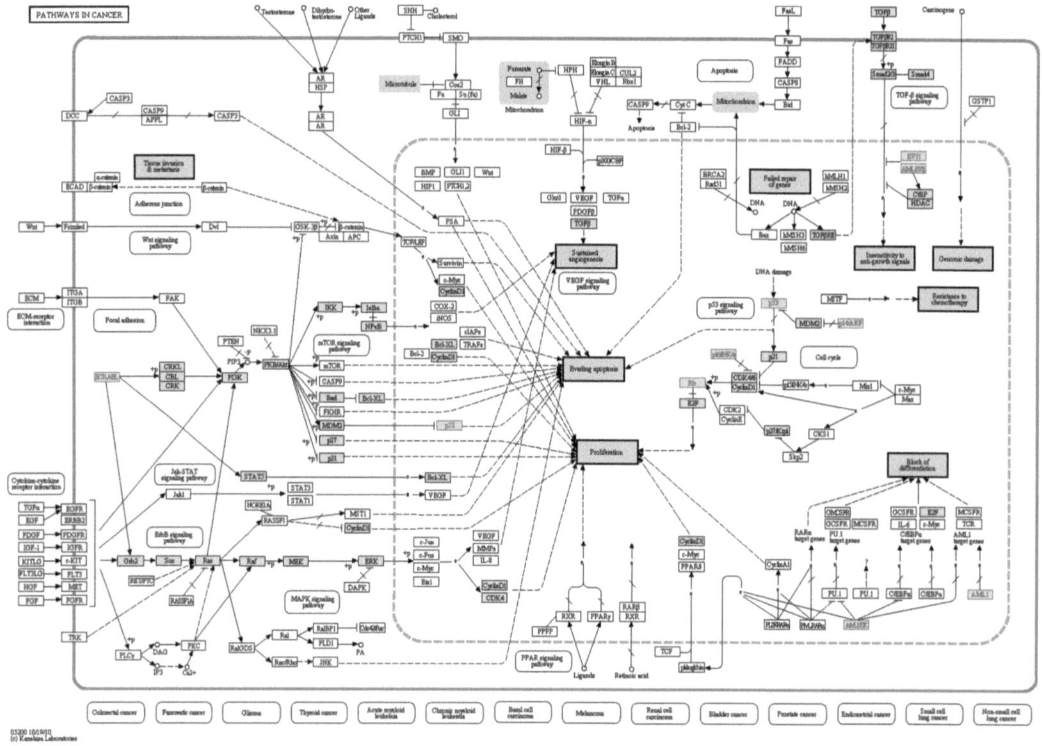

Fig. 6. Chronic myeloid leukemia pathway displayed on the global cancer pathway map.

Let us examine one particular disease/drug mapping for the pathway map of Alzheimer's disease (hsadd05010) shown in Fig. 7. The original map (hsa05010) contains four known disease genes (marked red) for Alzheimer's disease (H00056): APP (amyloid beta protein), APOE (apolipoprotein E), PSEN1 (presenilin 1), and PSEN2 (presenilin 2). In the disease/drug map (hsadd05010) there are additional genes with pink coloring, namely, genes associated with other diseases. Among them are: SNCA (alpha-synuclein) associated with Parkinson's disease (H00057) and Lewy body dementia (H00066), MAPT (microtubule-associated protein tau) associated with amyotrophic lateral sclerosis (H00058), progressive supranuclear palsy (H00077), and Pick's disease (H00078), and other genes associated with spinocerebellar ataxia (H00063) and Leber optic atrophy (H00068). All these additional diseases are neurodegenerative diseases, suggesting common mechanisms of neurodegeneration (7). Thus, the disease/drug mapping allows a holistic approach to understanding molecular mechanisms of diseases.

3.3. Adverse Drug Interactions

Coadministration of multiple drugs is known to sometimes cause serious adverse effects. Adverse events are described in the package inserts (labels) of prescription drugs, but the description is not necessary complete or up-to-date, and interactions with OTC

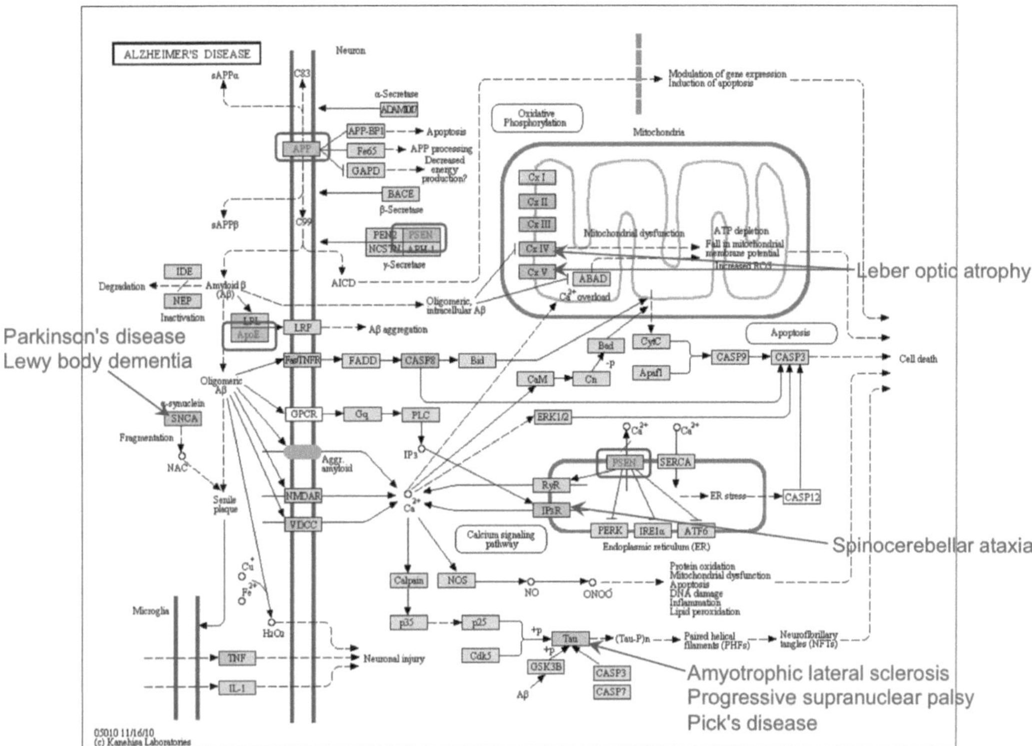

Fig. 7. Disease/drug mapping reveals relationships between Alzheimer's disease and other neurodegenerative diseases.

drugs and food are not well documented. In the KEGG DRUG database, there is an option (DDI button in each drug entry page) to search against known drug–drug interactions and to predict possible interactions. Known drug–drug interactions were extracted from the package inserts of all prescription drugs marketed in Japan. The extracted data were then processed to characterize and classify different types of drug interactions, such as those involving CYP enzymes or target molecules. Figure 8 shows the result of searching drug–drug interactions for Gleevec, molecular target drug for CML. The search result can be displayed on top of the WHO's ATC (Anatomical Therapeutic Chemical) classification. This reveals a group of drugs with some members known to interact with Gleevec, leading to a plausible conclusion that other members in the same group may also interact with Gleevec.

The ATC classification (br08303) in the KEGG BRITE database is manually constructed within KEGG by making correspondences between D numbers and ATC codes. During the daily update procedure of the KEGG DRUG database, the D number to target relationship and the D number to metabolizing enzyme relationship are stored in binary relation files. Then, by using the procedure similar to the Join Brite tool in KEGG Mapper, the BRITE hierarchy file and the binary relation file are combined. This is like adding a new column

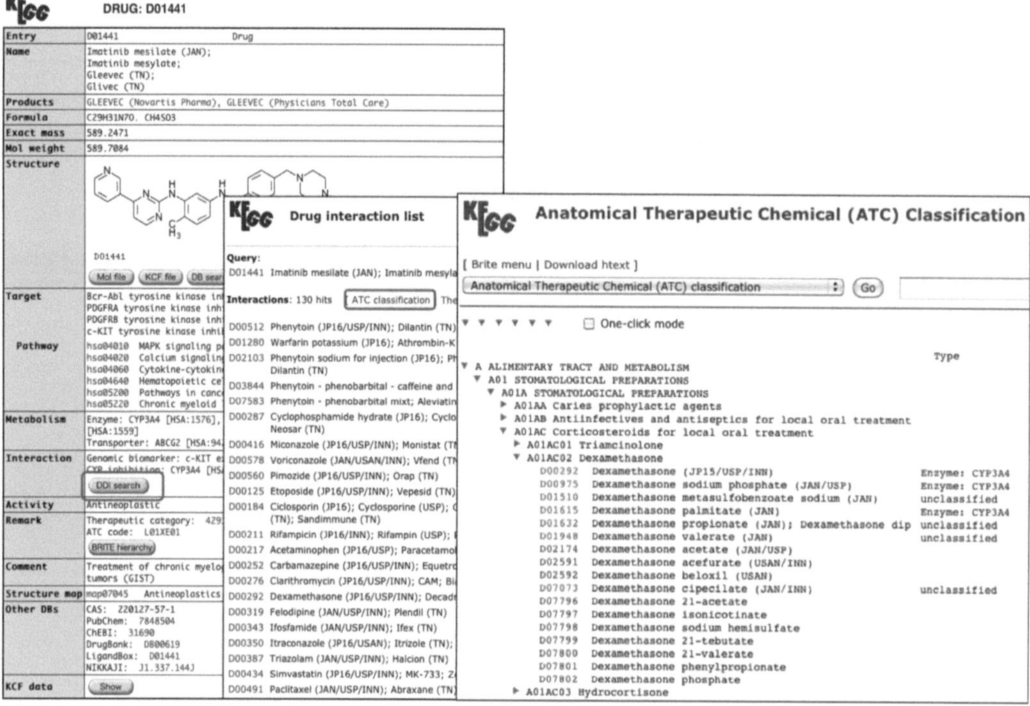

Fig. 8. Drug–drug interaction search for Gleevec. The result is shown on the ATC classification hierarchy.

to the existing BRITE hierarchy file, again revealing the target information (br08303_target) and the enzyme information (br08303_enzyme) associated with groups of similar drugs.

3.4. Cancer Pathway Analysis

The KEGG CANCER resource currently consists of 55 KEGG DISEASE entries, 14 KEGG pathway maps, one KEGG pathway map for "Pathways in cancer" (hsa05200), which is a combined map of 14 cancers, a BRITE hierarchy file for "Cancer stages" (hsa05201), a BRITE hierarchy file for "Antineoplastics" (br08308), and a BRITE hierarchy file for "Carcinogens" (br08008). Thus, information about genes, environmental factors, signaling pathways, and drugs is well organized, reflecting our view on diseases and drugs shown in Fig. 3.

Thus far, we have shown examples of using only the KEGG resource. As an attempt to integrate with outside resources, somatic mutation data obtained from Sanger Institute's COSMIC (Catalogue of Somatic Mutations in Cancer) database are mapped against KEGG cancer pathway maps using the Color Pathway tool of KEGG Mapper. The result of this mapping, such as shown in Fig. 9, can be examined from the "Cancer stages" hierarchy. Generally speaking, the disease genes, which have been identified from literature and whose gene names are marked red in the KEGG cancer pathway maps, well correspond to the actual observations of somatic mutations in cancer genomes.

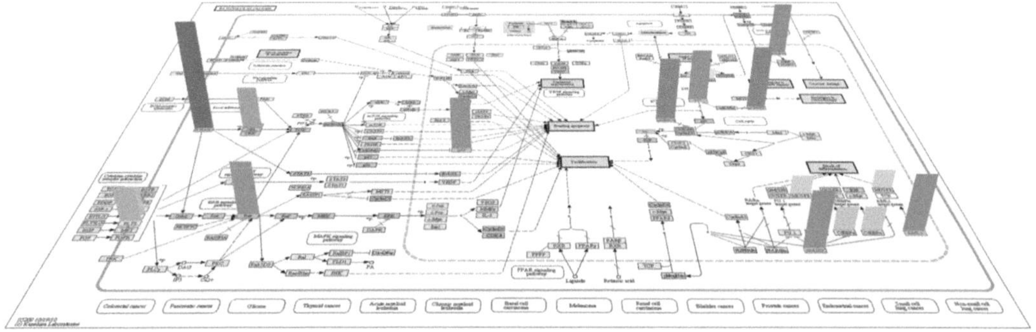

Fig. 9. Somatic mutation frequency in chronic myeloid leukemia (CML) displayed on top of the CML pathway.

4. Notes

1. KEGG Mapper is based on a set operation to match identifiers in the query data and the target data. The target data of KEGG pathway maps and BRITE hierarchies are created using the KEGG object identifiers (Table 1). Therefore, it is recommended to use KEGG object identifiers, such as K numbers, C numbers, and org:gene identifiers, in the query data as well. The use of EC number (Enzyme Commission number) (4) is highly discouraged, for it is treated as an attribute of K number rather than an identifier. The KEGG GENES entry identifiers mostly correspond to NCBI identifiers: locus_tag for prokaryotic genomes and Gene ID for eukaryotic genomes. If necessary, conversion tables are available for download from the LinkDB database at GenomeNet (http://www.genome.jp/dbget/linkdb.html).

References

1. Kanehisa M, Goto S, Furumichi M, Tanabe M, Hirakawa M (2010) KEGG for representation and analysis of molecular networks involving diseases and drugs. Nucleic Acids Res 38:D355–D360
2. Kanehisa M, Goto S (2000) KEGG: Kyoto Encyclopedia of Genes and Genomes. Nucleic Acids Res 28:27–30
3. Pruitt KD, Tatusova T, Maglott DR (2005) NCBI Reference Sequence (RefSeq): a curated non-redundant sequence database of genomes, transcripts and proteins. Nucleic Acids Res 33: D501–D504
4. McDonald AG, Boyce S, Tipton KF (2009) ExplorEnz: the primary source of the IUBMB enzyme list. Nucleic Acids Res 37:D593–D597
5. Takarabe M, Shigemizu D, Kotera M, Goto S, Kanehisa M (2009) Characterization and classification of adverse drug interactions. Genome Inform 22:167–175
6. Shigemizu D, Araki M, Okuda S, Goto S, Kanehisa M (2009) Extraction and analysis of chemical modification patterns in drug development. J Chem Inf Model 49:1122–1129
7. Limviphuvadh V, Tanaka S, Goto S, Ueda K, Kanehisa M (2007) The commonality of protein interaction networks determined in neurodegenerative disorders (NDDs). Bioinformatics 23:2129–2138

Index

Hiroshi Mamitsuka et al. (eds.), *Data Mining for Systems Biology: Methods and Protocols*, Methods in Molecular Biology, vol. 939, DOI 10.1007/978-1-62703-107-3, © Springer Science+Business Media New York 2013

H

I

K

L

M

N

MIX
Papier aus verantwortungsvollen Quellen
Paper from responsible sources
FSC® C105338

If you have any concerns about our products,
you can contact us on
ProductSafety@springernature.com

In case Publisher is established outside the EU,
the EU authorized representative is:
Springer Nature Customer Service Center GmbH
Europaplatz 3, 69115 Heidelberg, Germany

Printed by Libri Plureos GmbH
in Hamburg, Germany